BEHIND THE MASK
OF CHIVALRY

BEHIND THE MASK
OF CHIVALRY

▲ ▲ ▲

The Making of the
Second Ku Klux Klan

NANCY MACLEAN

New York   Oxford
OXFORD UNIVERSITY PRESS
1994

Oxford University Press

Oxford New York Toronto
Delhi Bombay Calcutta Madras Karachi
Kuala Lumpur Singapore Hong Kong Tokyo
Nairobi Dar es Salaam Cape Town
Melbourne Auckland

and associated companies in
Berlin Ibadan

Published by Oxford University Press, Inc.,
200 Madison Avenue, New York, New York 10016

Oxford is a registered trademark of Oxford University Press

Library of Congress Cataloging-in-Publication Data
MacLean, Nancy.
Behind the mask of chivalry :
the making of the second Ku Klux Klan /
Nancy MacLean.
p. cm.
Revision of thesis (doctoral)—University of Wisconsin-Madison.
Includes bibliographical references and index.
ISBN 0-19-507234-0
1. Ku Klux Klan (1915)—Georgia—Athens. 2. Athens (Ga.)—Race relations.
3. Athens (Ga.)—Social conditions. I. Title.
HS2330.K63M225 1994 322.4'2'0975818—dc20 93-27548

4 6 8 9 7 5 3

Printed in the United States of America
on acid-free paper

To my sister, Mary Anne,
my parents, Jack and Ann,
and the memory of my mother, Jeanne

▲ ▲ ▲

Acknowledgments

Sex has shielded me from some of the illusions and compulsions of my subjects. The fantasy of the "self-made man," for one, was never an option. So many people and institutions helped with the writing of this book in so many different ways that a full enumeration of their contributions would itself fill a book. But some deserve special mention.

Institutional support came from several sources. An Alice Freeman Palmer Fellowship, a Dissertation Fellowship from the University of Wisconsin-Madison Graduate School, and a Charlotte Newcombe Doctoral Dissertation Fellowship gave me the time for research and writing. Archivists and staff members at many places aided the research process, particularly those at Special Collections, the Russell Library, and the University Archives at the University of Georgia Libraries in Athens; the Georgia Department of Archives and History in Atlanta; the Atlanta Historical Society; the Library of Congress; the Robert W. Woodruff Library at Emory University; the Southern Historical Collection in Chapel Hill; and the William R. Perkins Library at Duke University. I owe special thanks to Dale Couch and Virginia Shadron for sharing with me their vast knowledge of the resources at the Georgia Department of Archives and History in Atlanta and their expertise in Southern history. Their generosity and good humor made research a pleasure.

Other debts are more difficult to describe, since the more I reflect on them the larger they loom. Singular thanks are due to Gerda Lerner. Her commitment to promoting women's history and developing women's minds helped bring together and sustain a re- markable group of women in the Graduate Program in American Women's History at the University of Wisconsin–Madison. As a forum for thinking and learning about gender in history, it was exceptional—and hindsight teaches, precious. My greatest intellec- tual debt is to Linda Gordon, who was the advisor for my disserta- tion, which this book comes from. Her formidable intellect, politi- cal good sense, personal warmth, and devotion to her students enriched this book and its author in immeasurable ways. Gerda Lerner, Florencia Mallon, Tom McCormick, and Steve Stern taught me much and gave me invaluable advice at various stages of this work. For their incisive comments on parts of this work and for years of challenge, engagement, and solidarity—spiced with ample doses of much-needed diversion—Bob Buchanan, Maureen Fitzger- ald, Joyce Follet, and Leisa Meyer also deserve thanks. At North- western University, Jim Campbell, Jim Oakes, and Bob Wiebe gave me the benefit of their careful readings and thoughtful criticisms. Thanks also to Steve Hahn for encouragement and very helpful suggestions at two critical stages of this project.

Others, outside the field of history, contributed in their own distinctive ways. Over the last decade, I have learned as much about the issues at stake in this work from activists and organizing as from scholars and research; political work schools an awareness of connections and a kind of rigor hard to acquire elsewhere. I owe a special debt to the people involved in the International Socialist tendency, whose ideas and activism taught me a great deal and whose commitment to a better world continues to inspire, even where particular arguments no longer persuade. Finally, there are the family members to whom this book is dedicated. Their years of faithful support, their belief in me, and their love made it possi- ble for me to do this work. For this, and much more, I am deeply grateful.

▲ ▲ ▲

Contents

▲ ▲ ▲

Introduction

Back of the writhing, yelling, cruel-eyed demons who break, destroy, maim and lynch and burn at the stake is a knot, large or small, of normal human beings and these human beings at heart are desperately afraid of something.

—W. E. B. DuBois[1]

This book is about the most powerful movement of the far right that America has yet produced: the Ku Klux Klan of the 1920s. The story it tells is disturbing. For it has to explain, among other things, how it was that sane, ordinary men came to believe that Catholics were stockpiling weapons to take over the country, that a cabal of Jewish bankers controlled world affairs, and that white people must ready themselves for an imminent race war with people of color. This study also has to explain how men who believed such things were able to persuade others. For, by mid-decade, well over a million—perhaps as many as five million—white, native-born, Protestant men had paid their dues and pledged their loyalty to the order's leaders and its program.[2] The Ku Klux Klan of that decade recruited more members and amassed more power in communities throughout the United States than any Klan before or since.

Yet it was neither the first nor the last movement of its kind, as former Grand Wizard David Duke's recent success in using elections to build a mass following for his ideas attests. On the contrary, since its first incarnation after the Civil War, the Ku Klux Klan has occupied an enduring place in American politics. Like the proverbial phoenix, the Klan has died in one setting, only to be reborn in another.[3] The key themes of Duke's recent campaigns in fact echoed the appeals of the second Klan: a form of populism that combined hostility to established élites with dedication to white supremacy, support for conservative family values, enthusiasm for "old-time religion," and antipathy to welfare recipients, trade unionists, immigrants, liberals, and leftists. This book is written in the hope that understanding the drawing power of such themes in the past can help us see why they still pull in our own time.

To make sense of the Klan, however, one must first surrender some comforting illusions. Above all, one has to give up the notion of the essential otherness of the kind of men attracted to it. In the 1920s at least, Klan members were not the deranged outcasts of popular imagination. A score of historians have now painstakingly researched the membership and activities of Klan chapters in localities across the nation. And they have found that most often the men who donned the order's robes and assembled beneath its flaming crosses were, as one contemporary put it, "if not the 'best people,' at least the next best . . . the good, solid middle-class citizens." Not only did the Klan draw from the broad middle of the nation's class structure, but it most commonly mobilized support through campaigns waged on the prosaic theme of upholding community moral standards.[4]

These campaigns illustrate how concerns about gender and sexuality animated this movement, as they would David Duke's efforts. The involvement of local Klans in Prohibition enforcement, for example, grew directly out of fears that alcohol consumption endangered conventional family life. Without attention to how notions of proper manhood, womanhood, and parenting infused Klan thought and action, no analysis would be complete. For the Klan's conservative ideology was a deeply gendered phenomenon. Klansmen could not discuss issues of race, class, or state power apart from their understanding of manhood, womanhood, and sexual decorum. This fusion of private and public imparted to Klan prejudices much of their peculiar force. Yet, as the classic example of the plantation mistress and the female slave illustrates, and as black feminists have argued most eloquently, it is impossible to understand ideas about gender or the sexual politics they inform without attention to their class and race moorings.[5] Guided by that principle, this study will show how the Klan's hostility to such things as teenage sexuality and birth control both emerged from and contributed to the racism, anti-Catholicism, and opposition to labor struggle it is conventionally and rightly known for.

That sense of connection between the Klan's respectability and its malice—of "the banality of evil" as Hannah Arendt once put it—leads this study to part company with the trend in recent historical writing about the Klan to de-emphasize the racial hatred of its politics and the violence of its practice. Just as many reporters assumed that David Duke's stylish suits and born-again Christianity somehow nullified his neo-Nazi convictions, so many scholars of the second Klan seem to assume that church-going, civic-minded, middle-class men would never have espoused the views or conducted the deeds the Klan is commonly associated with. Focussing on the organization's activities to the neglect of its ideas, and generally studying it in Northern settings where its members were more constrained, many of these studies concur that the Klan, as one recent synthesis put it, addressed "real social problems, not symbolic or imaginary fears." According to this interpretation, "the Klan was a popular social movement, not an extremist organization."[6]

This study seeks to move beyond such false polarities, which have dominated thinking about the Klan for too long. Scholars have dug in their heels over whether the movement was rural or urban, whether its causes were local or national, whether its members were civic crusaders or vigilantes, populists or racists, and whether its grievances were "real" or chimerical. As the blatant race-coding of ostensibly neutral public policy issues like welfare, crime, and teenage pregnancy in recent years indicates, this style of thought acts as a barrier to understanding.[7] The pitched battles over specious dichotomies have distracted attention from the real challenge: that of recognizing and accounting for how the Klan was all of these things. It was at once mainstream and extreme, hostile to big business and antagonistic to industrial unions, anti-élitist and hateful of blacks and immigrants, pro–law and order and prone to extralegal violence. If scholars have viewed these attributes as incompatible, Klansmen themselves did not.

This study aims to demonstrate the basic consistency of their motives and positions. The source of that consistency was a world view and politics best characterized, in my view, as reactionary populism. In it, the anti-élitism characteristic of populism joined with the commitment to enforce the subordination of whole groups of people.[8] The appeal of this politics was rooted deep within American society and culture: in the legions of middle-class white men who felt trapped between capital and labor and in the political culture they inherited from their forebears. Fearful for the future, Klan leaders drew from the wellsprings of American politics to fashion an ideology that would enable them to hold on to their basic values, make sense of rapidly changing social relations, and fend off challenges to their power. They drew from classical liberalism their ideas about economics, and from republicanism their notions of citizenship and the commonweal, in particular its long exclusion from the right to participate in political affairs of economic dependents—whether slaves, free women and children, or propertyless men.[9] The synthesis Klan leaders fashioned extended and modified, but by no means contradicted, values widely held in American society. It proved compelling enough to attract millions.

This book describes that synthesis and explains its appeal. My concerns are more cultural than chronological, more analytic than narrative. The emphasis is on how Klansmen understood their world, why they thought the way they did, and what moved them to action. Such an understanding can only be developed through a sustained examination of Klan ideology. Yet, when studying a social movement like the Klan, seeing the interaction between ideas and lived experience is crucial to understanding their sources and import. Hence this work employs a local case study to anchor its analysis of the Klan's ideology and practice nationwide.

The site of the study is Clarke County, Georgia, home of the University of Georgia and, in the 1920s, of Athens Klan Number 5. Once described by W. E. B. Du Bois as the "Invisible Empire State," Georgia was the birthplace and national headquarters of the second Klan.[10] A Georgia case study can thus illuminate the movement at its source. By and large, however, the reader will find no brief here for the special nature of either Athens or its Klan. On the contrary, it is their very ordinariness that attracts attention: a middling city and an average Klan chapter.

Yet in one respect the Athens Klan was unique in the South, and rare in the nation. Unlike most of their peers, its leaders failed to hide or destroy their chapter records. They left behind a rich cache of materials that found their way into the archives. The collection includes membership rosters, minutes, propaganda tracts, and correspondence—some of it from local women and men who asked the Klan to discipline relatives and neighbors whose conduct they disapproved of. Used in conjunction with other kinds of local, regional, and national sources—published materials, manuscript collections, public records, oral histories, and so forth—these Athens chapter records make possible a fine-grained portrait of the Klan world.

Readers will no doubt question whether conclusions drawn

from the study of a Southern Klan chapter can illuminate the movement in other areas of the United States. The South, after all, possessed a singular regional heritage. Slavery shaped its political economy and culture in ways that survived the war that buried the "peculiar institution." If white Southerners' racism was less unusual than most white Northerners like to admit, other aspects of Southern life at the turn of the century were distinctive: the prevalence of sharecropping, tenant farming, disfranchisement, and lynching, to name but the most obvious. These regional traditions influenced the Klan's development in the South. Their sway was most obvious in Southern Klansmen's more frequent indulgence in vigilante violence and in the tacit consent, if not outright support, that violence gained from regional élites. Ironically, Southern upbringing also probably steeled commitment to shared national Klan goals, such as curbing the power of the federal government, protecting "the purity of [white] womanhood," and, more generally, fortifying white supremacy.[11]

Yet, in the end, what proved most striking in the research for this study was less the differences between Southern Klansmen and their counterparts elsewhere than how much they all shared. The assumption of Southern distinctiveness with which the author embarked on this project gradually had to be shed in the face of compelling evidence to the contrary. In their basic values, as in their targets, the area of agreement between Klansmen in different parts of the country proved wide. Where they diverged, the differences tended to be of degree, not character, or in focus or tactics, not principles. In key ways, in fact, like the election of the Southerner Woodrow Wilson to the presidency not long before, the Klan of this era was both effect and cause of the reconciliation of North and South. Its very spread promoted a peculiar kind of national integration as Klan leaders worked to bury divisions among the nation's Anglo-Saxon men. The reasons for the convergence are not so hard to identify.

Local variations notwithstanding, the developments that agitated Klan members were not confined to any single locality. The labor unrest of 1919, for example, expressed class conflicts that respected no geographical boundaries. The expanding powers of the federal government and the changes in the structure of power at all levels of the state affected citizens in every area. The plight of agriculture in the 1920s ruined farmers in the Midwest and West as well as the South. The powerful image of the "New Negro" resonated among racists nationwide as Southern blacks moved north in record numbers, while Harlem's radicalism filtered outward. And the youthful pioneers of modern morality drank, danced, drove, and necked from one end of the country to the other. So it is not surprising that white men around the country rallied to Klan appeals with common core elements. Although the setting of this study is both local and Southern, then, the core of the argument has wider applicability.

To emphasize common purpose is not to deny that local experiences diverged. The history of each Klan chapter, in the South as elsewhere in the country, was in some respects unique. The configuration of an area's economy, the character of local politics, the talents or shortcomings of individual leaders, accidents of timing, and so forth, all worked to shape local Klans and to help or hinder their prospects. Yet these particularities, so important to the texture and excitement of history, should not blind us to larger patterns suggested by the remarkable synchronicity of the Klan's rise and decline in all corners of the nation. The method of this study should make it possible to have both: an understanding of larger patterns enriched by the drama of the particular.

After describing how the Klan was built, the soil it took root in, and the kind of men it attracted, the chapters that follow seek to re-create Klansmen's world view through thematic analysis of its core elements. Focussing on the acutely contested relations between classes, between sexes and generations, and between "races," respectively, the chapters open with cases of local Klan practice. They then go on to tease out the core ideas animating these efforts and analogous ones in other parts of the country. Chapter 7 shows how all of these ideas came together to make vigilantism the emblematic expression of Klansmen's way of thinking. The conclusion then steps back from the particulars of the story to explore some of the national and international contexts that give it added meaning.

I have assumed throughout this work that it is a mistake to take for granted, as so many observers do, the prejudices of Klan members. To say, for example, that the Klan harassed blacks or hated Jews because its members were racists is to confuse tautology with explanation. Another common fallacy is the attempt to deduce members' motives from readings of the census—to assume, say, that if few Catholics lived in a particular area, then anti-Catholicism must have ranked low among the local Klan's purposes. Such reasoning misconstrues the dynamics of prejudice— which often runs deepest where contacts are fewest—and also has the effect of naturalizing it, of making conflict appear to flow inevitably from difference.[12] The historian's task is more demanding: to interrogate even odious and seemingly irrational thought so as to grasp its inner logic and understand its sources and role in a specific time and place.[13]

The goal of this study, then, is to situate Klan members in the world of their day, to take seriously what they did, and to listen carefully to what they said. In this way, we can learn a great deal about what made them tick. Part of taking people seriously involves letting them speak in their own words and their own way. In quoting, therefore, I avoid using the intrusive and condescending *sic* to flag errors. Where the original grammar and spelling do not obscure meaning, they have been retained; where they distort, corrections are rendered in brackets.

I

▲▲▲

CONTEXT AND MOTIVE

1

▲ ▲ ▲

Mobilizing the Invisible Army

If there were such a thing as a typical Klan meeting, the klonklave held by the Athens Klan on the night of September 15, 1925, would qualify. Exalted Cyclops *J.P. Mangum*,* a fifty-two-year-old police-man, called the meeting to order at 8:30 in the Klan's klavern (meeting hall). Presiding over the evening's events with *Mangum* was a full complement of twelve "terrors" (officers). In many re-gards, the meeting resembled one any other organization might hold: minutes read and approved, new members voted in, dues col-lected, plans laid for a recruitment campaign, an educational dis-cussion, and even niceties: members received thanks from *Man-gum* for having visited him when he was sick and from the board of stewards of a local church for having attended its recent revival meeting with a contribution.[1]

Yet, mundane as the proceedings were, a few signs indicated that this club differed from others—notwithstanding the order's policy of not allowing discussions in meetings of "any subject, which, if published, would reflect discredit upon our great move-ment." Among the humdrum bills paid, for example, was one for labor and materials for a "fiery cross." Then there were the applica-tions to join, some from previous members, that the Klansmen in attendance voted to reject. The chapter had recently reorganized

---

* Names in italics have been changed. See Appendix for explanation.

3

due to a public scandal over the use of extralegal methods to combat vice, and it seems these men were viewed as possibly disloyal—"loose-mouth," "weak-kneed," or "traitors," in Klan parlance. Finally, one brief item in the minutes hinted at why absolute loyalty was so necessary. *L. S. Fleming*, the chapter Klokan (investigator), reported the case of a man who had been brought to the Klan's attention for failing to support his family. Not a few such delinquents found themselves kidnapped and flogged by crews of masked men in the 1920s.[2]

Such blending of the ordinary and the extreme was common in the Klan of the 1920s; indeed, the blurring proved a source of strength. The order's overlap with the mainstream made it possible to win the enthusiasm of men like *Chester D. Morton*, a local Mason, Shriner, Boy Scout leader, and member of the Booster Club and the board of stewards of the First Methodist Church. An ambitious young mortician who would soon become president of the Athens Lions Club and vice-president of the Georgia Funeral Directors' Association, *Morton* was not the type to belong to a fringe group. But then again, neither was he likely to pay monthly dues to an organization that merely replicated what he enjoyed through his other affiliations. In *Morton's* case, a clue to the Klan's special attraction comes from his rivalry with Jake and Mose Bernstein, first his employers, and later his competitors for trade and for position on the State Board of Embalmers.[3] Since the Klan admitted only white, native-born, gentile, adult men who believed in Christianity, white supremacy, and "pure Americanism," at the least it would keep out Jews such as the Bernstein brothers—unlike the Athens Elks, for example, who three times had elected businessman and civic leader Moses Gerson Michael to their highest office.[4] But the Klan might even manage to drive the Bernsteins and other Jews out of business altogether.

▲ ▲ ▲

The second Klan's founder, William Joseph Simmons, had not explicitly included such things among the Klan's goals when he established the order in 1915. The son of a poor Alabama country physician, Simmons was a man chronically on the make. Having tried his hand at farming, circuit-riding as a Southern Methodist Episcopal Church preacher, and lecturing in Southern history at Lanier University, by 1915 he had settled into a mildly lucrative position as the Atlanta-area organizer for the Woodmen of the World, a fraternal benefit society. Unsatisfied, Simmons dreamed of reviving the hooded order his father had served in as an officer after the Civil War.

For years, he thought about creating a new Ku Klux Klan. By October of 1915, he was ready to unveil the plans to a group of like-minded friends. Together, the group petitioned for a charter from the state. Then, on Thanksgiving night, they met atop Stone Mountain, an imposing several-hundred-foot-high granite butte just outside Atlanta. With a flag fluttering in the wind beside them, a Bible open to the twelfth chapter of Romans, and a flaming cross to light the night sky above, Simmons and his disciples proclaimed the new Knights of the Ku Klux Klan. Their passion for ceremony was not matched by a talent for organizing, however. Unclear about exactly what their message was, Simmons and his partners floundered over how to spread it. By early 1920, they had only enrolled a few thousand men.[5]

That would soon change. In June of that year, Simmons signed a contract with Mary Elizabeth Tyler and Edward Young Clarke, partners in the Southern Publicity Association. Having organized support for the Red Cross, the Anti-Saloon League, the Salvation Army, and the War Work Council, the two had mastered the art of modern propaganda. Hiring a staff of seasoned organizers, they set to work to amass a following for the Klan and a small fortune for themselves. Within a few months, membership jumped to an estimated 100,000. A wife at age fourteen and a widowed mother at fifteen who went on to make a career as a businesswoman, Tyler had a knack for turning adversity to advantage. When in 1921 the *New York World* set out to destroy the Klan by documenting over one hundred and fifty separate cases of vigilante violence charged to it—an exposure so damning that it prompted a congressional investigation of the order—Tyler turned both into recruiting opportunities. In the four months after the *World*'s exposé, the Klan chartered two hundred new chapters; overall membership leapt to some one million.[6]

Seasoned promoters, Tyler and Clarke knew not only how to sell, but what would sell. To Simmons' initial blend of white supremacy, Christianity, and the male-bonding rituals of fraternalism, they added elements geared to tap the fears of many white contemporaries in the anxious years after the Great War. Declaiming against organized blacks, Catholics, and Jews, along with the insidious encroachments of Bolshevism, the order put itself forward as the country's most militant defender of "pure Americanism." It stood for patriotism, "old-time religion," and conventional

morality, and pledged to fend off challenges from any quarter to
the rights and privileges of men from the stock of the nation's
founders. The message took. Although Tyler and Clarke had expected only Southerners to respond, men from all over the country
did. "In all my years of experience in organization work," Clarke
told Simmons, "I have never seen anything equal to the clamor
throughout the nation for the Klan."[7]

One month after the Klan's founding in Atlanta, Imperial Wizard Simmons spoke at a meeting called to promote the new order
in Athens. Yet, for several years, no more was heard of it in Clarke
County. No doubt it was outflanked by the officially sponsored
hysteria of the war years. With the state government enacting
"work or fight" laws, Athens schoolchildren compelled to sing patriotic songs and buy thrift stamps to avoid ostracism, "slackers"
and government critics branded as "traitors," and civic leaders
preaching that only the United States Army stood between local
residents and the "German Horror" of rape, pillage, and slavery—
and with cotton prices high all the while—the Klan lacked a distinctive appeal. At any rate, not until well after the Armistice did
the Klan reappear in Athens. Then, following a much-touted return
engagement of *Birth of a Nation* in January of 1921, the Klan renewed its efforts to win local men.[8]

Following a strategy devised by the Atlanta-based national office, Athens Klan promoters worked existing networks in the community to accumulate members. They looked to two areas in particular where it seemed their message might be well-received:
fraternal orders and Protestant churches. The Klan presented itself
to prospective members as the active embodiment of "the principles of the better class of lodges." Simmons, a member of fifteen
other fraternal organizations himself, rallied men with odes to "the
united powers of our regal manhood." To enhance the Klan's mys-
tique, he designed a special alliterative lexicon for the movement.
And he painstakingly worked out the details of elaborate rituals
whereby members advanced in the order by obtaining "degrees" as
they did in other fraternal orders. When he first came to Athens to
advertise the new order in 1915, Simmons in fact emphasized "its
unrivaled degree work." Many local Klansmen took the bait; they
delighted in impressing their fellows with their mastery of Klan
ritual.[9]

In presenting their order thus, Klan organizers staked a bid for
the loyalties of participants in the long tradition of fraternal association. The country had over six hundred secret societies by the

mid-1920s; together, they enlisted over thirty million people. The Klan curried support from a number of these groups, especially those of common mind. It endorsed the *Fellowship Forum*, an anti-Catholic publication that claimed a readership of one million white, Protestant fraternalists. The Junior Order of United American Mechanics (JOUAM), an anti-Catholic, nativist fraternity whose better-known members included populist leader Tom Watson and President Warren Harding, was also known as "a close ally" of the Klan. Indeed, the Georgia JOUAM shared its weekly Atlanta-based publication, *The Searchlight*, with the Klan until the Klan formally took it over in October of 1923. The distinction between the two was moot in any case, since Klan leader J. O. Wood edited the newspaper.[10]

Klan leaders cultivated their common ground with fraternal orders to reap a bumper crop of recruits. Almost all the traveling organizers (kleagles) hired by the Imperial Palace were Masons, an affiliation they used to meet prospective Klansmen in new communities. Soon after arriving, they approached leaders of societies such as the Masons, Elks, Odd Fellows, and Orangemen. Often these officials would allow the Klan to meet in their lodge halls, as the Masons did in Clarke County. Indeed, the Athens experience conformed to the larger pattern. When Simmons arrived to advance the new order in 1915, he came as the guest of C. A. Vonderleith, an organizer for the Woodmen of the World, whom Simmons already knew from his work with the Athens Woodmen. *I. A. Hogg*, an officer of the local womens' auxiliary of the Woodmen, would soon add the Klan to his roster of associations.[11]

The strategy worked. Even the meager records available for local fraternal orders reveal that a minimum of 120 Athens Klansmen, or twenty-nine percent, belonged to at least one. Among those that shared members with the Klan were the Woodmen of the World, the Elks, the Masons, the Odd Fellows, the Knights of Pythias, and the Shriners. Several Clarke County Klansmen also held office in these groups. Local Kligrapp (secretary) *Roy P. Yarborough*, for example, was a twenty-five-year veteran of the Odd Fellows and reportedly its most popular member ever. Known to his fellows as "Uncle *Roy*," *Yarborough* also served as clerk of the Woodmen of the World, and belonged to the Knights of Pythias.[12]

But the most ardent fraternalist in the chapter was *Wiley Frank Doolittle*, a Mason, Shriner, and Woodman. Enchanted by the magic of organized manhood, *Doolittle* had accumulated over the years an unrivalled library on fraternalism. "No one in Athens,"

the press eulogized upon his death in 1930, "possessed such knowledge of Masonry"; members of a half dozen local fraternal orders
paid him tribute as honorary pallbearers. Nationwide, the Klan
boasted that 500,000 Masons had joined by 1923. Along with members of other fraternal organizations, they often formed the backbone of local chapters.[13]

As fertile a harvest, organizers found, could be gleaned from
Protestant churches. Imperial Wizard Hiram Wesley Evans, who
assumed the helm from Simmons in 1922, described his organization as "a recruiting agency" for Protestant churches. Whether or
not this was true, the reverse was. Throughout the country, evangelical Protestants in particular flocked to the Klan, primarily Baptists, Methodists, and members of the Church of Christ, the Disciples of Christ, and the United Brethren. Men in more élite or
liberal denominations, in contrast, such as Unitarians, Congrega-
tionalists, Lutherans, or Episcopalians, appeared less likely to join.
In Clarke County, Klansmen also tended to be religious enthusi-
asts. Even the patchy church records available showed that at least
forty-three percent of Athens Klansmen belonged to a church—
about the same proportion as that of all white county residents. Of
these, thirty-seven percent were Baptists; thirty percent, Methodists; and smaller proportions scattered among other denominations.[14]

Many Athens Klan laymen helped lead their churches. At least
forty-six held positions such as deacon, elder, steward, committee
member, usher, or Sunday School participant. Twelve Klansmen
took part in the Men's Sunday School class at First Methodist
Church alone. Some members advanced the cause in other ways.
Klansman *L. T. Curry* served as Treasurer of the Businessmen's
Evangelistic Club, while *N. O. Bowers* championed "personal
evangelism" among young people through the Christian Endeavor
Society.[15] The wives and mothers of many local Klansmen, for
their part, participated in the women's missionary societies of
their churches.

Like laymen, many clergymen cooperated with the Klan. Of
the thirty-nine national lecturers working for the Klan at one
point, two-thirds were said to be Protestant ministers. Each Klan
chapter, meanwhile, had its own kludd (chaplain). By 1924, the
Klan boasted that it had enrolled 30,000 ministers. In that year, the
Klan also claimed as members three-quarters of the 6,000 delegates
to the Southeastern Baptist Convention.[16] In Clarke County, most
of the white Protestant churches had some connection to the Klan.

Either their pastors belonged, or they allowed announcements of Klan meetings or robed visits of Klansmen during services, or they accepted Klan aid in evangelistic efforts.

At least ten ministers belonged to the Athens Klan; several helped lead it. The Reverend M. B. Miller of First Christian Church, for example, served in the mid-'twenties as the Exalted Cyclops (chapter president) of the local Klan; his assistant, the Reverend Jerry Johnson, acted as Kligrapp (secretary). The Klan so valued the work of their member the Reverend B. Postell Read of Young Harris Memorial Methodist Church that, when he left his Athens congregation, Klansmen and women attended his last service in a body to express appreciation for his "service" to the community. The Reverend *B. B. Couch* of West End Baptist Church, also a Klansman, likewise won fulsome praise from his entire congregation upon his departure.[17]

Through such channels, the Klan built up its numbers. By 1923, its ranks included "three hundred of the finest men in Clarke County." Confident of their future, they began building a new klavern to hold their meetings. Members proudly announced that the hall would sport a forty-foot-tall electric cross. With its membership hovering around three hundred the next few years, the local chapter was a "baby Klan," as *The Searchlight* put it. The chapter drew in approximately one in ten of the native-born, Protestant white men eligible for membership—a considerable proportion, but small relative to some of its counterparts.[18]

Statewide, the Klan also thrived in the first half of the decade. Since Atlanta hosted the Klan's national office, or Imperial Palace, Georgia always played a significant role in Klan affairs. The Atlanta Klan enrolled upwards of fifteen thousand members and boasted the largest fraternal hall in the city. By the mid-'twenties, chapters blanketed the state. Cities like Macon, Augusta, and Columbus yielded larger absolute numbers, but towns like Pelham and La Grange, and hamlets like Dewy Rose and Tallulah Falls, held their own in ardor. Large and small, urban and rural, Klan chapters cooperated to achieve their common goals and to build numbers at each other's parades and rallies. A 1926 internal report maintained that Georgia was now second in membership in the South.[19]

In the nation as a whole, Georgia ranked eighth among states in estimated membership. Among regions, the North Central and Southwestern states enrolled the most members, followed by the Southeast, the Midwest and Far West, and, finally, the North At-

lantic states. By mid-decade, the total reached perhaps as high as
five million, distributed through nearly four thousand local chapters. Yet the numbers barely suggest the reach of the Klan's tenta-
cles. If its membership claims were true, the order enrolled as
many members as the American Federation of Labor at the peak of
its strength.[20] "Outside business," reported one sympathetic contemporary journalist, "the Ku Klux Klan has become the most vigorous, active and effective organization in American life." Indeed,
in the five months after the order established a formal national
lecture bureau, its speakers addressed audiences of well over
200,000 people. Those who missed the lectures could stay abreast
of issues through the order's press. Supplementing the publications
issued by the national office, state and local Klans published some
forty weekly newspapers.[21] Numbers like these created a legitimacy of their own.

But their effect was enhanced by the kind of men the Klan was
able to attract. The typical member, in Athens as elsewhere, was
not the uprooted angry young man one might expect; he was
middle-aged, married, and probably a father as well.[22] Ninety-two
percent of Athens Klan members were married men; more than
two-thirds were fathers, with an average of between three and four
children. While most local Klansmen were family men, not a few
were civic leaders. Klansman *G. M. Harris*, the county tax receiver,
had served two terms as mayor in the 1880s. The wealthy *Wiley
Doolittle* was described by the Athens press as among the "most
influential citizens." The roster of offices he had held included
mayor (three terms), president of the Chamber of Commerce, president of the Booster Club, president of the Kiwanis Club, and chairman of the Clarke County Democratic Executive Committee,
which four other Klansmen also served on. One of them, local Kligrapp (secretary) *Roy Yarborough* was also a county commissioner,
a notary public, and a justice of the peace. Joseph Kenneth Patrick,
for his part, was a charter member and later president of the Athens Lions Club, a member of the Board of Directors and the Rural
Committee of the Chamber of Commerce, and a future state
senator.

Just as the Klan recruited men from the mainstream, so it
boosted members' morale with the kinds of family and community
activities that clubs and churches also sponsored. Although excluded from the Klan itself, Klansmen's wives and sons could join
parallel orders: the Women of the Ku Klux Klan, created in 1923,
and the Junior Klan, created in 1924. Here, without distracting at-

tention from the leading roles of their menfolk, family members might work for shared ends. Athens Klanswomen and men thus cooperated to reward a visiting minister with an automobile for his leadership of a successful revival at East Athens Baptist Church in 1926, winning themselves the gratitude of the church's chairmen and deacons. The following year, they collaborated on a fund-raiser whose end was "to place a Flag and Bible" in the city high school.[23]

Klan chapters promoted sociability and mutual aid as well. At Klan picnics such as the "Great Klan Barbecue" hosted by Athens Klansmen in 1928, members gathered with their families and friends for afternoons of music, sports, and swimming, along with speeches. Sometimes, Klan rallies featured weddings of members, public rituals that interwove personal and political commitments. More important, fellow Klansmen were on hand in times of trouble. When a member fell sick, his brothers came to visit. When his family lost a loved one, they sent flowers and expressions of sympathy. And when he died, they stood ready to conduct a stately funeral. When crisis struck, Klansmen dug into their pockets for one another. When the bicycle of a local minister's son was stolen, for example, his fellow Klansmen voted to buy the boy a new one. After *E. D. Todd* lost his home in a fire, the chapter established a committee to inform his fellows of his needs. How much such courtesies meant was evident when the mother of two local Klansmen and grandmother of another passed away. Her six children and their spouses, "representing forty grandchildren and twenty great-grandchildren," joined to thank the Athens Klan and ask "God's richest blessings" for each of its members.[24]

▲ ▲ ▲

And yet Klansmen were not just Odd Fellows in robes and hoods. For all the ties that bound Klansmen to commonplace community networks and habits, the Klan was different. Leaders reminded members that their organization was "not a lodge," but "an army of Protestant Americans." As a *"mass movement"* to secure the alleged birthright of Anglo-Saxon Americans, it could achieve that goal only through "an aggressive application of the art of Klancraft." That required winning the confidence of the community by recruiting respected local men and making the Klan a *"civic asset."*[25] In short, breaking into church and fraternal networks was part of a larger strategy to accrue power. And that power would be used toward ends some people in these networks might balk at.

Signs that the second Klan would be more than just another

community organization were there from the beginning, not least in its name. The first call to re-establish the Klan came, not from William Joseph Simmons, but from Tom Watson. The foremost leader of Georgia's Populist movement in the 1890s, Watson had long since given up the struggle for interracial economic justice. Recently, he had turned his attention to Catholic and Jewish subversion. In August of 1915, he informed readers of his Georgia-based *Jeffersonian* magazine that "another Ku Klux Klan may have to be organized to restore Home Rule." Georgia's governor had just commuted the death sentence of Leo Frank, a Jewish factory supervisor convicted of the murder of Mary Phagan, a young white woman in his employ. The governor's mercy enraged those who believed Frank guilty, Watson among them. Four days after he issued his incitement, a body of men calling themselves the Knights of Mary Phagan kidnapped Frank from the state prison farm, took him to her home town, and hanged him from a tree.[26]

Three months later, Simmons resurrected the Knights of the Ku Klux Klan in Atlanta, where Frank's alleged crime and his trial had taken place. The Frank case has often been cited as a catalyst for the creation of the second Klan, whose founding members, according to popular myth, included some of the Knights of Mary Phagan.[27] In fact, no one has ever documented a direct connection between the two. The "truth" of the link lay less in personnel than in a common vigilante spirit. An appeal to that spirit would always be part of the Klan. Its promise of swift and secret vengeance, more than anything else, distinguished it from contemporary organizations with whom it shared ideas, rituals, and members.

Nothing in the early years helped more to make that promise come to life than D. W. Griffith's film extravaganza, *Birth of a Nation*, released in the same year as the second Klan's creation. In this racist epic of the Civil War, Reconstruction, and the restoration of white rule, Griffith harnessed all the emotive power of modern film-making technique to convince viewers that black men were beasts and white vigilantes were the saviors of American civilization. Given the right to vote and hold office, the film averred, African-American men dragged society into chaos; worse, they used such power to stalk white women. Griffith left no doubt about how this fate had been averted. In the final, climactic scene, the hooded and robed members of the Ku Klux Klan rode in to save his young white heroine from rape—by castrating and lynching her black would-be assailant. Their act ended sectional fratricide among white men and gave birth to a reunited America.[28]

For the Klan, the film proved a boon. When it came to Atlanta for a three-week showing, record-breaking white crowds packed the theaters to cheer on the white-robed crusaders. Recognizing an opportunity, Simmons ran newspaper advertisements for the revived order next to those for the film. Thereafter, the Klan routinely exploited showings of *Birth of a Nation* to enlist new members, for it sent the message the Klan wanted delivered. "No one who has seen the film," commented journalist Walter Lippmann in 1922, "will ever hear the name [Ku Klux Klan] again without seeing those white horsemen." Not surprisingly, the NAACP sought—in vain—to have the film removed from circulation.[29]

Black Americans in fact understood from the beginning that the second Klan was different, even from other racist organizations. In the view of many, it was an immediate threat. One month after the second Klan's founding ceremony, Georgia Republican leader Henry Lincoln Johnson begged the governor to make the order change its name, on the grounds that the Klan's re-establishment would encourage "mob outlawry." "My people (the colored people)," Johnson predicted, "will be the helpless, and often vicarious, victims." He was right. "Nobody knows," complained a Black Atlanta lodge officer to the NAACP in 1921, "the great destress" that this "great evil" had brought upon the black people of Georgia. The state's leading African-American newspaper, *The Atlanta Independent*, for its part, said of the second Klan that, like the first, its "aim and purpose is to terrorize helpless black men and women." "The epitome of race hatred and religious intolerance," it constituted "the most dangerous menace that ever threatened popular government."[30]

Outside Georgia, members of other groups the Klan pitted itself against made it clear that they, too, saw the order as outside the framework of ordinary politics. The editor of the *Catholic World*, for example, warned in 1923 that if the Klan were allowed to persist and the state failed to protect Catholic citizens from its provocations, they would employ "self-defense, even to the extent of bloodshed." In that year, in fact, Catholics were the leading force in organizing a militant anti-Klan group called the "Red Knights," or "Knights of the Flaming Circle." The group welcomed anyone opposed to the Klan and not a Protestant, and was said to be recruiting well in Pennsylvania and West Virginia, and later in the industrial cities of Ohio as well. In several parts of the country, often under the auspices of the Red Knights, Catholics responded to the Klan's provocations with mass, armed counterattacks so determined that the National Guard was called out on at least one occasion.[31]

Some radical farmers and unionists shared the Red Knights' assessment of the novel danger posed by the Klan. Iowa farmers charged a Klan meeting with pitchforks; their counterparts in the Arkansas mountains turned shotguns on Klan intruders, killing one and wounding several. The United Mine Workers of America (UMWA), a renegade in the contemporary craft-dominated labor movement because of its commitment to interracial industrial unionism, tried to close ranks against the Klan. In 1921, while leaders of the American Federation of Labor equivocated, refusing a black delegation to its national convention the right to introduce a resolution calling for the suppression of the Klan, the UMWA barred from its ranks miners who joined the Klan. Although unable to rid the union of Klan influence, many militants tried. Oklahoma UMWA members spied on a Klan meeting and expelled the miners who attended, while their Pennsylvania counterparts put union members suspected of Klan membership on trial.[32]

The imminent dangers perceived by the Klan's targets offer a clue to why the order did not take off until 1921. For one thing, an organized national vigilante movement would probably have seemed superfluous before that. The suppression of challenges to prevailing relations of power was official federal policy in the war years. When *Wiley Doolittle*, former mayor of Athens and future Klansman, proclaimed at a flag-kissing ceremony in June of 1918 that "hereafter disloyalists might expect to be branded on the forehead and on either cheek, and the rope would be the end of traitors, in legal process of law or otherwise," other civic leaders were not scandalized. On the contrary, Atlanta's leading newspaper quoted him approvingly. Similarly, an Alabama-based Justice Department official writing in the same year made clear his assumption that vigilante activity was legitimate. Having found that, so far, the mobs had confined themselves to attacks on black people, workers, and wartime dissenters, he described the night-riders as merely "*potential* sources of lawlessness and disorder."[33]

The Espionage and Sedition Acts of 1917 and 1918 backed community pressures for "one hundred percent Americanism" with censorship and possible prison terms for dissenters. Insulating the war effort from criticism was only part of their purpose; suppressing domestic labor struggle and left-wing radicalism was as important. These strictures had barely been lifted before the postwar Red Scare, the most extensive peacetime violation of civil liberties in United States history, began. Its climax was the Palmer Raids of January 1920, a nationwide dragnet against radicals named for the Attorney General under whose direction they proceeded.[34] As national leaders gradually began to breathe more freely after 1920 and favor a return to normal methods of rule, so did many local élites. Then the kind of methods they had recently endorsed came to seem excessive, even illegitimate.

In Athens, some became nervous when the Klan sought to keep the wartime spirit alive for its own purposes. When the Klan returned to Athens in 1921 and promoted itself as a force for "law and order," the implicit promise was that it would enforce its concept of order in a manner similar to its namesake's. Recoiling, some one hundred "good citizens" signed a petition against it. Their numbers included many prominent civic and business leaders. The signers declared that the "announced purposes of the Ku Klux Klan . . . have the approval of all good citizens." They took issue with the order on only one point: its usurpation of the powers of lawfully "constituted authorities." The Klan may have performed a necessary service during Reconstruction, they argued, but "no such necessity exists now." Hence its re-establishment was "ill advised," and its "self-constituted guardians of the peace, working at night and in disguise" were mistaken in their zeal.[35] However timid their criticism of the Klan, the signers did understand its penchant for violence. To silence them, Klan leaders rounded up more imposing voices.

Just as the Klan used burning crosses to sanctify its message, so it used ministers to sanctify its methods. So valued was their function that the Klan chased after them systematically and allowed them to belong without paying dues. The harvest such policies reaped distressed Klan opponents beyond measure. The Atlanta-based Committee for Interracial Cooperation (CIC) reported in 1923 that Southern ministers "who approve the Ku Klux Klan methods" far outnumbered those who objected to them; "this type of minister," the CIC concluded, "has made the Ku Klux Klan possible."[36]

Indeed, as the community leaders thought most capable of interpreting God's will, Klan clergymen seemed to give his blessing to the order's activities. By 1926, CIC director Will Alexander, a former minister himself, confessed privately that "the large number of [Methodist and Baptist] ministers who are in the Ku Klux Klan . . . renders me hopeless as far as the masses of ministers are concerned." Such despair was more than justified in Athens, where only one minister ever publicly condemned the Klan—and then quite late in the day and not by name. In 1927, the Reverend J. D. Mell, president of the Georgia Baptist Convention, condemned the vigilante activities of "masked mobs." Even then, he was careful to qualify his remarks by stating that "good men" participated.[37]

For their part, the men charged with responsibility for training the minds of Georgia's youth maintained a stance of benign neutrality toward the Klan. Asked by the *New York World* to take a stand in a campaign to push the 1924 Republican and Democratic national conventions to adopt anti-Klan planks, the chancellor of the University of Georgia had only this to say: "Organizations with secret membership have a tendency to produce reliance on group action which is not in accord with the American spirit of personal responsibility and independence." The university's community governing board, a Prudential Committee composed of leading businessmen, became more agitated over liberal dissent. It acted to censor student journalists who criticized some of the state's cities, and one of its members, newspaper editor Hugh Rowe, called for the expulsion of a campus YMCA staff member who organized interracial student discussion groups on the grounds that he was probably a paid Russian agent. Prudential Committee members insisted, calling upon university rules for support, "that there be no political criticism at the University of our government."[38]

Thus intimidated, even the few faculty members who despised the Klan hesitated to speak out. "The trouble with us," reflected English professor John Wade, "is that we have as little courage as we have voice. But with things as they are now in Georgia, *more* courage would likely mean martyrdom, not of the effective variety." In such a climate, it is perhaps unsurprising that more than a third of the freshmen trying out for the university's debating society in 1923 defended the proposition that "the activities of the Ku Klux Klan as now practiced are of the best interests to the United States."[39]

Just as the endorsement of clergymen and the equivocation of educators could shield the Klan from criticism, so could the backing of politicians protect it from hostile legislation or prosecution. The Klan, of course, needed the help of public officials to realize such elements of its program as immigration restriction, Prohibition enforcement, opposition to American participation in the League of Nations and the World Court, tax relief, prohibition of interracial marriage, exclusion of Catholic teachers from the public schools, the closing of parochial schools, and prohibition of property ownership by non-citizens.[40]

Yet the order went about electoral politics with a zeal beyond that of other contemporary organizations. It pushed members, not only to go to the polls themselves, but also to turn out family members, friends, and neighbors. Such prodding paid off. As Exalted Cyclops *J. P. Mangum* boasted to the Grand Dragon in November of 1925: "We have just had an election in the city and [won] out[;] they dont know how we did it but we did our stuff and sed nothing." "Our stuff" most likely referred to such Klan tactics as the mass distribution of model ballots; the "decade system" in which, after the order had decided on a candidate, each member would then go out and talk up him up with ten non-members; or the "poison squads" of Klanswomen who used ordinary gossip networks to spread malicious rumors about those opposed by the Klan.[41] The Klan also organized public shows of strength to sway wavering non-members and frighten opponents at key times. More importantly, the Realm office warned recalcitrant public officials to beware "the invisible eye." The threats were not idle. The Klan kept extensive files on public figures to blackmail them if need arose. Where evidence of impropriety was lacking, the order sometimes tried to manufacture it. Imperial officers thus deployed prostitutes and bootleggers in vain hopes of entrapping racial liberals like Will Alexander, and the ministers C. B. Wilmer, Plato Durham, and M. Ashby Jones.[42]

Through such varied means, the Klan steamrollered opposition and gained influence in Georgia. Broker of the votes of an estimated 100,000 of the state's 300,000 Democrats by 1923, the Klan held "the balance of power" in state politics, as even a reporter who sought to play down its domination had to admit. The order enrolled such well-placed officials as Governor Clifford Walker, Chief Justice of the State Supreme Court Richard B. Russell, Sr., State Attorney General George M. Napier, Atlanta Mayor Walter A. Sims, Solicitor General (district attorney) of Fulton County John M. Boykin, and Fulton Superior Court Judge Gus H. Howard, in addition to many less strategically placed men. Some evidence suggests that the roster also included Georgia's United States senators Tom Watson, Walter George, and William J. Harris; United States congressman and past president of the Anti-Saloon League W. D. Upshaw; and President of the Georgia State Senate Herbert Clay.[43]

Similar patterns prevailed elsewhere. In 1923, for example, at least seventy-five congressional representatives were said to owe their seats to the Klan; at the annual conference of state governors the year before, only one was willing to discuss, let alone condemn, the Klan. The reason was not hard to find. The Klan held sway in the political life of many states; it dominated some outright, such as Indiana and Colorado; and it swept anti-Klan governors from office in a number of others, most spectacularly in Oregon and Kansas.[44]

The ability to dispose of opponents so handily where it had the requisite numbers gives an indication of why Klan leaders put such a high premium on electoral politics. "It is of vital importance that our friends be placed in office," Georgia's Grand Dragon explained; "the life of our organization" might hinge upon the outcome of elections. The election of Klan enemies to the legislature, such as Athens resident Andrew Erwin, could not be tolerated and should be reversed. The Grand Dragon's own command of the votes of at least half the state's delegates to the 1924 Democratic Party national convention helped ensure that the convention would vote down a platform plank against the Klan; of Georgia's fifty-six delegates, only *one*—that same Athens resident—supported the plank. Indeed, wielding its power in the nonpartisan style of the Anti-Saloon League, the Klan managed to prevent either major party and all the nation's presidents in the decade from condemning it publicly.[45]

But it was back at home that the insulation mattered most. With it, the Klan could fend off measures that might have made its night-riding operations more difficult, such as a 1922 bill—aimed at the Klan—to prevent the wearing of masks on Georgia's public highways. The order went on to deliver one of the biggest electoral defeats in state history to the governor who proposed it, Thomas W. Hardwick. His successor, Clifford Walker, a Klansman himself, learned the lesson. As governor, he consulted Klan leaders before introducing new initiatives to the state assembly.[46] On the local level, prosecution of Klan violence was hardly likely when municipal governments, police departments, and courts were rife with Klan members and sympathizers. "Everybody in the courthouse belonged to the Klan" in Atlanta, recalled a local city attorney; "virtually every judge, the prosecuting officers . . . all the police and the mayor and the councilmen." If he exaggerated, it was not by much. With the cards thus stacked in its favor, the Klan could act with impunity.[47]

Indeed, newspaper editors in the South, like politicians, tended to quaver in the face of Klan's power. Clearly, they did not view the order as an innocent analogue of other fraternal lodges. While local papers boosted these, most maintained an eerie silence regarding the Klan's activities. Few offered outright support, yet neither would they investigate or expose it. With the notable exceptions of the *Columbus Enquirer-Sun* and eventually the *Macon Telegraph*, no Georgia newspapers condemned the Klan until the second half of the decade, when its power had begun to wane. The Athens press was no exception. On the contrary, its editor denounced Julian Harris, the anti-Klan editor of the *Enquirer-Sun*, for giving comfort to "the South-haters" with his coverage of the movement. "Nothing unpleasant must ever be printed" seemed to be the operating principle of most newspapers, observed one Athens educator and resident; another later recalled, "they put only nice things in the *Banner-Herald*."[48]

Night-riding and inciting hatred were not nice; neither were they "newsworthy" if their targets were blacks or poor whites. The topic was just too ticklish to touch. Coverage of the Klan's activities, after all, might deter outside investors, agitate blacks, and stimulate discord among whites—to say nothing of losing subscriptions. Yet, uneasiness about the Klan's methods remained. The Athens press thus gave editorial support to two area judges who came out against the Klan in 1926 for its "lawlessness," its efforts "to intimidate men and . . . dominate who shall run for office," and its habit of "trying men in secret." Throughout the South, in fact, the most commonly stated rationale for élite opposition to the Klan was, in the words of one Texas judge, that society could not abide "two systems of government for punishing crime," one "working at night with a bucket of tar and a sack of feathers."[49]

▲ ▲ ▲

The Klan's internal structure and methods of operation were well suited to such activities. All power ultimately resided in the hands of the Imperial Wizard, whose reign, in the words of one observer, was a "virtual dictatorship." Indeed, short of overthrow, no real checks on his power existed. All Klan officials at the national and state levels received their positions by appointment from above, not election from below. Local klaverns, once chartered, could decide whom to admit as members, participate in the selection of local officials, and plan their own activities. But their sovereignty was more apparent than real. Chapter decisions were always subject to the veto of the state Grand Dragon, who also had the power to withdraw charters and thus self-government.[50] Klan officials no doubt saw in this organizational structure a safeguard for the lucre of office.

Yet the command structure also matched the organization's ultimate mission: combat. Klansmen regularly described their organization as "an army" and "a fighting machine" and their projects as "battles." "Ours is a military system, requiring performance of duty and honor above all else," explained a manual for building local chapters. Describing the Exalted Cyclops as "the commander-in-chief," it went on to elaborate the duties of the rest of the "military machinery" in like vocabulary.[51] Democratic debate and decision-making might undermine the Klan's larger program for the imposition of social order. One Klan leader admitted as much in an apologia for its "complete dictatorship." "The only way in which the Klan can be protected," he said, from "demagogues who might be able to sway a portion of the membership . . . is to have a government strong enough to suppress all such attacks." Such discussion and voting as occurred within the national Klan, in fact, bore more resemblance to plebiscites than to democratic processes. In 1926, for example, the order boasted that Imperial Wizard Evans had been re-elected and changes made in the Constitution "without a single dissenting vote."[52] What national leaders proposed, the ranks disposed.

The Klan was not alone among voluntary organizations in its lack of democratic process; yet the commitment to secrecy that accompanied these procedures did appear singular. And it made sense for a movement that had so much to conceal. The order's internal publications often stressed confidentiality, particularly about "Secret Work." One advised chapters with members who "cannot keep their mouths shut" to expel them. Keeping one's mouth shut included telling lies under oath in court if necessary, something Klansmen routinely did on the rare occasions when members of their movement faced indictment.[53] Secrecy also served to intimidate potential opponents and stifle public discussion, since no non-members could be sure of whether or not their white acquaintances belonged to the Klan.[54]

Public silence was but one element of the absolute fidelity the order demanded. Prospective members had to pledge their loyalty to the Klan and promise to willingly endure any penalty devised by their brethren if they proved "untrue" to this vow, among them "disgrace, dishonor and death." Members who might consider violating this pledge were forewarned of the danger. "The Klan is cruel to those who betray it," an internal document reminded the faint-hearted; "pause to consider the status of those who have betrayed it." The Klan punished, sometimes brutally, men who betrayed their oaths. W. S. Coburn, an aide to then-deposed Imperial Wizard Simmons, lost his life for indiscretion in 1922; Philip Fox, a member of Imperial Wizard Evans' staff, confessed to the assassination.[55]

The stress the Klan placed on internal hierarchy and obedience was part of a more general militarism. Militarism, as historian Alfred Vagts has shown, is associated with war and armies, yet it goes beyond military purposes and may even interfere with them. Historically, the term has connoted "a domination of the military man over the civilian." Enthralled with martial rank, prestige, and custom, militarists exalt "caste and cult, authority and belief." The principal exponents of modern militarism have not been military men, but civilians disenchanted with the banalities of bourgeois politics and the quest for material gain, yet also inimical to liberalism, labor, and the Left. For inspiration, they have looked to the ceremony, discipline, and mystical nationalism of militarism, through which they expressed their desires for a society organized along corresponding lines.[56]

Klansmen demonstrated their fealty to martial values in numerous ways. Most visibly, they chose to present themselves to the public in elaborate ceremonies, hidden beneath masks and uniforms that wiped out their individual identities. These outfits, in turn, varied according to members' places in the internal hierarchy. Rank-and-file Klansmen donned simple white robes and hoods; Exalted Cyclopses and Grand Dragons, more ornate ones; and the Imperial Wizard the most ostentatious and colorful regalia of all. The salutations required for the Imperial Wizard likewise showed infatuation with castelike warrior traditions: "His Majesty," "His Excellency," even "Emperor of the Invisible Empire." Expressing its aspirations to the powers nationhood conferred, the Klan referred to the society outside itself as "the alien world." To be admitted to the Klan, an "alien" had to undergo an initiation ceremony described as "naturalization." Not surprisingly, the movement became a magnet for men imbued with martial values. Around the country, law enforcement personnel and military men joined the Klan in large numbers.[57]

Klansmen's discomfort with civilian values and their embrace of a lock-step hierarchy within their own movement suggested a profound uneasiness with the direction they saw their society heading. What they meant when they pledged to defend "pure Americanism" will become clearer through an examination of the time and place that produced the movement. For the bonds solidified by fraternalism, by common religious feeling, and by family involvement and mutual aid did not simply enhance members' loyalty to one another. These bonds also steeled them for battle against "alien forces."

2

▲ ▲ ▲

"Where Money Rules and Morals Rot": The Vise of Modernity

Soon after he reached the Paris Peace Conference, Woodrow Wilson received a communication from his secretary in Washington. "If America fails now," Joseph Tumulty warned, "socialism rules the world."[1] That sense of a world poised on the brink of a precipice reached well beyond the halls of Versailles. Back home, the events of 1919 defied assumptions that millions of small-holding white men had constructed their lives on. As black Americans stood their ground and fought back, undaunted, against their white assailants in the nationwide race riots of that year, white supremacy appeared vulnerable. As the woman suffrage movement won its seventy-year-long battle for the right to vote, male prerogative no longer seemed assured. As one in every five American workers walked off their jobs to go on strike, the rights of property seemed less clear. And, finally, as Wilson set out to build a global League of Nations, the levers of power moved farther from the hands of non-élite white men than ever before. A middle-class man inclined to fear, in fact, could see in the events of 1919 the nightmare of the republic's founders come true: growing economic inequality had bred concentrated power above and below a great mass with little stake in society.

Some of the men in between, like Athens barber *A. J. Boyd*, would soon enlist in the Klan in order to reimpose their notions of

order on this topsy-turvy world. *Boyd* was no novice in encounters
with change. Born soon after the end of Reconstruction, he had
lived though a profound transformation in local, regional, and national life. But that is precisely what made the immediate postwar
years so ominous: the concatenation of challenges made new sense
of alterations long under way in the economy and society. The
message was both clear and shrill. The nineteenth-century world,
a world in which a white man could still reasonably hope to become his own boss if he harnessed himself to the task, was fast
passing. Without the self-sovereignty that small-holding promised,
maintaining his authority over African Americans and immigrants—let alone over his own wife and children—would be more
difficult than ever before. So, too, would be making himself heard
by the most powerful men in society. For as their enterprises and
interests grew to national and international proportions, the input
of men like *Boyd* came to seem irksome to them.[2]

Athens had a different look and feel to it now than when *Boyd*
grew up back in the 1880s. Founded astride the Oconee River in
1801 to host the state university, the town in time spilled out over
the red clay hills undulating away from the river's east and the
west banks. By the early twentieth century, several rail lines and
major highways linked Athens to markets elsewhere in the country
and helped it become an important hub of cotton trade and manufacture. By 1920, its varied enterprises had attracted nearly seventeen thousand residents. Automobiles and streetcars now vied with
horses and buggies for its city streets.[3]

As Athens grew after 1880, its internal divisions also became
plainer. The wealthiest residents congregated along Milledge and
Prince Avenues in the white-columned antebellum mansions that
earned Athens the name "Classic City." To their immediate west
was one of many black neighborhoods dispersed through the city.
Almost twenty percent of the city's African Americans lived in
its congested, ramshackle, usually unpainted houses, considered by
white landlords to be "one of the best investments for small
amounts of money." Denied the municipal services that white residents enjoyed, their yards held, alongside their chicken coops, outdoor privies and wells and piles of trash that the city failed to
collect.[4]

East of this community was the city center. Anchored on the
south by the university campus, it hummed with customers and
the residents who owned and staffed its shops and offices. Here,
too, could be found the nascent commercial leisure industry of

movie houses, poolrooms and soft-drink parlors that superseded older forms of recreation like the cock-fighting *Boyd* had once practiced. Among the main patrons of these pool halls and picture shows were white youth from the largely working-class neighborhoods of East Athens and West Athens. Those who lived in these communities and labored in the area textile mills knew they were looked down upon by the better-off inhabitants of the city as "more or less poor white trash."[5] In short, Athens's placid appearance masked antagonisms no less potent for being largely mute.

▲ ▲ ▲

Perhaps, if the turbulence of the late 1910s had remained confined to distant places like Harlem, Versailles, and Petrograd, it might not have so upset men like *Boyd*. But the tumult could not be contained. Dramatic national and international events found echoes in communities like Athens. Klansmen were hardly alone in seeing such connections. "The world is seething in social unrest and disquietude," lamented a group of Athens ministers in 1921; "anarchy overwhelms whole sections." The local press, for its part, asserted connections between the Bolshevik revolution in Russia, the postwar upsurge of rebellion in the United States, and lenient parenting and crime in Athens. The "world," said editor Hugh Rowe, was "reaping just about what it sowed."[6]

Workers' newfound strength attracted some of the most anxious attention, at least from those in a position to employ. The insatiable demand for labor during the war enabled Southern manufacturing workers to more than double their average yearly wages between 1914 and 1919, in the process narrowing the North-South wage gap. By 1920, wages in Georgia hit an all-time high, inflation notwithstanding. Athens' black workers made particularly spectacular gains; wages in the main occupations open to them generally doubled and in some cases tripled. Employers felt keenly the slippage of their power in this labor-scarce market. Where only a few years before they had sat firmly in the saddle, now they had to tolerate insubordination for fear of losing the workers they did have if they tried to discipline them as they would have in former days. Unable to control the situation themselves, planters and businessmen looked to the state legislature to pass compulsory work laws and punitive measures against vagrants and labor emigration agents.[7]

Among white workers, the most obvious sign of newfound confidence was a spate of strikes and union-organizing efforts in the southern Piedmont from 1918 to 1921. In Georgia, textile workers built unions in Columbus, Macon, Griffin, and Atlanta. Other, traditionally organized, groups also caught the strike fever: machinists, railway workers, streetcar workers, and building trades workers. In four cities, the strikers proved so determined that the governor had to deploy the state militia to overpower them. Meanwhile, the number of union members and locals in Georgia mounted continuously from 1915 to 1920.[8]

In Athens, skilled workers established or reactivated several locals in the 'teens: painters, barbers, typographers, and carpenters among them. These locals then formed a city central federation in 1914. Although no evidence survives of union organizing among local textile workers, operatives at the Southern Mills conducted an impromptu work stoppage when management tried to introduce piecework. "They would starve first," they said. Giving larger import to these local events, the Athens press bombarded readers with sensational, front-page accounts of battles between labor and capital elsewhere in the South and the nation.[9]

The insurgence of white labor could scarcely be separated from the disgruntlement spreading among African Americans, even before the war. Young people especially resented the curbs on their freedom. Many white planters in the Clarke County area thus complained in 1911 that "the younger generation is rapidly becoming unmanageable," attributing their recalcitrance to "a deep-seated dislike of control and discontent with farming life and conditions." An expert in agricultural economics concurred, pointing to "the growing aversion on the part of the negro to supervision. He desires his movements to be absolutely unrestricted," even if it meant less income. That desire was palpable in black families' growing rejection of wage labor in favor of rental arrangements that allowed more autonomy. In town, black women wrested more freedom for themselves and more time for their own families by choosing laundry work over domestic service where possible.[10] Such efforts, urban and rural, led many white employers to suspect that black workers were secretly organized. One planter thus cautioned Governor Joseph Mackey Brown in 1913 that they "must be crowded back by some means."[11]

The war brought these antagonisms into the open. Finding Jim Crow in Georgia unions, black workers expressed their aspirations in other ways. Some moved to escape the South's caste system, undaunted by the knowledge that they would face new ordeals in the North. "Negroes are leaving here by the hundreds," marvelled an Augusta resident in 1917. "They know where they are going; they know what they are up against."[12] Those who stayed behind were hardly the docile folk of New South propaganda. Clarke County white employers, both urban and rural, fumed over the boldness of their black employees in 1919. One hotel manager described "such gross indifference . . . [as he] had never encountered before." A fertilizer plant manager maintained that his black workers "were absolutely uncontrollable." Before long, he and other local white employers believed, such black workers would "begin to organize." White nerves became so sensitive that they registered changes in blacks' spending habits as political statements. High wartime wages enabled even some poor people to buy such things as silk shirts or automobiles. This "unusually extravagant buying" irked establishment whites. Some griped that blacks spent twice as much on cars as they did on war bonds.[13]

But the bravest statement of black aspirations came in politics, as some pushed to open a second front in President Wilson's "war for democracy": below the Mason-Dixon line. In March of 1918, for example, over a hundred African-American Atlantans signed and put into mass circulation a letter vigorously demanding the rights they declared due them as "sovereign American citizens." They denounced lynching as "worse than Prussianism" and condemned the "discrimination," "humiliation," and "segregation" their people were daily subjected to as "a violation of the fundamental rights of citizens of the United States." Most dramatically, they attributed these "brutalities and indignities" to the way Southern whites had "filch[ed]" the votes of black men in an "effort to re-enslave us." Finally, the signers vowed to continue to "exert our righteous efforts until not only every eligible black man but every eligible black woman shall be wielding the ballot proudly in defense of our liberties and our homes."[14]

In their focus on the Great Migration and the Harlem Renaissance, in fact, historians of African-American life in these years have slighted the surge of resistance to white supremacy by Southern blacks.[15] Contemporaries did not. "Did you ever know a race to awake as our race has awakened in the last year or so?" exclaimed a member of the Augusta NAACP. "Augusta is almost a different town. The old spirit of humble satisfaction, of let-well-enough-alone is fast dying out." In the three years after 1916, the number of NAACP branches in the South jumped from six to 155. Together, they amassed a dues-paying membership of over 42,000. For the first time, Southerners now dominated the organization's rank and file. Georgia blacks organized scores of these chapters, not only in cities like Atlanta and Augusta, but also in tiny towns and hamlets around the state.[16]

In 1917, the same year the NAACP came to Georgia, thirty-one Athens residents chartered a branch in their community. The following year, they brought NAACP leader Walter White to town to speak. G. C. Callaway, a member of the local executive committee, gave voice to his co-workers' aspirations. He informed the national office that Athens had a reputation as "the best town" in the state for blacks. But that relative comfort was no longer enough. Callaway looked to the NAACP "to force in to the nation liberty fre[e]dom [and] equality." Everyone, he hoped, would "orginice and join the fight."[17]

Branches in cities took up the cudgels for their rural brethren. The Atlanta NAACP, whose membership reached 1,700 in 1919, prosecuted cases of debt peonage and defended two blacks who had killed whites in self-defense. Harking back to the abolition crusade, Atlanta members constructed "a system of underground railroads for . . . persons fleeing from the cruelty and oppression of the rural communities and small towns." The chapter also undertook a massive, successful voter-registration drive in 1919. In one month, they bought over a thousand new black voters to the polls in Atlanta— more than double the number who had taken part in some past elections. This campaign caused panic among leading whites, who hauled out their white employees to offset black votes. The "solidity" of the African-American vote in the election, as much as the NAACP's forthright insistence that it expressed "definite and long standing grievances," was perhaps behind the legislation proposed in the state assembly the following year to prohibit blacks' voting or holding office at all.[18] The proportion of blacks involved in such outspoken protest was tiny; the challenge their undertakings posed to the status quo was great.

Even more fearsome to racist whites than the NAACP were black veterans of the Great War. The mere vision of an African-American man in a uniform, a symbol commanding respect, could arouse white fire-eaters to violence. But the threat posed by black soldiers was not merely symbolic. Once having experienced an alternative to Southern life, most would never be the same—nor would the communities they returned to. Seventy percent of the Clarke County planters with black employees in the Army reported in one study that the veterans left the farm soon after returning from the service. A majority said that the remaining workers then became dissatisfied as well.[19]

Even federal officials became disturbed about black veterans moving north. "They are inclined to put what they understand to be their rights as American citizens above every other consideration," observed one official in the War Department. Others were not so circumspect. A 1918 military intelligence report described as "a potential danger," not white vigilantes, but the black soldier "strutting around in his uniform," particularly if he was "inclined to impudence or arrogance." If these men tried to act on "the new ideas and social aspirations" they had acquired in France, the author declared (in allusion to rumors of romantic liaisons with white women), "an era of bloodshed will follow as compared with which the history of reconstruction will be a mild reading, indeed." So alarmed was the Division of Military Intelligence over "Negro subversion"—defined as black veterans' fighting "any white effort, especially in the South, to reestablish white ascendancy"—that it undertook a secret investigation to find out whether they had a collective organization to promote their goals.[20]

Here, it seemed, was brewing the black rebellion whose specter haunted the white establishment. Not only were black soldiers trained in combat, but it appeared their civilian peers might no longer turn the other cheek, either. This, at any rate, was the message of the race riots of 1919. In them, African Americans fought back en masse, for the first time, against white assailants. Certainly local racists noticed that, according to Athens merchants, black purchases of firearms skyrocketed in these years, restrained only by limited supply. The newly formed Federal Bureau of Investigation became so worried that in 1920 it initiated investigations throughout the South into the extent of gun purchasing by blacks. The reports that came back often indicated either a noticeable increase or that they were *already* almost universally armed with good weapons—hardly a comforting prospect to panic-ridden white supremacists.[21]

The unrest among blacks was serious enough to prompt a small group of liberal whites to action. Several met in January of 1919 in Atlanta to form what would come to be known as the Commission on Interracial Cooperation (CIC). World War I had "changed the whole status of race relationships," an internal account of the CIC's origins later explained; blacks became determined to obtain "things hitherto not hoped for." The CIC identi-

fied three different groups among Southern blacks: the "openly
rebellious, defiant and contemptuous" leaders to whom "one talks
. . . in vain as to the need for patience"; "below . . . the great
mass of uneducated Negroes" who gravitated more and more toward the radicals' positions; and finally, those "thoughtful, educated Negro leaders" who counseled a need for "patience" yet had
a "tendency to despair" at its failure to produce results. White CIC
founders sought to isolate radical leaders and raise the credibility
of more conservative ones with the mass of skeptical poor blacks
by alleviating some of the most onerous aspects of white supremacy. Limited as the goals of the CIC were, its very existence
marked a sea change in some quarters of Southern white society.
In a culture in which, as CIC director Will Alexander rued, many
whites still found "killing a Negro less reprehensible than eating
with him," CIC committees in hundreds of local communities involved leading blacks and whites in ongoing discussions on how to
stop mob violence and improve race relations.[22]

Indicative of the indivisibility of the challenges to prevailing
relations of power in these years, adult women and college students proved among the most avid of the CIC's white supporters.
Thus, some intrepid young men and women from the YMCA at the
University of Georgia met with students from the Knox Institute, a
local black private school, in an interracial discussion group—until
their leader was driven from the university.[23] From the first meeting of leaders of women's organizations to discuss the war's impact
on race relations in 1920, according to Alexander, "the most effective force in changing southern racial patterns has been the white
women." Their efficacy came from the way their very participation
in interracial work challenged the myth of the "Southern lady" so
central to policing the lines between black and white in the
South.[24]

Involvement in interracial work after the war betokened a
metamorphosis among Southern white women. Among its other
markers was a growth in feminist agitation by middle- and upper-
class women in the 1910s. The Athens Women's Club joined with
society women's groups across the state to demand the admission
of women to the University of Georgia in Athens. The Clarke
County Equal Suffrage Association, established in 1912, held public rallies and debates to promote woman's rights.[25] Some men,
such as local newspaper editor Hugh Rowe, feared that the suffrage
movement was breeding "sex antagonism and prejudice." Anti-
suffragists went further: votes for women would overturn the so-

cial order. Still, the "world-wide pull of the feminist movement" seemed irresistible; "neither race, nationality, nor the hostility of man," it seemed to Rowe, "[could] stop it."[26]

Feminism's spread both reflected and fueled changes in the everyday lives of young, middle-class white women. Despite the opposition of a majority of the male faculty and the "skepticism" of male civic organizations, the University of Georgia finally admitted women as full-time, regular students in September of 1918. Many male students resisted the change. "Boys did everything to embarrass the co-eds," one contemporary recalled, from boycotting them socially, to swearing in front of them, to denouncing "the evils of women" in public meetings. Yet their tantrums failed. Female students stood their ground. One even publicly defended the recent gains of her sex. She insisted that college training was women's "just right rather than a high privilege bestowed upon them." She pointed out that "most twentieth century girls do choose to enter a profession or industry." They did so in part because they were "unwilling to become economic burdens or social parasites" now that so much work had moved outside the home. "The modern girl," she warned, "will not submit" to the desires of men "who insist on girls being dolls to be flattered and entertained. . . . She demands recognition and opportunities for her capabilities."[27] That the writer chose to remain anonymous indicates the opposition such ideas still confronted; that she wrote it at all indicates the willingness of some women to challenge a gender ideology inherited from an older social order.

In this context of redefinition, simple gestures came to denote larger agendas for both sides. Perhaps because new standards of female dress were the most visible marker of change, they served as a potent symbol in the renegotiation of female roles. Casting off the long skirts and high-necked blouses of their mothers' generation, "business girls" adopted styles at once more relaxed and more flamboyant. Impatient with polite conventions about female modesty, girls from the posh Lucy Cobb Institute in Athens followed their lead. They endured the ritual inspection of their dress length each Saturday morning before they could go to town—only to hike their skirts above their knees once they were out of the matron's sight.[28] The significance of young women's determination to shed Victorian attire, denoting as it did also aspirations toward sexual self-determination, was not lost on defenders of female domesticity. Reports circulated in 1921 of a bill pending in the state legislature to fine or imprison women whose skirts ended more than

three inches above their ankles. Even the press joined in the out-
cry. The failure of all such efforts to dissuade young women from
baring their knees and bobbing their hair merely confirmed their
unruliness in the eyes of those who believed that the "new
woman" jeopardized the social order.[29]

Like rising skirts, the rise of smoking among young white
women became a controversial issue. Whereas the new dress styles
indicated a decline in female sexual modesty, female smoking
threatened to dissolve gender distinctions altogether. "French
Women," a banner headline in the local press thus proclaimed,
"Get Mustaches from Smoking." Although only about thirty female students at the University of Georgia smoked in these years,
they were also the first generation of co-eds. As such, they appeared to many to be the wave of the future. Their insistence on
their right to smoke and their defiance of university rules forbidding it no doubt made that future appear ominous to those interested in preserving clearly demarcated gender roles and unques-
tioned submission to authority.[30]

The very newness of such behavior, adopted as it was in an
already turbulent social context, made it seem seditious. Many
contemporaries could not believe that a woman who smoked could
be a reliable mother, or that a girl who wore short skirts would
ever heed a husband's wishes. Only experience could calm their
fears. And, in fact, it would be a decade before the fearful realized
that these changes did not have the apocalyptic potential they had
imagined. In the 'twenties, "girls was shipped off for just any offense," recalled a maid at the State Teachers' College in Athens in
the 1930s. "Now they does pretty much as they pleases. [The
school] even provides smokin' rooms for 'em."[31] Until then, smoking, like so much other new female behavior, seemed to pose a
grave threat to the social order.

Continuities notwithstanding, young women did expect more
from men than their mothers had. When disappointed, they proved
more willing to buck the Victorian middle-class convention of
marriage as a permanent union devoted to child-rearing. Nationwide, the divorce rate increased by two thousand percent between
the Civil War and the Great Depression, when one in six marriages
ended in divorce. While the total number of divorces per capita in
Georgia was about half the number nationally in 1916, the *rate* of
increase after the turn of the century was more rapid than in the
United States as a whole—notwithstanding the fact that state's
courts granted alimony to fewer than one in ten women. In Clarke

County, the number of divorces leapt from fifteen in 1916 to forty in 1922.[32] Faced with this surge, the local press recoiled from its earlier liberalism. Having defended "the divorce blessing" in 1914 as many women's only means of "escape" from "oppression," after the war the Athens *Banner-Herald* deplored the "divorce evil," called Athens "a little Reno," and referred to its court as "the Divorce mill."[33]

Its turnaround issued from an astute intuition. Such things as the rise of divorce, feminism, black radicalism, white racial liberalism, and the postwar strike wave were not isolated, random occurrences. These instances of insubordination to old masters were the birth pangs of a new kind of social order, one whose relations of power and culture differed from those of the nineteenth-century world men like *Boyd* and Rowe had been born into. In their eyes, it appeared to eviscerate discipline, stability, and predictability—in short, to undercut the kind of hierarchy from which men like themselves had derived security.[34]

▲ ▲ ▲

As disturbing as the many challenges to authority was the awareness that they had not emerged from thin air. Rather, they realized potentials created by long-term changes that together weakened the foundations of the nineteenth-century world. By altering the ground families constructed their lives upon, for example, economic development after 1880 began to open up to question previous relations between men and women. On one side, employment apart from the family, more control over their fertility, and easier access to divorce offered women the prospect of greater autonomy. On the other side, economic insecurity, geographical mobility, and the market's takeover of much domestic labor made men more willing to desert wives and children. Taken together, such developments made families less permanent, more contested, and in many ways, different institutions.

Of course, the changes should not be overstated. In hindsight, some of the continuities seem as impressive. Cotton continued to dominate the economy, for example. Although agriculture now occupied a minority of the local labor force, it was still the leading occupation of Clarke County men, as of most Southern men, just as cotton manufacture was the area's leading industry. Most rural people, moreover, still strained to make ends meet. The average annual income for a farm family in a typical Georgia Piedmont county in 1924 was $591, from which expenses for farm operations

had to be paid. As late as 1930, fewer than one in ten Clarke
County farm households enjoyed electricity, telephones, or running water. Like their grandparents, most rural people still lived in
unpainted houses on dirt roads, drew their water by hand, and trav-
elled by wagon or buggy when need arose.[35]

Whether they lived in the countryside or in town, nearly every
group in Georgia still relied on kinship networks for economic survival. Among the wealthy, endogamous marriage practices concentrated economic resources so that they could be mobilized to greatest effect. Among the less well-off, children still counted as net
assets rather than debits. Small-farm owners and tenants alike
counted on the labor of all family members. The more the hands
to tend it, the better the crop—or the bigger the plot, for a tenant
household.[36] Like rural landlords, mill managers favored large families and penalized small in their employment and housing policies. "The size of your house," as an Athens mill villager recalled,
"depended on the size of your family." From the children who
earned wages, to the grandparents who looked after toddlers whose
parents worked in the mills, household maintenance required collective effort.[37] Kin also furnished aid no one else would in times of
unemployment, illness, or incapacity. Indeed, the support of one's
children made it possible to avert what a Savannah unionist described in 1922 as "the terror of Old Age": "POVERTY, and the
POOR HOUSE."[38]

Nearly all social relations, in fact, still bore the stamp of an
older patriarchal model that subordinated individual needs and
rights to the welfare of hierarchical collectivities. That model sanctioned private violence in the service of public order. Most important, in the South, was the force used to bolster white supremacy. Lynching illustrated the legitimacy of such force among
whites. Between 1882 and 1934, lynch mobs murdered more than
five thousand people, the vast majority of them Southern black
men. In Georgia, at least 549 people, 510 of them black, were
lynched over roughly the same period. Yet, between 1885 and 1922,
the state prosecuted only *one* person for lynching.[39]

Violence against African Americans short of killing was still a
routine feature of Southern life. The whipping of workers common
in slavery persisted into the twentieth century, particularly in rural
areas, where physical compulsion remained an accepted tool of labor control. The NAACP received numerous complaints in the
1920s about the killing of black farm hands by white employers or
overseers. One Albany, Georgia, minister, in reporting the murder

of a black man whose only offense was to have "cursed" his cheating boss, concluded forlornly, "often things like this hapen in this county."[40]

Labor relations in the countryside influenced state penal practice. A few complaints notwithstanding, Georgia prison camps still used whips to discipline their charges in the 'twenties. Some contemporaries—including the editor of the Athens *Banner-Herald* and the American Bar Association's Committee on Law Enforcement—even called for the re-establishment of the public whipping post. A speaker at an annual convention of Georgia sheriffs and peace officers advocated it for "petty criminals," in particular for "men who neglect their families."[41]

The paternalistic power wielded by planters was also copied by industrialists. Since many mill villages were unincorporated, they had no democratically constituted public authority. Mill officials owned them and ran them, and hired, paid, and controlled their police forces. Workers who violated the employers' codes of moral conduct, even during their off-hours, stood to lose their jobs and homes. "You didn't have no private life at all," one mill worker later complained. "You could come home and take your pants off and leave one leg on, and they'd tell you about it at the mill."[42] Of even more concern to mill owners and their political supporters than illicit sex, drinking, or gambling were strikes and union activity. Especially in the turn-of-the-century South, employer-sponsored vigilantism against labor organizers was extensive and unabashed. It included warnings out of town, beatings, floggings, tar-and-featherings, and occasionally, outright murder.[43]

Public social control, in turn, derived legitimacy from the power relations of family life. Most whites accepted male dominance as necessary to maintain family order. Wife-beating, although publicly frowned upon in the 'twenties, nonetheless appeared common. Courts treated it lightly when it came before them. Similarly, parents took for granted their right to whip their children. Most also continued to back corporal punishment in Georgia schools, especially in rural areas and small towns.[44]

Yet, for all the continuities, as productive property became more concentrated and the number of people engaged in wage labor grew, Southern society changed in fundamental ways. Where once the class structure of Southern white men had bulged in the middle, now it looked more like the pyramid that its Northern counterpart had become. At the pinnacle of the emerging order stood the economic moguls of the nation, distant figures such as Henry

Ford and John D. Rockefeller. Their imitators in Piedmont cities and towns like Athens were industrialists (generally textile manufacturers), bankers, and large-scale merchants. Often travelling in the same circles as this regional élite, if not with the same resources, were college-educated white professionals. At a lower, albeit still respected, rung stood small business owners, managers, salaried clerks, and skilled tradesmen. Finally, near the bottom of the pyramid were unskilled white operatives. Their paltry wages nearly matched those of the black laborers and service workers who filled the very bottom tier. Although far over-represented at the bottom of the heap, African Americans existed in smaller numbers in the middling ranks as well.[45]

The countryside featured a comparable hierarchy. Here, too, prospective white smallholders found old avenues to independence impeded by economic concentration, even in the once yeoman-dominated Piedmont. A landlord-merchant class stood atop the rural class structure, presiding over a mass of propertyless blacks and whites. Yet the forms in which the landless sold their labor varied in ways that corresponded to class fractions in town. Racism ensured that the lowliest positions—sharecroppers and wage laborers—would be occupied almost exclusively by African Americans. Lacking farm animals and equipment, sharecroppers constituted, in both fact and law, a rural proletariat akin to the unskilled wage laborers in the region's mills. Renting tenants, in contrast, shared some of the attributes of skilled craftsmen, such as ownership of their own tools, greater freedom from supervision, and better, if diminishing, prospects for acquiring land of their own. More likely to be white, renters were also more likely to identify their interests with those of small farmers.[46]

Southern society had not always divided on such lines. Farm ownership had remained widespread among whites throughout the nineteenth century. Yet a series of circumstances led the rate of independent ownership to plummet between the Civil War and the World War. Credit being scarce in the South, farmers in need of cash turned—or were pushed by furnishing merchants to turn—more and more to commercial crop production. Reflecting that growing orientation to the market, an orientation made possible by railroads that began to connect non-plantation areas to Northern markets, Clarke County farmers in 1920 planted three times as many acres in cotton as they had in 1880, at the expense of food and other crops. In Georgia as a whole, cotton came to account for two-thirds of the value of all crops by 1920.[47]

Meanwhile, acquiring a farm of one's own became a more monumental task for young white men. As descendants multiplied while the land supply held constant, a trend to smaller and less viable farms developed. Young men, in particular, now had to hire themselves out in growing numbers as tenants in hopes of acquiring the wherewithal to purchase farms of their own. Many never would. The proportion of Georgia farm operators who were tenants thus grew from forty-five percent in 1880 to sixty-seven percent in 1920. In Clarke County, the ratio was starker: almost three of every four farm operators in 1920 worked someone else's land.[48]

Developments in town and country were closely connected, since, as one study put it, "the impoverishment of farmers was industrialization's driving force." Manufacturing acted as a magnet for refugees from hardscrabble farms. Athens' population practically doubled between 1890 and 1920, making it one of Georgia's leading mid-sized cities. By then, two-thirds of the county's residents lived in the city proper or in smaller satellite towns. The number of wage-earners and of manufacturing establishments in the county had also more than doubled since 1880. Among the largest firms were the Climax Hosiery Mill, the Athens Manufacturing Company, and the Union City Thread Mill in East Athens, and in the outlying areas of Princeton and Whitehall, the Southern Manufacturing Company and the Mallison Braid and Cord Mill. As the manufacturing labor force grew to more than one and a half thousand people, the aggregate value of the products they produced grew over fifteen times, to reach over ten million dollars in 1919.[49]

While country life had its drawbacks, mill villages left more to be desired in the view of many whites, especially older men. In place of the self-paced and seasonal if hard work of farming, manufacturing workers endured monotonous workdays and grave risks. Every three to four working days in Georgia in 1921 a worker died in an industrial accident. In Athens, only one in twenty manufacturing workers worked fewer than fifty-four hours a week in 1919; three in four put in sixty or more hours. "In my young days," an Athens mill worker later recalled, "the life of a mill worker wasn't very long. The close confinement, long hours, lint and dust that they had to breathe, all worked together to shorten their lives."[50]

Whether one lived in town or country, in any case, was becoming less important than what one did. A widening chasm separated mill villagers from their "uptown" contemporaries. The latter included not only the self-styled "better people" as one might expect—the mill owners, bankers, large merchants, and professionals who dominated public life—but also small business people, white-collar employees, and supervisory personnel. Even craftsmen took care to distance themselves from unskilled workers. Having themselves usually hailed from landowning families, large or small, uptown whites joined planters and independent farmers in disdain for mill operatives. Scorned as "lintheads" and "white trash," their failure to succeed in farm life, according to uptown people, stemmed from laziness, immorality, even genetic inferiority.[51]

Mill villagers tended to reciprocate the dislike of uptown whites and express a strong "them and us" consciousness and values of their own. A man born in a mill section of Athens later recalled what it was like to grow up "on the wrong side of the tracks." "The people who had money" in the city were "kowtowed to as if they were kings," while the people in his neighborhood incurred disdain from other townspeople. To reclaim their dignity, he and his friends mocked rich boys, whom parents warned them not to play with, as "sissies." According to one study, mill workers also spurned as virtual traitors those who moved up the ladder to become foremen. Some mill operatives (almost half in one 1930 study) were in fact favorable to unions, although convinced by the disastrous results of earlier efforts that their employers would never tolerate them. Others resented company housing and "welfare" programs, which used money they believed theirs by right. Still others consistently voted against the wishes of town élites in politics.[52]

The segregation between mill and uptown people was extreme, in some ways more unremitting than that between blacks and whites. Mill and town rarely associated at all. They almost never intermarried, they belonged to mutually exclusive clubs and organizations, and they avoided social encounters. The friction between the two groups was so great that the few institutions involving people from both, such as churches or schools in mixed communities, often could not hold social events. In Athens, according to one contemporary teacher, the split between mill children and other children was already unbridgeable by the fifth grade.[53]

More than any other group of white wage-earners in the local population, in fact, mill workers most resembled the bogey of Jeffersonian republicans: the Old World proletariat. Their pitiful incomes ranked among the lowest of all manufacturing workers in the United States in the 1920s. Most mill households had to rely on employers for their roofs as well as their wages. Even their churches were usually not their own, but were built and backed by management to mold and subdue them. Moreover, unlike white farm tenants who might acquire land, or craftsmen who might start their own businesses later in life, mill operatives' station was usually permanent. Although many parents desired better lives for their children, escape from the industry was rare. Paltry education and intensive labor ensured that most would lack the resources for mobility. In the elementary school attended by mill workers' children in west Athens in the late 1910s, for example, sixty-five to seventy pupils competed for the attention of the first-grade teacher. Were the overcrowding not enough to discourage learning, the lack of a fifth grade in their school barred these mill children from secondary education even if their families could have afforded the loss of their wages. There was thus much truth to the common saying of mill parents that "there is no chance for the children of such as us."[54]

Yet, as economic development closed off some old options for white men, it opened some new ones for African Americans. A step down from landowning, tenancy marked a step up from sharecropping. As whites moved down and blacks moved up, race and class ceased to converge as neatly as they had in the nineteenth century. "The Negroes in Georgia," as one Athens contemporary put it, "can no longer be divided from white people by a sharp line of economic cleavage." While tenants now outnumbered owners among whites, more African Americans owned their farms in Clarke County in 1920 than ever before. For the first time, the number of white and black owners was almost equal: 180 to 163.[55]

Another sign of the old system's unravelling was the mounting numbers of blacks who left the country for the city, where they felt less pressure to defer to whites. By 1920, a quarter of a million black Georgians lived in urban areas. Like other Georgia cities in the decade, Athens became a magnet for blacks fleeing rural areas. And here also, black property-holding grew after 1875. By 1913, sixty-three black households owned over one thousand dollars of taxable property. Mary Wright Hill was an emblem of their achievements. The principal of an East Athens school, Hill lived in a ten-room house, married in a posh church ceremony, put all her children through college, and treated her daughter to a tour of Europe. Some of Hill's peers rejected the servility demanded by Southern whites more overtly. At the 1914 Atlanta convention of the Woman's Christian Temperance Union, Athens black women refused to sit in the segregated area the organization's white leaders had designated for them.[56]

The growing self-assertion of the black middle class in fact lay behind the emergence of top-down racial reform efforts such as the CIC. On one hand, the pool of educated and relatively privileged black leaders that now existed in most sizable communities in the South shared many of the values held by their white counterparts, a convergence that made cooperation possible. On the other hand, the very economic restructuring that aided some African Americans in their efforts to move up the ladder also decreased the region's reliance on plantation labor. In this setting, some middle- and upper-middle-class whites began to imagine, for the first time, a racial order based less on coercion and more on consent. Coming from universities, churches, newspapers, and some large enterprises in the region's cities and towns, these men and women recognized that without reform blacks would quit the region, which itself could become a backwater. "The danger," explained M. Ashby Jones in one CIC missive, came "from the loss of labor on our farms and from the condemnation of the outside world." Thus, both sides could agree on a strategy that brought together the self-styled "best elements" of each race to achieve greater harmony through a process of gradual reform.[57]

Economic change altered the foundations, not only of relations between blacks and whites, but also of men and women and parents and children within white society. Although families remained important economic units, their character changed. Where the quintessential nineteenth-century white family labored under the direction of its male head, now sons and daughters, sometimes wives as well, earned wages in their own right. White women and children had in fact pioneered the movement into the mills in the 1880s, with adult men following only as the farm economy constricted. By 1919, adult women accounted for almost one in five manufacturing workers in the state. In addition to manufacturing, which usually involved them as contributors to family coffers, more and more young white women took clerical jobs and moved out of their parents' homes. By the end of the decade, white women in professional and semiprofessional service in Athens in fact outnumbered those employed in cotton mills. The number of "business girls" living on their own so multiplied by 1909 that the Athens YWCA established a boarding house for them.[58]

More generally, the decline of independent proprietorship undermined fathers' ability to control their children through the prospect of inheritance. Save the minority of farmers who still owned land and the small-business owners and craftsmen with a shop or trade to pass on, by 1920 most non-élite parents in Georgia lacked resources with which to win their children's obedience. Even their houses were rarely their own. Freed from agriculture by the shrinking pool of available land and eager to achieve independence, young people proved the likeliest to leave the countryside for the city. As child-labor and compulsory-schooling laws began to take effect after 1914, moreover, urban adolescents enjoyed unprecedented leisure.[59]

Young women and men in these years in fact helped fashion a new cultural constellation aptly depicted by one historian as "filiarchy." As industry eclipsed agriculture and technological innovation more and more drove the nation's economy, deference to youth increasingly supplanted the veneration of age that had described classic patriarchy. Willingly or not, parents and churches ceded cultural authority to their children's peers and a commercial mass culture. The new pattern, described by some as the displacement of Victorianism by modernism, became plain by war's end.[60]

As it did elsewhere in the country, a heterosocial youth culture began to take form in Athens in these years. Like their peers in New York and Chicago had before them, Clarke County adults witnessed its advent with apprehension. Movie-mania, "dance madness," "joy-riding," and even the newly discovered phenomenon of juvenile delinquency all seemed to express disregard for the authority of parents and disdain for their gender roles.[61] Students at the University of Georgia, once quiescent, also became more defiant in the 1920s. Several petitioned the administration to abolish mandatory chapel attendance—on the grounds that it was "boring." Others started a newspaper, *The Iconoclast*, whose contents lived up to its name. Like adults, students began to polarize over issues of gender, sex, and culture. One student leader thus complained to his friends and his diary about how his male peers were "unable to think above their belts." Disgusted by their obsessive talk of women and their "obscene jokes," he was also perplexed by his own fantasies.[62]

White men's loss of power over their own children and wives was accompanied by a loss of leverage in public life. In early nineteenth-century Georgia, as elsewhere in the country, politics was notable for the breadth of participation of white men. Even after the defeat of Reconstruction and the restoration of white rule, non-élite white men—and black men—continued to turn out to the polls in large numbers. If a Republican sweep was ruled out, the Democratic Party was nonetheless far from united. Under its auspices a planter-dominated old guard struggled against challenges from both small farmers and city-based proponents of an industrialized New South. In the late 1870s and early 1880s, these rivalries erupted in the so-called Independent revolt against Bourbon Democratic "machine rule." Later in the decade, the region's rulers faced a challenge from a different quarter, as the Knights of Labor, with its vision of an end to "wage slavery" and a "cooperative commonwealth," made a pitch for the loyalties of America's direct producers. In Georgia, the Knights signed up some nine thousand members in thirty-one assemblies, three of them in Athens.[63]

By far the most momentous struggle, however, was that waged in the 1890s by the People's Party. Prompted by a catastrophic decline in cotton prices and the devastation wrought by the Long Depression, Southern farmers began to question the direction of American society and politics. When their attempts to organize producers' cooperatives through the Farmers' Alliance ran aground, large numbers turned to politics. Convinced by talented organizers like Tom Watson and by their own experience in state politics that the Democratic Party was beyond repair, they joined the national third-party movement on a wide-ranging anti-monopoly platform. Worship of the almighty dollar, said the Populists, had corrupted politics and dehumanized society. But their most radical position was an insistence that black and white small farmers shared a common interest in reform.[64]

Within a few years, the People's Party mounted the most significant electoral challenge ever faced by the region's rulers. At least one in three Georgia voters resisted bribes, intimidation, and violence to vote Populist in 1892, an act that cost fifteen men their lives. Two years later, the third party took more than forty-four percent of the state's votes and one-third of its counties. Nearly half of all Clarke County voters went Populist in 1894; some of the faithful even established a Farmers' Alliance Warehouse and Commission Company and a Workingmen's Cooperative Store. But enthusiasm was not enough; by 1896, the People's Party was on the wane. Bribery, fraud, threats, violence, and cooptative reforms by Democratic élites had combined with internal divisions among "the people" themselves to seal the third party's fate.[65]

The defeat of Populism shaped the future of politics in decisive ways. As pragmatic as its white leaders' proposal for alliance with black farmers had been, their willingness to extend it marked a watershed in Southern politics. Bourbon Democrats' understanding of the profundity of this challenge to the building blocks of their political economy was evident in the lengths they went to to defeat the Populists and to prevent such a challenge from occurring again. Across the South, the resurgent Bourbons pushed through legislation to so separate whites and blacks that they might never again recognize common experiences and needs. Over the next two decades, Southern states and localities issued a veritable avalanche of Jim Crow laws and ordinances. Some went as far as to segregate the dead in cemeteries. Even more devastating to black communities and to the prospects for interracial social movements was the related campaign to disfranchise black men—to guarantee the political quiescence that other forms of intimidation had failed to.[66]

Yet, in Georgia at least, planters and Democratic party bosses bent on suppressing challenges from below were not the only proponents of disfranchisement; former Populist standard-bearer Tom Watson also clamored to exclude black men from the electorate. Watson's reasoning presaged the kind of convoluted thinking about class and race that would later characterize the Klan. On the grounds that the Democratic élite used black votes to deter challenges from a disaffected white majority, Watson promised in 1904 to deliver his following to any Democrat who would support a constitutional amendment to disfranchise blacks. Hoke Smith did; with Watson's backing he trounced his competitor by a four-to-one margin in 1906. Two years later, an amendment that took suffrage away from most of the remaining black voters became law. While Watson's campaign to deprive black men of voting rights marked an about-face from his interracial appeal for economic justice in the 1890s the different stages of his career were unified by a common core: his devotion to the interests of middling whites as he understood them. Whereas in the 1890s, the primary threat to them appeared to come from above—from robber barons and conservative planters—as time went on, the challenge from below— from propertyless labor in town and country, especially blacks and immigrants—grew, and Watson turned more and more attention to it.[67]

Yet things did not work out the way he had planned. With black voters pushed to the margins, white politics became less, not more, democratic. Even drastic restrictions on who could vote failed to satisfy the South's governing class. On the contrary, throughout the region a commercial civic élite, led by organizations such as the Chamber of Commerce and made up of substantial proprietors and their allies in the press and professions, sought to limit the range of issues to be decided by the remaining voters. In Athens, beginning in 1913, an élite reform coalition involving upper-class residents and members of the university faculty repeatedly sought to shift the city government away from the mayor-council form inherited from the 1870s to a city commission or city-manager system. Proclaiming the latter "more efficient," they also looked to it to dilute the voting power of working-class and lower-middle-class wards and to ease out politicians without college educations.[68]

By the 1910s, Southern white men outside the seats of power in the economy and government were understandably skeptical about their ability to influence the occupants through the old channels. Regular elections notwithstanding, politics could hardly be called democratic. Rather, planters, industrialists, and their urban commercial and professional allies together rode herd on the excluded majority, black and white, and the less powerful voters who remained.[69] One resident of Athens, a shoe salesman, thus explained later that he never bothered to vote because "the little man's vote don't count for nothing nohow." In the county as a whole, the proportion of the population who went to the polls dropped from nineteen percent in 1868 to eight percent in 1920—the passage of woman suffrage notwithstanding.[70]

Growing constraints on popular control of public affairs were not unique to the South, of course. On the contrary, the accession of Woodrow Wilson to the presidency in 1912 culminated a process of sectional reconciliation under way since the Compromise of 1877. Like their Southern brethren, Northern élites were also working to insulate governance from the unpredictability associated with popular participation. Indeed, they invented the new forms of city government that soon spread across the country, and they acquiesced when Wilson imported Jim Crow into federal government offices. Throughout the country, in fact, the central attribute of American politics in the first two decades of the twentieth century was a mounting élitism: the two main parties atrophied, voter participation dropped precipitously, and a rising national administrative bureaucracy filled the vacuum in political decision-making. The "chief function" of the party system inaugurated by the election of 1896, in the words of political scientist Walter Dean Burnham, was "the substantially complete insulation of élites from attacks by the victims of the industrializing process."[71] As the victims lacked means of redress, their frustrations festered.

▲ ▲ ▲

Yet, by and large, these frustrations rarely erupted in Georgia in the years between the demise of the People's Party and the Great War. If the economic upturn after the turn of the century did not eliminate anxieties, it did assuage tempers. Such flare-ups as occurred could be put out before they ignited the combustible mixtures at hand. But then wartime economic boom gave way to serious recession in the summer of 1920. Cotton farmers encountered the steepest price dive in cotton history: from 40 cents a pound in July to 13.5 cents in December. Other problems exacerbated the strain. The boll weevil, in the words of Athenian Harry Hodgson, "played havoc with all of Georgia." Over the years from 1920 to 1925, cotton production in Clarke County and its northern neighbors dropped by half to three-quarters; the counties to its south sometimes lost their entire crop. Not for a decade would production recover its former level. In the meantime, the human toll was steep. The county suffered a net loss of over three hundred farms during the 1920s; every fourth farm went under.[72]

Cotton being the centerpiece of the state economy, the crisis soon infested other areas. "The truth is," one Georgian concluded a treatise on the spin-off effects of the farm disaster, "we are all as good as busted." "The last three years have been tough ones," his Athens counterpart confided in 1922. "Our businessmen are in a bad plight[,] with the farmers discouraged and in many cases labor leaving the farms." "The South is in great distress," wrote another man; "where the farmer cannot function the merchants, banks, hotels, etc. all go down."[73]

Without cotton, the mills could not run full time. In any case, they had their own problems in the 'twenties, as the introduction of synthetic fabrics compounded the industry's surfeit of national and international competition. Managers responded by cutting production, laying off workers, and demanding more from those who remained. Never again would Athens mill operatives enjoy the wages they had during the war. On the contrary, throughout the Piedmont, anxiety became chronic about what "these hard times" would do to families.[74]

Governor Thomas Hardwick in 1922 described the preceding few years as "one of the periods of most profound depression" in all of Georgia's history. It cut a wide swath, as a hardware merchant from Winder observed. "The world is [a]ffected by this panic," he said. "The rich and strong are losing fortunes daily and the poor is hungry and homeless." Never before in his three decades of work with farmers, Georgia's Commissioner of Agriculture maintained, had they appeared "so depressed and in such an alarming financial condition" as they were by the summer of 1921. "It is pitiful," bemoaned a South Carolinian; even returning white veterans were "glad to work for 10 cents per hour." In Athens, men who had once had jobs, particularly in the building trades, were by 1922 spending their days roaming the streets "begging and pleading for work."[75]

Some kind of retrenchment was perhaps to be expected in the circumstances, but there was no consensus about who should bear the brunt. For a time it seemed that, South as well as North, labor might win a larger say than ever before. Or so the tens of thousands who walked the picket lines had hoped. But that possibility was foreclosed in short order by the combination of government persecution of the Left and a robust open-shop drive. Southern employers, particularly in the textile industry, had a reputation for violent suppression of labor unrest, one borne out by their response to the strikes and organizing drives of these years. From 1920 to 1922, they fought hard to roll back the gains organized workers had made, as did their counterparts in the nationwide effort to impose the "American Plan."[76] The number of union members in Georgia plummeted, and most of the smaller city central labor bodies collapsed. Nationwide, union membership dropped from its 1920 high of over five million to just over three and a half million in 1923; by 1930, the proportion of nonagricultural workers in unions would be about half what it had been in 1920. With labor defeated, radicals driven underground, and liberals demoralized, the government was free to pursue policies supportive of big business, often at the expense of small.[77]

Even the *New York Times* acknowledged in 1920 that the restrictive monetary policies of the Wilson administration and the Federal Reserve Banks had "created a bitter feeling" across the South. "The hand that is feeding the world," one Georgia farmer complained, "is being spit upon." "The people of the whole country are distressed beyond measure with the present situation," complained an Augusta insurance agent. "In a land of plenty . . . yet the people are almost starving for a lack of money and credit to keep business and trade moving . . . [while] J. P. Morgan and the International Bankers are governing the country in the interests of big business." A state envoy of the Farmers' Union reported by 1923 "unrest among the farmers . . . as wide as Georgia's boundaries"; it was so profound as to pose "a menace to Georgia's security."[78]

By late 1920, the confidence of non-élite Southern whites in the government was at a low ebb. An Athens lawyer hoped for some measure that would "restore the confidence of the people in Congress, and give them more courage for the future." "Since the propaganda of the war," an elderly minister from Virginia observed, "the folks have lost faith in the 'Powers [that] Be' until their is a state of unrest that is close to the danger line." Even some of those who might be expected to be most loyal to the government grew mutinous. "Tell them to try another war and see where they will land," warned the Commander of the Huntsville, Alabama, post of the American Legion after the defeat of the Soldiers' Bonus Bill. "The flagwaving patriots . . . can go where it is hotter than it is in Alabama before we will lift a hand again for J. P. Morgan, Standard Oil, and other big interests."[79]

The trouble went deeper than the recession or big business's political influence. The government itself seemed out of control. In so enlarging the power of the executive, Woodrow Wilson had excited time-honored republican fears of concentrated power. One Georgia farmer complained that the people had endured sufficient "autocratic encroachments o[n] our liberties . . . to nauseate to extreems during the late war." Thomas Hardwick, one of Georgia's United States senators, likewise condemned the "Beauracracy which has grasped this government by the throat under Mr. Wilson." "I am deeply alarmed," he told an ally, "at the tendency to centralize this government, to enthrone an autocrat, to abandon, one by one, the great fundamentals that underlie and protect our liberties." "It is high time," warned a Single Tax advocate whose ideas would later be appreciated by Atlanta Klansmen, "to stop this temporizing with Wilson or we will be in far worse [shape] than if we had a hereditary 'ruler.' "[80] Even after Wilson's death, Washington, D.C., presented to Tom Watson a "loathsome" specter of Old World corruption. Evoking "Paris at its worst" in the days of Louis XV, Watson voiced disgust at "the waste, greed, graft, thievery [and] harlotage" in the nation's capital. A Georgia editor of the *National Farmers Magazine* spelled out the logic: "no people may remain free where money rules and morals rot."[81]

Feeling afraid and excluded, tens of thousands of white Georgians turned for leadership to the old Populist standard-bearer, Tom Watson. Himself a rich planter and lawyer, Watson still had a knack for addressing the concerns of middling men. Persecuted by the Wilson administration for his opposition to the government during the war, Watson now appeared a martyred hero to many non-élite whites. In his race for a United States Senate seat in 1920, he thrashed his establishment rivals at the polls. Once in Washington, Watson acted as a faithful outsider. He thundered against the executive office's usurpation of power, against the imperial designs of American foreign-policy makers, against the tight-fisted Federal Reserve Bank, against the machinations of monopolies—even against the imprisonment of socialists for antiwar activity and against American intervention against the Soviet government in Russia. At the same time, he fulminated against the Catholic menace and fought the appointment of blacks to federal jobs.[82]

Watson was hardly alone in his understanding of the problems facing the country. Indeed, it is impossible to understand the Klan's rise without recognizing that vast numbers of people, all over the United States, embraced views like his. White farmers, small merchants, and others wrote to Watson to express gratitude for his efforts and to denounce those they held accountable for their plight. Their primary grievances with the economy, like Watson's, centered on monopolies and high finance. "The farmers are being imposed upon by every class of speculative enterest," complained a representative of the Brooks County Farmers' Union; "this must be curbed if we expect this republic to live." Grocers from Valdosta sought Watson's help against the "meat trust," which they blamed for farmers' problems and their own. Others voiced more directly the old Populist theory of economic crisis; their target was the government and the banks. J. S. Dean of Buchanan, Georgia, thus agreed with Watson's attacks on the Federal Reserve. Dean, too, believed that "the money questian . . . was the most vital questian." Like the old People's Party, he found the root of the problem not in the economy itself but in "law making since . . . the laws [are] being made by those who are to be benefited by said laws." Dean complained that "this set of robbers" was using the Federal Reserve to make farmers "poorer year by year and . . . more & more dependent on those whom the laws protect."[83]

Even some outside Watson's ranks shared these convictions. J. S. Hale of Barnesville, Georgia, made the same complaints to Watson rival Hoke Smith. Capitalism per se was not the problem; like Watson, Hale believed that "money is a blessing when properly used." But "the hoarding" practiced by the "big interest[s]" was something else. It was "causing wreck and ruin throughout this whole country" and making a mockery of democratic government. The rule of "the money power" had become "more cruel" than that of "the German K[a]iser." Hale hoped for "a Moses to lead the people from under the yoke of bondage." "By the help of God," he concluded, "the people will not crouch and cower to the will of the money kings."[84]

This kind of malaise would find an outlet in the second Klan, whose leaders seized on the old Populist analysis and remolded it to their own ends. In Georgia, the connections were not merely ideological; they were personified in Watson.[85] Whether he actually joined the Klan cannot be determined from the evidence available; he did endorse it as "a worthy organization."[86] But that the Klan embraced Watson's vision and he theirs, albeit a more pro-labor and civil liberties version (he had, after all, denounced "the '100 percent' idiots"), there can be no doubt. Watson did not merely promote reactionary causes—couched in populist language—that the Klan would later take up, including the disfranchisement of African Americans, attacks on the Roman Catholic Church, tirades against socialism, and campaigns against finance capital tinged with anti-Semitism. He also maintained close and amicable relations with the Klan. He defended the order from congressional investigators in 1921, he helped Klan candidates for public office, he shared his subscription lists with the organization, and he supplied the Klan's official national lecturer with material on how Jews and Catholics endangered the country.[87]

The Klan, in turn, worshipped Watson as a hero of "the common man." One national Klan representative and past ally of Watson described him as "the political genius of our age." Rank-and-file Mississippi Klansmen praised Watson "as the most active proponent of true Americanism" for his defense of "liberty and freedom" and his fight against the "papists." After Watson's death, Klan leader E. Y. Clarke offered the organization's sympathies to Watson's widow for the loss of this "champion of right and courageous defender of the downtrodden[,] suffering and oppressed." At least one Klavern of the Georgia Klan was named in his honor, while the state's Grand Dragon described Watson as "beloved" by the members. Perhaps most indicative of the connection, however, were popular perceptions among Klan sympathizers that Watson was a representative of it.[88]

Also indicative of the way the nascent Klan movement would

perpetuate and deepen earlier cleavages was the way previous enemies of Watson went on to oppose the Klan. The CIC thus counted
among its leaders several men who had fought against Tom Watson
in the Leo Frank affair of 1915, such as the ministers M. Ashby
Jones, C. B. Wilmer, and Plato Durham. Some representatives of
the Democratic political establishment also worked against first
Watson, then the Klan. In Athens, Mayor Andrew Erwin urged
party loyalists in 1920 to "fight against Watsonism" in the surrounding counties; in the 1924 Democratic Party national convention, he was the only Georgia delegate to support an anti-Klan platform. Less courageously, a few other prominent local figures
followed the same trajectory.[89]

Such opponents of Watson and the Klan tended to have more
cosmopolitan economic interests than their peers, particularly in
attracting outside investors to Georgia. University of Georgia president David Crenshaw Barrow, for example, the first to sign a 1921
Athens anti-Klan petition, was also a supporter of Woodrow Wilson, the Federal Reserve, and the League of Nations. "A liberal attitude is required to attract new residents," explained Atlantan Walter Taylor, one of the components of which was "a willingness to
stop trying to regulate others' lives."[90] These frictions were not
only longstanding, but political in the most basic sense of the
term: conflict over who should wield power and how. When some
of those who felt they had a right to power found themselves ignored, they resorted to more Machiavellian tactics of getting their
way, violence among them.

That in the circumstances of the late 1910s and 1920s some
men would turn to force should not surprise us. European historians and sociologists have long recognized the use of collective violence as a tool to readjust relations of power. "Violence flows from
politics," writes Charles Tilly; "more precisely from political
change." One can expect outbreaks, he suggests, at "those historical moments when the structure of power is changing deci-
sively."[91] Clearly, the postwar years constituted such a moment.

In adopting collective violence to achieve their ends, Klansmen
could draw support from indigenous American vigilante traditions
that began even before the Revolution. Foremost among these
models were the White Cap bands who periodically came forward
to police social relations in the nineteenth-century Southern upcountry, in the Midwest, and in frontier communities. Named for
the hoods they wore on their night-riding raids, these bands of upstanding white community residents terrorized deviants from their

collective sense of right and wrong. Whether the victim of their masked floggings was an adulterous wife, a hard-drinking father, a rapacious businessman, or an ambitious black sharecropper, the object of the White Caps' visit was the same: to enforce the private and public conduct the world of white proprietors like themselves depended on. Rarely were they prosecuted.[92]

The night-riding members of the second Klan would operate with a similarly holistic world view, against a similarly broad range of perceived threats, and with similar indemnity. Yet their movement differed from its predecessors in a fundamental way. It was the first national, sustained, and self-consciously ideological vigilante movement in American history. No other White Caps operated on such a scale, for so long, or with such a propaganda apparatus. Such novel coordination and promotion appeared necessary to participants because the paternalistic social relations earlier vigilantism was associated with had so weakened. As individual wage-earning supplanted the petty production of households organized by their male heads, and as class differences grew among both whites and blacks, it became more difficult to present the ascribed hierarchies of race, gender, and age as natural and inevitable. Indeed, many African-American women and men, white women, and youth of both races took advantage of this uncertainty and the resources now available to them to claim new rights. The Klan's scope and frenzy were thus the measure, less of members' power, than of the distance separating them from the provincial, patriarchal world of their dreams.

3

▲ ▲ ▲

Men in the Middle:
The Class Composition
of the Klan

"It has worried me to think," Klansman *S. B. Yarborough* would muse in the 1930s, "that I've worked hard all my life and just can't seem to make no headway." Surveying his years of effort, *Yarborough* concluded forlornly, "It's right down disheartening to try so hard and never git nowhare." Like his older brother, *Roy*, Kligrapp of the Athens Klan, *Scott Yarborough* had ample reason for frustration with his life. The *Yarborough* brothers had grown up in a mill community outside Athens in a family with ten children. Only six survived infancy. Their father, *B. F. Yarborough*, had lost an arm in the Civil War. Relegated to teaching in a rural public school, he developed a reputation for bitterness and cruelty toward his students and his own children. The pay from school-teaching being too paltry even to buy shoes for all of his family, *Ben Yarborough* sent his sons and daughters to work early "doing everything that come to hand," from hired labor in the fields to mill work.[1]

Roy and *Scott* dreamed of escaping from the mills—and no doubt from their father as well. In time, both managed to. Beginning on a few dollars of borrowed money, *Roy* acquired his own farm and established a small grocery business. Over the years, he accumulated some $10,000 worth of personal assets and was able to provide jobs for some of his seven children. Yet industriousness proved a feeble shield against the vagaries of the economy and unexpected family illness. By 1910, *Roy* had to mortgage his farm; by 1927, he had lost almost three-quarters of his 1910 assets. For his part, immobilized by personal tragedy, *Scott* remained in the Princeton mill long after his brother had struck out on his own. When his first wife died in childbirth a year after their marriage, it "'most nigh killed" the seventeen-year-old husband she left behind. The loss so depressed him that he quit work for a time and did not remarry for six years. Once back in the mills, however, he moonlighted in the evenings and on weekends as a barber in his home. Ultimately, he put away enough from haircuts to open his own "Red, White and Blue Barber Shop." "We eat three meals a day, all my taxes are paid, and I don't own a cent to nobody," *Scott* said. "I don't have to call on nobody for nothing. That's what the barber business has meant to me."[2]

The *Yarborough* brothers' dream of independence—and the modesty of its fruition—bound them to fellow Klan members. Like their occupations and living conditions, the social standing of local Klansmen varied considerably. Yet, in general, these were middling men: neither élite employers and brokers nor, as today's popular conceptions of the Klan would have it, "poor white trash." While few had the resources to hire others, most exercised more control over their labor than their working-class contemporaries: the operatives who tended the cotton looms and the sharecroppers who tilled the fields of luckier men. "If not the 'best people,'" as one observer put it at the time, Klan members were "at least the next best . . . the good, solid, middle-class citizens." Some sixty-three percent of Athens Klansmen, for example, lived in homes or on farms that they or their parents owned. Perhaps simple, these homes were nonetheless public badges of relative economic independence; only thirty-two percent of Clarke County families could claim this distinction.[3]

Yet, what attracted men to the Klan was not simply their relative standing. It was the changes they experienced in that standing over the years leading up to and following 1920, as their expectations were first raised, then abruptly dashed. By and large, these were men who had climbed the economic ladder, if only by a rung or two. The Protestant work ethic had paid off for them, most dramatically during the wartime bonanza. Then, suddenly, just when their prospects had appeared most promising, they confronted unforeseen obstacles—if not disaster. Being on the edge to begin with, they reeled under the wave of the hard times that washed across the land. Already feeling vulnerable, Klansmen-to-be then looked

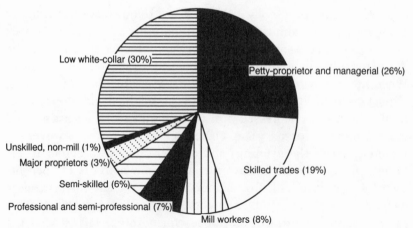

Figure 1 *Occupational Distribution of Athens Klan Members (See Table 1 for occupations within categories, precise numbers, and definitions)*

on as labor and capital locked heads. Whether unskilled workers pushed up wages or capital beat them back and expanded on the proceeds, middling men feared being crowded out. Large numbers joined the Klan in hopes of warding off that fate and reclaiming their cherished independence.

▲ ▲ ▲

The portrait that emerges from Klan membership rosters at mid-decade indicates that the vast majority of men in the Athens Klan stood between capital and unskilled labor. In their occupations, as in their assets, local Klansmen hovered at the midpoint of the spectrum. The greatest proportions clustered in the three categories of lower white-collar employees; petty proprietors, managers, and officials; and skilled tradesmen (see Figure 1). Klansmen's patterns of affiliation also indicated their relative status: middle-class, not "society." Large numbers of Klan members and their families belonged to the Baptist and Methodist churches, yet only a few to the more élite Presbyterian or Episcopal churches.[4] Similarly, scores of Klansmen participated in fraternal organizations, generally dominated by the lower middle class, and they counted heavily among the activists and leaders in the Chamber of Commerce and Lions Club. Few, if any, however, belonged to the more select Rotary Club, and only one Klan wife was included among the leaders of the Athens Women's Club or the League of Women Voters.

The single most common occupation among local Klansmen

was owner or manager of a small business. It accounted for double the number of its nearest rivals (see Table 1) and contributed many local leaders such as *Roy Yarborough*. The rural counterparts of these small businessmen were the small farmers who made up six percent of the chapter's membership, a figure that no doubt understates their actual participation.[5] Nonetheless, even in a data base that underestimates rural participation, at least 136 local members either grew up on farms, lived on them in the 'twenties, or derived some income from them. Of the farmers whose tenure could be determined, just over one-fifth were tenants, as compared to three-fifths of the Clarke County white farm population at large. The other four-fifths of Klan farmers owned the land they worked; about one-third of them, enough to hire a tenant or two. Numerous white-collar workers, members of the "new middle class" in this period, also joined the Klan.[6] Lower-level white-collar employees—

Table 1 Dominant Occupations of Athens Klan Members

Dominant Occupation	Number	Percentage
Broker	1	
Manufacturer	1	
Major merchant	4	
Major planter	1	
Combined major proprietor (e.g., planter and merchant)	3	
Total major proprietor[a]	10	(3%)
Teacher	2	
Lawyer	2	
Minister	10	
Physician	2	
Pharmacist	3	
Total professional	19	(5%)
Chiropractor	1	
Embalmer/mortician	2	
Optician/optometrist	1	
College student	2	
Total semi-professional	6	(2%)
Superintendent, foreman, or overseer	13	
Owner or manager of small business	57	
Minor government official	2	
Farmer, tenure unknown	4	
Farm owner[b]	14	
Farm tenant	4	

Table 1   *(Continued)*

Dominant Occupation	Number	Percentage
Combined working farmer and merchant	1	
Total petty-proprietor & managerial	95	(26%)
Public employees (post, police, fire, etc.)	31	
Salespeople	29	
Agent (real estate, insurance, etc.)	12	
Clerk	23	
Bookkeeper, accountant	7	
Cashier, bank teller	1	
Collector	1	
Railroad conductor	1	
Other low white-colar	5	
Total low white-collar	110	(30%)
Carpenter/woodworker	22	
Blacksmith	4	
Barber	5	
Plumber	2	
Electrician	7	
Baker	5	
Mechanic	9	
Machinist	4	
Other skilled trades	10	
Total skilled trades	68	(19%)
Non-mill factory worker	3	
Blue-collar transportation	15	
Presser	1	
Other semi-skilled	3	
Total semi-skilled	22	(6%)
Skilled mill workers	4	
Unskilled mill workers	16	
Mill workers, unspecified	10	
Total mill workers	30	(8%)
Laborer	1	
Drayman	1	
Sawyer, lumberman	2	
Total unskilled, non-mill	4	(1%)
Total members in list	364	(100%)

Note: "Dominant occupation" refers to the one the member remained in for most of the period under consideration, or he held at the end of the period under consideration (1900–1927).

a "Major proprietors" were defined as those with over $7,000 in assets, hence with the resources to employ extrafamilial labor.

b Two sons living with landowning fathers were included in this category.

public employees, salesmen, clerks, and agents—in fact made up the largest general occupational category in the Klan. They included 110 men, or 30 percent of the chapter.

Interestingly, even at mid-decade when the Klan's status was declining, unskilled men did not join in the same kinds of numbers. Although several hundred local white men earned their livings as mill operatives, only eight percent of chapter members—thirty individuals in all—had mill work as their dominant occupation.[7] The work experience, living conditions, and life opportunities of these men differed in significant ways from their better-off fellow members. The five *Hess* brothers, for example, all Klansmen, were sons of an immigrant German baker who died in 1908, leaving behind a widow and eleven children. The brothers began working in cotton mills before they had reached adolescence. Two of the five eventually made it into the skilled trades, but neither acquired much in the way of worldly goods. Similarly, the brothers *Stephen* and *Tom Gilmer* went into the mills as adolescents. While *Stephen* remained in the mills, *Tom* found employment as a clerk for the Athens Coffee Company in the 'twenties, acquiring more status and a home, but few other tangible assets. The humble condition of most other poor Klansmen meant that they usually left no paper trail for historians to follow; nor did they tend to play the leading roles in the movement that small-business and white-collar members often did. Craftsmen, in contrast, joined in larger numbers than mill operatives and won more recognition from their fellow members as leaders. Skilled workers in fact made up eighteen percent of the Athens Klan, the largest general category after lower white-collar workers and petty-proprietors and managers.

Men of substantial property and social power—the manufacturers, planters, large merchants, bankers, and leading attorneys who made up the local economic élite—were even less likely than unskilled workers to join. Only ten such men belonged to the Athens Klan at mid-decade, and none was among those recalled by contemporaries as the richest and most powerful in town. Nor did their assets qualify them as such.[8] Among the few from "blueblood" backgrounds who did belong was *R. D. Moore*. In 1914, his ownership of a 647–acre plantation in neighboring Morgan County and two other farms put him near the very top tier of the rural population. Described by the press as "one of the most progressive farmers in Georgia" and a man "prominent in all business and financial circles," *Moore* also served as president of the Athens Foundry and Machine Works, vice-president of both the Athens

Mattress and Spring Bed Company and the Citizen's Pharmacy, and director of various Georgia banks. He was among the handful of local Klansmen with a live-in servant. Another member, *R. R. Ransom*, grew up with three servants. His father, a planter and merchant, had accumulated over $40,000 in assets by 1921. Klansman *Wiley Doolittle*, for his part, owned one of the largest furniture companies in the state, directed the Athens Mattress Company, and served as chair of the board of directors of the Athens Gas Light and Fuel Company, among other business laurels. Yet such moguls were the exceptions rather than the rule. Wealthy Clarke County residents were more likely to sign up with the opposition. The names of premier businessmen thus studded a 1921 petition to discourage revival of the Klan—although most of the signatories fell silent as the breadth of its appeal became clear.[9]

To explain the order's appeal, however, the profile of Klan members needs refinement. The typical Klansman was not simply petit-bourgeois; he appeared less economically secure than the norm for his class. When a proprietor, he was often a newcomer, and the value of his holdings was less than the average among owners at large. Such small farmers and lower-middle-class townspeople, according to contemporaries, were the most "acutely conscious of the Town-Mill distinction." They were especially anxious to distinguish themselves—particularly through their moral codes—from the ordinary workers they viewed as beneath them.[10]

Although local members' assets varied widely, the modal figures are significant (see Figure 2). In 1921, members' mean assets were valued at $2,031, in the range of comfortable landowning, working farmers. Yet the presence of some wealthy men biased the figures upwards. Klansmen's median assets were significantly less. At $540, they were on a par with renting farm tenants, more independent than croppers or mill workers, but hardly secure. Indeed, they just squeezed past the $500 minimum qualification for voting rights.[11] Sixty percent of local Klansmen never amassed more than $1,000 in holdings over the years 1900 to 1927.

To some degree, the low mean assets reflected the youth of the Klansmen still dependent on their parents. When the mean is adjusted to include only those over thirty years of age in 1920, 53 percent had over $1,000, the baseline for landowners; thirty percent enough to be renting farm tenants; and only eighteen percent as little as sharecroppers. Still, only nineteen percent of the older men had enough assets to hire tenants, so the age-adjusted figures do not signify a different position in the class structure.[12] In short,

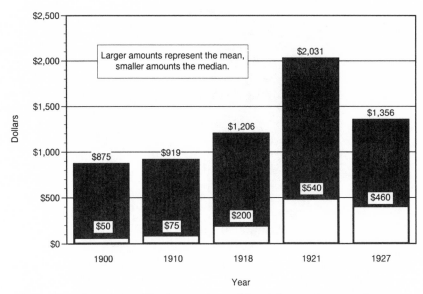

Figure 2 *Value of Taxable Assets Owned by Athens Klan Members*

while the typical Klansmen was better off than the vast majority of blacks and a large proportion of whites, he was vulnerable. His standing was unstable, and he knew it. Indeed, since numerous Klansmen lived in or near the largely working-class communities of East Athens and West Athens, they were daily reminded of the precariousness of their own positions.

Such success as Klan proprietors and white-collar workers had, moreover, was often recently acquired. Almost half of Athens Klansmen moved up the occupational ladder after 1900 (see Figure 3). Klansmen's improvement of their condition was also evident in their acquisition of assets. Both the mean and the median assets of members grew substantially between 1900 and 1921. Their mean assets jumped from $875 in 1900 to $2,031 in 1921; median assets from $50 to $540 over the same period. Like the *Yarborough* brothers, numerous other Klansmen had accomplished the unusual feat of moving up in or out of the mills. Whereas in 1900, thirty-eight percent had been mill workers, by 1926–27, that proportion had dropped to six percent.[13] The parents of Klansman *F. S. Gray*, for example, were illiterate mill workers, yet he eventually became a policeman. The *Carlton* brothers, *P. S.* and *Claude*, also came from unlettered parents and entered the mills at early ages. Yet *Paul* managed to become a salesman and Henry a superintendent. Klansman *H. G. Walker's* rise came late in life; at age thirty-one, he still

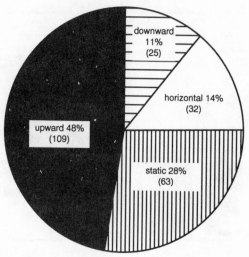

Figure 3 *Mobility of Athens Klansmen, 1900–1927 (Total number on whom data are available = 229)*

worked as a common mill operative. But by age forty, he became a mill superintendent, and by fifty-seven, a partner in a loan company with leading local citizens, including fellow Klan member *G. M. Harris.*

Examination of the life histories of members in particular occupational groups further illuminates the often hard-won character of their standing. Among the owners and managers of small businesses, a few prospered, like *I. W. Sheats,* who managed to acquire over $19,000 in assets by 1921 from his musical instruments business. Yet most Klan proprietors ran "mom and pop" operations. The frailty of their enterprises is suggested by the owners' mean assets. They rivaled those of the lower ranks of working farm owners with no tenants.[14]

The modest successes of these proprietors, moreover, usually required years of exertion. *K. L. Nash* joined his siblings in the mills as a boy to support the family after his father died. In time, *Nash* was able to get a job as a policeman, and later to open a family-run café, increasing his personal assets from $75 in 1918 to $1,145 in 1927. Like *Scott Yarborough, A. J. Boyd* moonlighted his way out of the mills. Having begun work at age eight as a water boy, he later joined with partners to acquire the first white barbershop in Athens. Through frugal living, Davis later obtained his own shop. The tenuous standing of many Klansmen is further underscored by the way, like *Boyd,* some had to push blacks out of the way to get where they were. By 1913 whites held six of Athens' eight barbershops, formerly a black trade; five of these white businesses belonged to Klansmen. Similar racial skirmishes took place in the pressing trade and the postal service; these occupations, too, yielded several Klan recruits.[15]

Not all the Klansmen who had experienced mobility came from mill backgrounds, of course. *P. W. Spurling*, for example, grew up on a dirt farm in a nearby county. After leaving home, he drifted from one occupation to another until a stint in the Mallison Braid and Cord Mill persuaded him to save up for a business of his own. It was a plain grocery store serving a poor black clientèle, but it was his, and he could "make a living" from it. *H. B. Bailey* went from being a railroad ticket agent to managing a bus station with his wife. Although they worked fifteen-hour days, *Bailey* said he vastly preferred this to his former employment because "we are our own bosses." Most Klan farmers would make the same boast, but they had to scramble to hold on to their prized positions. In all the years surveyed except the boom year of 1921, their mean assets hovered below the mean in the general farm population for owners with no tenants.[16]

The professionals and semi-professionals in the Athens Klan were in a position analogous to that of small businessmen and farmers. Although masters of their own labor, few had extensive education, and none commanded enough wealth to employ. Their mean assets were $3,751 in 1921, dropping to $1,559 by 1927, which ranked them with the lower echelons of the rural petite bourgeoisie. The individual stories of some of these men are suggestive. Klansman *Chester Morton's* father was a furniture salesman; *Chester* trained for a license as an undertaker. For some fifteen years, he worked for other men until, in 1927, he was able to set out on his own with a partner. James Willie Arnold, one of only two lawyers in the Athens chapter at mid-decade, began as a cotton trader, but entered law school at the age of thirty-six. By 1927 he had $2,850 in assets—less than when he was admitted to the bar in 1921.

None of the ten ministers who joined the Klan was even as secure as *Morton* or Arnold. The Reverend *C. G. Wilson*, pastor of East Athens Baptist Church, had served his congregation for decades by 1927, yet his assets amounted to a mere $250. Rather than attending college as the children of more prominent professionals did, *Wilson's* daughters and sons helped support the household through such jobs as telephone operator and travelling salesman for an overalls company. Among University of Georgia professors, in contrast, a more élite lot culturally if not economically, the Klan could claim only two members.[17]

Of the members who were fairly wealthy by contemporary standards, few gained their comforts from inheritance. Joseph K. Patrick moved to Athens in 1908 with few assets and little cash to his name. He worked as a carpenter to finance his pharmaceutical education at the University of Georgia. In 1913, he opened the Citizen's Pharmacy along with two partners, one of them his future co-member *R. D. Moore*. Working fourteen hours a day, or so he reminisced, Patrick raised his assets from $1,300 in 1918 to $16,140 in 1927. *N. S. Rich* also amassed his wealth via independent entrepreneurship. *Rich* moved to Athens in 1919 and used his modest savings to buy out a small bakery with only seven employees. By 1925, the payroll of *Rich's* Bakery was over $44,000, more than six times its original amount, and *Rich* was operating plants in four different towns, all financed through earnings.[18]

Athens Klans members were thus, by and large, men whose own experience prepared them to accept economic individualism. They found in the Klan an organization that upheld their beliefs and applauded their efforts. Indeed, the order showed special regard for those who profited from the application of the Protestant work ethic. Klan leaders trumpeted the rewards that accrued to self-made men, and again and again emphasized "self-control" and the forgoing of "immediate gratification" in favor of long-term economic security.[19] "Man is made or unmade by himself," as one Klan leader put the prevailing faith. Many national Klan leaders, Simmons and Evans included, were themselves "self-made men" and self-consciously so. They had risen with much effort from humble beginnings—some from large, hardscrabble farm families—to positions of relative wealth and comfort, and they sympathized with other humble men on the make. The organization clearly aimed at upward-striving men, moreover. The *Searchlight* implied this when it asserted that "only men who amount to something or who want to amount to something" were admitted to the Klan.[20] But perhaps most important, the Klan offered explanations when things went awry.

Indeed, a third, and crucial, factor in the Klan's appeal was the economic crisis of the early 'twenties. It cut short the climb of men on the make and defied their dreams of being their own bosses. Whether Klansmen suffered losses severer than their non-Klan counterparts' is impossible to say with the data available. The argument here does not posit members' atypicality in any case, or assume a mechanical relationship between economic condition and Klan membership. Joining the Klan marked but one possible response to the crisis, albeit a common one. Just as not all petit-bourgeois white men gravitated to the Klan, neither did all those hard hit by the depression. Class standing and economic insecurity created a potential among white men for openness to the Klan's message. Whether a man responded positively to it depended on other factors.[21]

However they came to terms with the experience, most Georgians got a rude shock when the biggest sustained boom in the state's history abruptly gave way to its worst depression. Property assessments and tax returns in Clarke County as elsewhere registered losses for several years running. The most devastating was 1921, when the Klan's meteoric rise began. The State Tax Receiver, whose task of organizing collections from begrudging citizens had become a nightmare, was moved to comment that Georgians were "in the worst condition financially that they had been in for many years." "This financial depression," he continued, "was accompanied by the most deep-seated, far-reaching feeling of dissatisfaction that possibly had ever existed."[22]

That malaise was aggravated by the unevenness of the losses. As hundreds lost the fruits of years' effort, a few managed to expand their holdings. While Georgia's small-town and country merchants suffered, for example, "department stores and other large mercantile establishments in the cities" held their own, even prospered. In Clarke County, where hitherto almost all stores and services were operated by their owners, the 'twenties proved a critical turning point. Local and state merchants complained of inability to compete with newly arrived national outlets and chain stores. Their volume of sales and refusal to extend credit or make deliveries enabled them to undersell local shops. In time, they priced many out of business.[23] Others, such as barbers and filling-station proprietors, suffered a surfeit of local competition, customers who could not pay their debts, and, sometimes, government regulations that hit small business hardest. "We are all tied up with red tape now," Klansman *A. J. Boyd* thus complained, after costly new state hygiene and licensing requirements drove his brother out of the barbershop business in 1924.[24]

The records of members' assets offer grim testimony to the toll that postwar recession and economic reorganization took of them: nearly half suffered economic losses between 1918 and 1927 (see

Figure 2).[25] The better their starting position was, it seemed, the worse their losses. Thus, Klan leader Bela Dunaway had acquired $30,100 in assets in 1921; by 1927, the family assets had plunged to $2,700. In effect, he fell from the local capitalist class to the petite bourgeoisie. For Dr. *S. M. Turner*, the fall was even more painful. *Turner* had made himself a relatively rich man between 1918 and 1921—only to see his holdings plummet from $13,920 to $505 by 1927. Several other wealthy members incurred less spectacular but substantial blows, ranging as high as a third of their 1921 holdings. These men lost between a few and several thousand dollars, yet they did not hit bottom. Among the humbler, such losses hurt more. The Reverend *B. B. Couch's* assets dove from $3,500 in 1921 to $850 in 1927; *S. S. Aycock's*, from $1,200 to 100; *Llewellyn Daw's*, from $200 to $25, and so on.

To say that almost half suffered losses, however, is to understate the damage. Some Klansmen too young to own much themselves witnessed their parents' living standards plunge. From the time *A. B. Page* was ten, his widowed mother rode an economic roller coaster. She lost $2,375, or thirty-four percent of her assets, between 1910 and 1918; and regained approximately half by 1921, only to lose again thereafter. *Hiram Brantley* and his two sons, *R. N.* and *G. K.*, all joined the Klan after a distressing decade. The elder *Brantley* had climbed from tenant farmer to merchant over the years 1910 to 1918, and built his assets up from $45 to $2,085 as he went. In the next three years, however, he lost almost half of this hard-won bundle. *Kyle* and *Robert Odell* saw their father endure a similarly dramatic rise-then-fall, while the brothers *B. C.* and *H. O. Thorndale*, whose father, *Dirk*, was also a member, saw him lose thirty-six percent of his 1921 assets by 1927. For *Grady Thrasher*, the impact was perhaps especially personal. His father's hardware business, "*Thrasher & Sons*," the would-be inheritance of his six sons, experienced a steady fall in assets, from $50,000 in 1900 to $925 in 1918; the family's personal assets dropped from $22,415 to $3,825 over the same period.

For young men such as the *Brantleys* and *Thrasher*, downward mobility was hardly an abstract threat. Indeed, eleven percent of Athens Klansmen had already experienced it; by and large, they were wage-earning sons of farm owners or other small proprietors, or men who lost their own land or position. One such member was *P. F. Nelms*. After years of work as a boss spinner and then foreman in a local cotton mill, *Nelms* was laid off in 1911 at age forty-nine.

He was soon so broke that he had to file for tax exemption. By the 1920s, he was making a bare living as an ordinary operative.

▲ ▲ ▲

It might appear that the Athens Klan simply attracted a cross-section of the white male population. To some extent, this is true. In Piedmont communities such as Clarke County, the petite bourgeoisie made up a large proportion of the economically active white male population. Hence its dominance in the Klan is not remarkable. Still, as noted earlier, one is struck by the relatively small number and proportion of chapter members contributed by those highest and lowest in the class structure. Moreover, a similar clustering of members in middling occupations within a profile that was otherwise a cross-section of the population eligible for membership occurred in other parts of the country as well, in large manufacturing cities as well as in smaller towns. Although the exact proportions varied, those most likely to belong to the order included white-collar employees, small-business owners, independent professionals, skilled workers, and farmers. In contrast, unskilled workers, landless rural people, and the very wealthiest residents were almost everywhere under-represented.[26]

The significance of the clustering of Klan membership in the middle ranks of society emerges more clearly when it is viewed in transnational perspective. As some historians and social critics have pointed out, the common tendency to confine class analysis to labor and capital has tended to obscure the very existence of the petite bourgeoisie as a class with its own distinctive relationship to capital and labor, its own internal dynamics, and its own modes of thought. This difficulty is compounded by the diffuse character of the petite bourgeoisie, or, in more colloquial usage, lower middle class. Its members run the gamut from owners of small businesses to artisans and salaried clerks. Most petit-bourgeois occupations also shade off at the bottom into the upper ranks of the working class and at the top into the lower ranks of the capitalist class. These varied occupations and the internal pecking orders within them result in differences, sometimes striking, in living standards, social status, and perspectives.[27]

Significant as they are in the experience of everyday life, these differences mask underlying commonalities that justify treating the petite bourgeoisie as a class. Foremost among these commonalities is its members' relationship to the means of production and to other classes. These are clearest in the case of the "old middle class": landholding farmers, shopkeepers, and artisans. All own the means by which they earn their livelihoods and rely on their own labor—often on that of other family members as well—to survive. Rarely do they employ extrafamilial help, and if so, only for short duration or in small numbers. In contrast, members of the "new middle class"—managers, paraprofessionals, and salaried white-collar employees—lack the economic independence that character-ized the old. Yet they, too, exercise more autonomy in their own labor than members of the working class proper, whom they some-times direct.[28]

This structural position helps account for the characteristic ambivalence of the lower middle class vis-à-vis the capitalists above them and the unskilled workers beneath them in the social order. Like capitalists, small proprietors have a vested interest in private ownership. Yet, unlike the actual bourgeoisie, members of the petite bourgeoisie live for the most part off their own and per-haps family members' labor, and often harbor suspicion of those who live off the work of non-kin and can out-compete small rivals. Indeed, the small-holder is vulnerable to being crowded out by larger, more resourceful units of capital. These latter features make the petite bourgeoisie, like the working class proper, potentially anti-capitalist. Yet, here again, full identification is often compli-cated by fear that gains for workers will come at the expense of small-holders less able to pass on the cost of higher wage bills or taxes. In short, the very placement of the petite bourgeoisie means that it is perennially pulled and pushed in two directions: towards capital and against labor, towards labor and against capital.[29]

In trying times, this buffeting tends to become particularly pro-nounced. Economic crisis and intense struggle between labor and capital aggravate the ambivalence imbedded in the structural posi-tion of petit-bourgeois people. Faced with the prospect of disaster, they can feel pushed first from one direction, then the next, shuf-fled back and forth by forces beyond their control. "It is in such moments of extreme crisis," observed historian Arno Mayer, "that the vague sense of negative commonality—of being neither bour-geois nor worker—is transformed into a politicized awareness or consciousness of economic, social, and cultural identity." In these circumstances, the petite bourgeoisie "loses self-confidence and be-comes prey to anxieties and fears which," Mayer notes, "may well predispose it to rally to a politics of anger, scapegoating, and atavis-tic millenarianism." This volatility has repeatedly found expression in mass anti-Semitism, whose foremost proponents and front-line troops have come from the petite bourgeoisie.[30]

How this Janus-faced perspective will be resolved in any particular case cannot be ascertained in advance. Which way particular sections of the class incline depends, not simply on the economic situation, but on culture and politics as well. Some analyses suggest that where a labor movement actively fights for a universalistic agenda, it can rally much of the lower middle class behind its banners. Conversely, a labor movement with merely sectoral or ill-formed goals, or one that has suffered critical defeats, may forfeit that magnetic power.[31] Whether this holds true as a general proposition is not clear, but the latter situation undoubtedly characterized American labor in 1921. Thinly based, internally divided, and sorely defeated in the postwar contest with employers, it had lost the momentum that might have enabled it to attract those wavering in the middle.

These general considerations make possible a deeper understanding of the class composition of the Athens Klan. The structural position of the petite bourgeoisie, first of all, helps explain the otherwise curious lines of demarcation that struck contemporary observers of Piedmont communities like Athens. Both white-collar employees and craftsmen tended to identify with the "uptown" urban population and with landowning farmers in the countryside. In effect, middling men lined up against mill operatives and landless farmers—even though they distrusted the planters, manufacturers, and leading merchants and professionals who made up the élite proper.[32] This tradition, in turn, affected the way lower-middle-class men responded to the trials of the postwar years.

Faced with a shrinking pie, many small businessmen joined large employers in rage over what they saw as the impudence of workers, who by early 1920 had pushed wages in Georgia to an all-time high. So concerned were propertied Clarke County residents, some future Klansmen among them, that in 1918, they formed the Athens State Guard. For over two years, the sixty-odd citizens in this "local vigilance corps" conducted military training several times a week in case Athens should "have occasion to suppress outbreaks . . . [or] disturbances." Similarly, the local chapter of the Elks, which many Klansmen belonged to, boasted of their order's "aid in stamping out . . . the multiple-headed Red menace," including "exercising vigorous censorship over shop and office conversion" to catch any "Red utterances." Even after the onset of economic trouble in late 1920 helped employers cut wages, fear remained prominent through 1925 that the "exodus" of labor would strengthen workers' bargaining power and cut employers' profit margins. Despite the magnitude of the crisis, in fact, non-farm employers found themselves unable to drive real wages back to pre-war levels.[33]

Farm owners and ambitious renters, for their part, bristled against the bankers and creditors whose maneuvers boded ill for simple "dirt farmers." Yet many of these same rural people looked askance at labor demands that might raise the prices of manufactured goods. "Union labor must not get it all," one man from Blakely, Georgia, thus complained in 1920; "the labor & capital used in growing cotton ought to be respected & paid as much as the labor tha[t] spins & weaves. There is a wide difference now." In fact, unions or not, manufacturing prices and wages had proved "stickier" than farm prices and wages.[34]

The numerous white-collar workers who joined the Klan lacked even the nominal economic independence of small farmers and petty shopkeepers. Still, by tradition, they had enjoyed greater independence on the job and more income and opportunities for advancement than manual workers. Those advantages, however, now appeared less secure. As blue-collar workers improved their incomes, they narrowed the wage gap salaried employees had prided themselves on. By 1919, one historian reported, "the professional classes, salaried clerks, civic officials, police, and others in a similar category were worse off economically than at any time since the Civil War." Among white-collar workers in the country at large, signs of resentment against manual workers and feelings of relative deprivation began to appear. Such feelings also surfaced in the Klan, whose leaders often assumed an audience of non-manual workers. "When we pay such high prices, it is time," complained one Klansman with spiteful exaggeration, that "the laborer be satisfied with a mere Ford."[35]

White-collar workers saw their prospects jeopardized from above as well as below. By the 1920s, the rationalization and degradation of office work had become a topic of national discussion. A surviving copy of contemporary "Office Rules" for the Athens Railway and Electric Company shows how the de-skilling of white-collar work affected Athens men. The rules prescribed the minutia of office conduct, going as far as to warn clerks not to leave their posts "to run out and see a dog fight or hear a band play." More seriously, the new rules informed subordinates that they "should never enter a private office" without permission, that they should not "offer any suggestion unless asked for," and that "your directing officer is your Superior and not your friend." Such rules could not but insult those expected to obey them. Perhaps it was not coincidental that the author of the rules, the firm's president, was among the first to sign an élite anti-Klan petition, while several of his employees joined the Klan.[36]

Like low-level white-collar workers, craftsmen had to answer to someone else in order to pay their bills. Yet skilled workers' ownership of tools and relative freedom from direct supervision allowed them a kind of independence that neither unskilled workers nor sharecroppers enjoyed. "The key to the carpenter's outlook," as labor historian Irving Bernstein put it, "was his chest of tools. He owned them and he alone had the skill to use them."[37] The artisanal heritage was, after all, not proletarian, but petit-bourgeois: as small producers, craftsmen owned their means of production and jealously guarded their independence. The fitful spread of mechanization in the South likely gave artisanal traditions, weak though they were, greater viability.

Certainly this was the case in Athens. Even the local craftsmen not self-employed seldom worked in the kind of large plants whose sheer scale and anonymity helped break down proprietorial consciousness. Most manufacturing concerns in Athens in 1919 employed between one and twenty wage-earners. Even when the mills are included in the calculations, the average number employed by any given concern was only twenty-eight. Face-to-face relations between workers and employers were still common in such facilities, as were owners who worked on the shop floor. Both patterns blunted class divisions. Klansman *N. S. Rich*, for example, labored alongside the men in his employ. At least some of these workers joined the Klan; the same was true of *Hammond's* Bakery and *L. T. Curry's* Garage. Similarly, the skilled craftsmen at the Athens Foundry and Machine Works, where a few Klansmen worked in the 'twenties, maintained friendly relations with the boss. He took "the boys" on fishing trips in summer and put his own sons to work in the foundry to "make men" of them. Economic concentration and mechanization would in time erode the privileges of craft workers and widen the gap between them and their bosses. But in the 1920s, these developments were only beginning in Athens.[38]

Craft workers, moreover, still tended to think of themselves in petit-bourgeois terms. Even those in labor organizations, presumably the most class-conscious, often brought the individualism of small proprietors to their collective struggles. Most unions at the time—with important exceptions like the United Mine Workers and the Industrial Workers of the World—were craft unions peopled largely by native-born white men. They had their own version of the ideology of a commonwealth of self-made men. AFL unions' proprietorial approach to their members' skills and jobs often led them to collaborate with employers out of fear of unskilled labor. This was particularly true in the South, where the bogey of black competition served to discipline the unruly. Railroad work, among the best organized trades in Georgia, was a case in point: the railway unions fought nearly as hard against black competitors as against their employers.[39]

The dead hand of the artisanal past weighed on the Georgia Federation of Labor (GFL) as well. Notwithstanding its noteworthy left-wing positions on scores of issues, the GFL in the first decades of the twentieth century also demonstrated condescension toward unskilled workers, racism toward African Americans and immigrants, strong patriotism, and sexual conservatism.[40] The state labor federation had supported immigration restriction in 1907 in terms that would have suited Thomas Jefferson. It denounced the "flooding of the South and Georgia with . . . the scum of Europe, a people nowise in sympathy with the spirit of our institutions and form of government, and whose presence . . . [will put] us on a plane with the Northeast, with its tenements crowded with unassimilative pauper labor." One Atlanta railroad unionist exhibited a similar contradictory consciousness in 1922. He maintained that the great rail strike of that year was a fight "for the same freedom . . . our fore Fathers fought for 60 years ago"—the Confederacy.[41]

Unions were few and weak in Athens relative to those in larger cities, South and North. A lack of records makes it impossible to determine whether any Klansmen belonged, but several did work in trades where racially exclusive craft locals operated, including carpenters, painters, and barbers. For its part, the Atlanta Barbers' Union, which campaigned to drive black barbers from the trade, was praised by the Klan as a model of cooperation between "capital and labor." Earlier, Atlanta barbers had opposed "the tipping evil," stating that it might be acceptable for Northerners or blacks, but for them it was humiliating. Presumably, as freeborn white sons of the South, they thought themselves above the fawning behavior tipping might foster.[42] Craftsmen could also still approximate the "respectability" of the town middle class. Rare was the skilled worker in the Athens Klan whose wife worked outside the home, or whose daughters worked in other than white-collar occupations where they could still be considered "ladies."[43]

Yet, unskilled laborers, urban and rural, enjoyed no such advantages. In class terms, their membership seems difficult to explain. One might simply attribute it to the pull of the Klan's positions on other issues and move on, since these groups made up a relatively small part of the chapter's membership. But closer scrutiny of their situation bears out the emphasis here on the ambiguities in the social position and life experience of the petite bourgeoisie. An exploration of how these affected propertyless wage laborers can offer added insights into the social and cultural roots of the second Klan's rise.

In the white farm population, relations between different tenure groups were so complex as to undercut the emergence of clearly differentiated, much less actively opposed, classes. For many rural white men in the 1920s, propertylessness was but a phase in the life cycle. Beginning as croppers, they might work their way up to being renters, and perhaps even buy their own farms in mid-life. A detailed 1924 study of conditions in a nearby typical Piedmont farming county, for example, revealed that most farm landlords had purchased their acreage with earnings from years of farming on their own. Of the owners, two-thirds had worked as renters, and just under a fourth as croppers. The age hierarchy of the agrarian ladder was further evident in that only nine percent of men under thirty-five owned their farms, while more than half of men over fifty-five did. Agrarian class relations baffled even the trained investigators: twenty-three percent of white renting farmers were related by kinship to their landlords, and seven percent of tenants had tenants of their own, some more than one.[44] No doubt family ties bound a still larger number to other owners and planters. Under such circumstances, petit-bourgeois ways of seeing could thrive for reasons as much structural as cultural.

The membership of mill operatives is more puzzling; they had little chance of acquiring economic independence or even moving up the occupational ladder. Their communities would seem barren ground for the Klan's message of economic individualism. Given the history of mutual antagonism between unskilled people and the "uptown" types who dominated the Klan, moreover, one might expect mill men to shy from the order. Most did, but some did not. The decision of these mill workers to join the Klan was conditioned by the way the mill labor force in the South of the 1920s was still a class in formation. Innumerable ties linked its members to the petite bourgeoisie of the countryside. Most adult mill workers hailed from farm backgrounds; some combined mill work with farming. Others maintained connections to their rural roots through kin and friendship networks or membership in country churches. The very geography of most mill villages—decentralized, quasi-rural operations—encouraged ongoing identification with farm life.[45]

The Princeton and White Hall mill villages, situated in the countryside beyond Athens city limits, yielded many former and current residents to the Klan, mill workers along with neighboring farmers, merchants, and craftsmen. "It was just like living in the country," recalled *Scott Yarborough's* second wife of her childhood in the Princeton Factory district, where her father ran a picker machine in the mill while her mother raised "her own cow, chickens, and garden." Movement between farm and town characterized the lives of many non-mill blue-collar and lower-level white-collar workers as well. Oral histories revealed a great deal of shifting back and forth between farming, factory work, and marginal small businesses.[46]

The rural heritage of mill workers helped old Populist ideas, such as the critique of finance capital, retain a resonance for them. The tenant farming father of Klansmen *Earl* and *Travis Nunnally*, for example, had subscribed to Tom Watson's publications. As former yeomen or tenants themselves, many older mill workers had suffered from restrictive monetary policies and usurious creditors. Indeed, many harbored deep suspicions of banks. Some viewed as allies populist demagogues such as the latter-day Watson and Cole Blease, notwithstanding their often deeply anti-labor attitudes. Blease attracted support from mill workers, according to historian David Carlton, because his obstruction of "Progressive" reform appealed to a community that "regarded the government as its enemy, as an engine of oppression controlled by a hostile class."[47]

Also suggestive of the Klan's appeal to adult male mill workers was the way gender and age shaped class perspectives. In the 1920s, older mill men waxed nostalgic about agrarian living and the relative autonomy it had provided them; of all groups, they were the likeliest to want to return to farming. Older men were also likelier than young men to join the Klan. Most younger men who had grown up in mill villages, in contrast, had no desire to farm; they enjoyed the sociability that wage-earning made possible. They were, in a sense, the first large cohort of lifelong white working-class people in the Piedmont. Indicative of the generational differences in class identity was the way that younger workers—women and men—pioneered both the labor struggles of the period and the revolution in morals that so disturbed, not only uptown residents, but many of their farm-born parents as well.[48]

Like age, sex made for distinctive experiences of class. Mill fathers who lacked control over any productive property could still deploy the labor of other family members. Notwithstanding the advantages mill wives and daughters may have gained from their employment, husbands and fathers wielded ultimate authority over their earnings and behavior. Such control enabled the men to maintain a sense of themselves as masters of their households in the way that farmers or family businessmen might. Popular mill expressions caught the dynamics well. "If you've got sense," one operative's wife warned a single woman, "you'll stay your own boss." Likewise, many mill fathers disapproved of child labor, but believed that "you can't have the law tellin' a man what to do with his children."[49]

Like the vision of independent male proprietorship the Klan offered, its commitment to white supremacy was also attractive to many mill workers. The deep roots of racism in the yeoman heritage of mill workers were replanted by mill owners' strategy of employing an overwhelmingly white labor force, whose fear of displacement by blacks then made them more tractable. Whereas the ethnic bonds of immigrant workers in the mass production industries of the North often enhanced their sense of class identity, the ethnic identity of Southern mill workers—overwhelmingly Anglo-Saxon and Protestant—cut against class consciousness. If anything, it lined them up with the South's ruling élite against blacks and Catholics. *Kate Yarborough,* the second wife of *Scott Yarborough* and a longtime mill worker in her own right, articulated the logic behind this tense alliance when she said, in reference to the expulsion of African Americans from the mills after Emancipation, "I'm sure glad niggers are free, for if they warn't this mill down here would be filled up with 'em."[50] In these circumstances, the membership of some tenant farmers and mill workers in the Klan is not surprising; the wonder is that so few joined.

▲ ▲ ▲

The clustering of Klan members in middle-rank occupations was thus not coincidental. Embattled as they were, petit-bourgeois white men saw ample reason to organize. Yet, while class can be isolated as a category with which to analyze data on membership, in lived experience it was enmeshed with other social ties. Petit-bourgeois white men's fear of losing class position was no abstraction. In concrete terms, it meant losing control not only over their own labor, but also over African Americans, male and female, and the women and children in their own households.

The fraternal tradition Klansmen laid claim to illustrates this complex interweaving of class with race, gender, and age. Like Klansmen, most were married, middle-aged white men. The class composition of fraternal orders varied somewhat: white-collar workers and independent proprietors dominated Masonic lodges, while less prestigious orders geared to mutual aid also enlisted significant proportions of skilled workers and smaller numbers of semi-skilled workers. Still, like the Klan, fraternal orders included few unskilled workers. In general, their members had some claim, however modest, to economic independence.[51]

Deeper ideological affinities underlay the common membership profile. Fraternal orders, as sociologist Mary Ann Clawson has shown, gave ritual affirmation to an historically specific and exclusive male identity. The values celebrated by fraternalists were those of early-modern artisanal culture, in which male producer-proprietors prided themselves both on their own economic independence and relative equality and on their difference from and patriarchal dominion over the women, youth, servants, and apprentices who lacked such independence, therefore the right to full citizenship. Membership in such organizations skyrocketed in the late nineteenth and early twentieth centuries—just as the take-off of industrial capitalism, with its growing economic inequalities and unprecedented class conflict, was undercutting the foundations of artisanal culture. Even as fraternal orders embraced the market and male individualism, they protested the passage of artisanal brotherhood and hungered for a social order in which individuals subordinated their desires to the well-being of the community.[52]

"Through its construction of ties based on images of masculinity and craftsmanship," Clawson concluded, "the mixed-class, all-male American fraternal order worked to deny the significance of class difference and to offer gender and race as appropriate categories for the organization of collective identity." The Klan shared this project, albeit a shriller, more menacing version. It, too, sought to banish class from white men's consciousness and to redivide the world along the imagined "natural" lines of race, sex, and age.[53] Still committed to economic individualism, Klan members looked elsewhere to explain what had gone wrong with their world.

II

▲▲▲

CONTENT AND PURPOSE

4

▲ ▲ ▲

Reactionary Populism:
The Politics of Class

"For the first time in the history of our country," E. D. Rivers, a Great Titan of the Klan, state senator, and now candidate for governor, warned in 1930, "we are faced with being ruled by an oligarchy [of] centralized wealth." Rivers joined other Klansmen as spokespeople in a campaign being waged in Clarke County against "the invasion of these minions of monopoly—the alien chain stores." Employing a traditional idiom of popular protest, Rivers identified chain stores with "the taking away of the freedom of government from the masses." "The Little Group of Kings in Wall Street," one campaign advertisement admonished, is "very deliberately wiping out your independence." Another prominent Klan speaker expressed concern for the "young men of the country who will become 'automatons' " with no choice but to work for such monopolies. "Are your Sons and Daughters for sale?" the Citizens' Protective League demanded of local parents. "Do you realize you are gradually selling them into slavery?" The League implored residents not to let "Wall Street [continue] . . . destroying the community life of America." It was vital, an Athens Klan lecturer had earlier warned, to "break up [the] MONOPOLY that is now RUIN[IN]G and CRUSHING DOWN ON THE ENTIRE POPULATION of the world."[1]

In the fight against chain stores, radical rhetoric roused popular support for restorationist ends. The critique of economic concentration aimed, not to promote radical democratic change, but to

avert it. The Speaker of the Georgia House, Richard B. Russell, Jr.,
warned that "if the monopolistic tendency is allowed to continue
unchecked, it will result in socialism or communism." The Citizens' Protective League agreed. If the mergers weren't stopped, "we
are going to face exactly the situation that has been gone through
with Russia." River's claim that this was the "first time" that concentrated capital endangered the welfare of the people made a
mockery of the very populism it evoked. When he said the trusts
aimed to take away popular sovereignty, he clearly had in mind
the middling groups the Klan represented, since blacks and many
poor whites had lost long ago what little sovereignty they had.
Most telling, perhaps, of the Klan's reactionary motives in the
anti–chain store campaign were its associations. Rivers attacked
"atheism, communism, chain stores and companionate marriage"
as though they were of a piece.[2]

Local Klan lecturers on "Americanism," for their part, blamed
Jews and Catholics for the chain-store peril. One speaker dared his
listeners to "find out who owns stock" in companies like the "A
& P Grocery stores." Jews and Catholics, it seems, hid behind the
initials. He further complained that "department stores, all of
which are principally owned by Jews or foreigners," were pushing
out "American" businesses. He raved against the inroads made
into Georgia by Sears & Roebuck, which he insisted was owned by
"JEWS. JEWS. JEWS." Its entrenchment would "spell ruination" for
the state's independent merchants. He told listeners to find out
whether their druggists, undertakers, grocers, butchers, and clothing and shoe merchants were "JEWS OR CATHOLICS," and if so,
to boycott them and organize others to do the same. National Klan
leaders concurred. If present trends continued, Imperial Wizard
Simmons warned, immigrants from Southern and Eastern Europe
and their children would soon crowd native-born whites out of
"the business class."[3] Such charges struck a chord with local members, several of whom operated in competition with Jews and immigrants from Italy and Greece. The Bernstein Brothers, for example, prospered in undertaking and furniture sales, where Klansmen
Chester Morton and Bela Dunaway and his sons struggled to acquire a footing; the Michael Brothers owned the local department
store, and Joseph Costa and family ran a flourishing ice cream and
soda business.[4]

The campaign against chain stores illustrates how impossible
it is to understand the Klan if one conceives of it as a simple conservative force. The Klan was indeed conservative, fiercely so, as

the anti-Semitism and nativism of the chain-store struggle make clear. Yet the order's politics were different from those of the usual standard-bearers of conservatism, "the better people" in Clarke County as elsewhere. The Klan put forward a populist critique of American society suited to the middling men who made up the core of its following. They resented, sometimes vociferously, "the silk hat crowd" and the social transformation their reign had wrought.[5] Yet the Klan's was no ordinary populism. While it gave voice to middle-class fears of economic concentration and political disempowerment, it also put up ferocious opposition to social reconstruction from the left.

In this dualism lay the appeal of the Klan's class politics to the lower-middle-class men who flocked to the order in such numbers. It articulated the animosity petit-bourgeois whites felt toward *both* capital and labor—and it spoke in idioms at the core of American culture. From classical liberalism, the Klan drew its anti-statist economics. From republicanism, the Klan drew many of its assumptions about the good society and the prerequisites of citizenship. From evangelical Protestantism, it drew a structure of feeling that expressed its members' feelings of being embattled from above and below and that sanctified aggressive self-defense.[6] The composite is best described as reactionary populism.

▲ ▲ ▲

Klan leaders prided themselves on their fidelity to the vision of the founding fathers. On one hand, they exalted the old liberal tradition of possessive individualism. That property was the basis of freedom was the grounding assumption of the Klan's political theory. In this line of reasoning, as C. B. MacPherson observed, the individual could only be "free inasmuch as he was proprietor of his person and capacities." Politics thus became "a calculated device for the protection of this property and for the maintenance of an orderly relation of exchange." Klan propaganda often manifested such assumptions. "The function of the government," wrote a Klan-recommended writer, "is to protect individuals in their right of person and right of property." The great merit of the United States Constitution was that it had "established individual property rights more securely" than any other form of government, guarding against the twin dangers of "feudalism" and "all forms of socialism or communism."[7]

But Klansmen also looked to the republican tradition for backing about who had the right to participate in politics and about

what ends it should promote. Each of the requirements the order
thought necessary for individuals to function in a democracy, Simmons explained, "necessitates a large degree of economic freedom;
for without this, the individual is enslaved and driven in things
political." The conviction that disinterested devotion to the commonweal required economic independence provided the Klan with
a way to oppose the substance of contemporary democracy while
paying obeisance to it in name. Republican thinkers, after all, historically had excluded from the polity all economic dependents,
whether slaves, servants, married women, or youth. Now, Evans
insisted that in the American political tradition "equality" denoted
neither political nor social rights, but simply the chance to enjoy
the fruits of one's labor. As evidence, he adduced the suffrage restrictions advocated by a long line of leading American statesmen.
Charles Gould, whose work the Klan recommended to its members, informed readers that universal suffrage amounted to "mob
rule." Gould applauded the restrictions the writers of the Constitution had instituted "to protect authority from the populace." As
had some of the nation's founders, Klansmen took pains to distinguish the United States as a "republic" from both monarchy and
democracy. The United States' Constitution, as one writer put it,
"provided a middle ground between the two extremes of autocracy
. . . and democracy, the golden mean between heredity and direct
government."[8]

Klansmen committed themselves to what they understood as
the social vision of Thomas Jefferson: a republic of small proprietors. According to Simmons, the "real America has always been a
country America." "The farmer is the wealth producer of the nation," concurred the *Imperial Night Hawk*, "the backbone of all
industry." Simmons saw urgent danger in contemporary population trends. In 1920, for the first time, most Americans lived in
urban areas, and city residents cast the majority of votes. "Ignore
the problem of the white small farming class yet a little longer,"
Simmons warned, "and we shall be driven into farming on a great
scale, with armies of stolid peasants doing the work." Simmons
found the prospect of the day "when the countryside, like the city,
shall have lost its free independent population" horrible to contemplate.[9]

Although Klansmen shared Jefferson's adulation of independent farmers, they modified his vision to suit a modern class structure. They extended their loyalties to "the middle class" as a
whole, among whom Klansmen also included small businessmen,

white-collar workers, independent professionals, and skilled craftsmen. Common to both the yeoman ideal and its broad petit-bourgeois variant was the belief that the future of the republic depended on those with a stake in society. Simmons asserted that the success of early American democracy was attributable to the homogenous interests of "the small property-holders and skilled workers" who made up the citizenry. "The importance of the middle class in history," declared Charles Gould, "cannot be overestimated." As a mediating force between ruling and exploited classes, the middle class had provided stability to hierarchical social orders, from the ancient slave states forward. When the middle class was "depressed" or "destroyed," the ruin of whole societies ensued. Another Klan author cited "the failure of the middle class" as the preeminent reason for the problems of all nations—most immediately, his own.[10]

Now, it seemed, sinister forces imperiled the fragile balance a republic depended on. Far more than their contemporaries, Klan representatives gloomily foresaw the end of the republic. The rhetoric of republican alarm and despair—"luxury," "corruption," and "decay"—and morbid analogies between the contemporary United States and the declines of ancient Greece, Rome, and Old World Europe regularly peppered Klan propaganda.[11] Simmons predicted "a steady drift toward monarchy," a "natural outgrowth" for "a decadent republic" that had spawned "a great class of the rich on the one hand and a great class of the poor on the other." Both of these classes tended "toward corruption"; neither could be trusted to serve the commonweal. *"As a people and as a nation,"* he warned, *"we are face to face with dissolution."* "Democracy [was] threatened from every side" in the contemporary United States, "by greedy and designing powers above, as by a great mass of incompetent, unprincipled and undemocratic voters from below." "Both plutocracy and Bolshevism," announced Simmons, "are new forms of tyranny" the Klan would combat. The Klan, said Simmons' successor, Hiram Evans, was a tool for "the common people . . . to resume control of their country," implying they had already lost it.[12]

The foremost threat to the republic, in the Klan's view, came from below. The order's press and many of its leaders, North and South, saw the "labor question" as *the* critical one in their society in the early 1920s. They were terrified about how it would be answered. According to an Arkansas Klan leader, the "grave industrial unrest" of the era had driven men into the Klan. "Look at the

list of our strikes," explained Klan propagandist and preacher
Charles Jefferson. "In no other country is the conflict between labor and capital so implacable and so bitter." "Everybody who reads
the newspapers or talks with his neighbors," agreed Imperial Wizard Simmons, "knows that the conflict between labor and capital
is drifting us into another civil war. . . . And how much more
deadly is disunity between classes than between sections." Creating "a closer relationship between capital and labor" was one of
the Klan's oft-stated goals.[13]

So, too, was fighting the left. Indeed, nothing else elicited from
Klan members quite the same distemper. Socialism was "without
a possible exception . . . the most destructive philosophy preached
by thinking men." "The 'Red' is the most dastardly creature in-
festing the earth," intoned the *Searchlight* in 1923, "the worst
menace to civilization." The activities of the Industrial Workers of
the World and the Communist Party proved that "America Needs
the Klan." In later years, Klan leaders would boast that it was their
organization that had first "discovered Communism in the United
States and which first assailed it."[14]

Despite the lack of a serious radical presence in the state, anti-
communism in fact animated the Georgia Klan throughout the decade. Clifford Walker campaigned on it in 1920, charging his opponent in the race for the governorship with helping the "Bolshe-
vists" "interfere with our business and invade our states' rights."
Once in office in 1922, Walker proclaimed the need to suppress
"the Demagogue" who pitted "class against class." If his measures
were enacted, "the voice of the radical . . . of the bolshevist"
would be ignored. Ed Rivers likewise made an attack on communism a central feature of his 1930 campaign speech in Athens as
the Klan's candidate for governor. Throughout the 'twenties, the
Georgia Klan proclaimed the severity of the "Red Menace" and
heralded the Klan as "the only organization that is endeavoring to
combat same."[15]

Such vehement anti-communism seems odd in light of the
weakness of the American left relative to its European counterparts, particularly after the Red Scare, and its virtual absence in
the Southeast. But the paradox is more apparent than real. Communism condensed into a single entity all the leveling influences
Klansmen perceived in the contemporary world—from economic
concentration to the organization of African Americans, immigrants, women, and youth. Hence the depiction of it as "the extreme of Democracy." Anti-communism became a sign expressing

Klansmen's belief that all these hierarchies were linked: tampering with one would unloose all the others. In order to rouse mass popular opposition to changes in any one area, Klan leaders situated them in a worldwide conspiracy. The threat of "anarchy" from Bolshevism was thus discovered "even in villages and hamlets where it would be least expected."[16]

Not surprisingly, then, combatting communism—in all its faces—appeared an urgent task to Klansmen, particularly in the early 1920s, but even thereafter. "Klan Declares War on Radical Forces in U. S.," one headline thus proclaimed, while the Klan press published a whole series on the theme "Bolshevism—Menace to America." After the first World War, socialism ceased to be only "a remote threat" in the United States; "Never was the Red Peril so real." "Labor strikes take on the nature of social revolutions," complained Simmons. "The advocacy of Bolshevism arouses mighty crowds to wild enthusiasm." If such subversion continued, he predicted, native-born Americans would "probably divide and civil war will result."[17]

Above all, the Klan responded to the ripple effects of the October revolution in Russia. "Not only has bolshevism been the terror of Russia," declared Alma White, a female Klan propagandist; "it is now threatening the whole civilized world." It was "scattering its political ruin, industrial desolation and governmental diseases among all the nations of the earth." Simmons pointed with despair to the revolutionary stirrings of workers in Germany, Italy, and France, and to the budding rebellions of colonized peoples in Turkey, China, India, and Egypt as omens of the worldwide collapse of capitalism. "We must at least keep the hydrophobia out of this country," argued another Klan author, "while the political revolution lasts in Europe."[18]

Klansmen viewed that revolution as the antithesis of the philosophy the United States was founded on. Bolshevik success confirmed time-worn republican fears that a class of propertyless labor would disregard private property and upend social order. Simmons cited Jefferson's fear of cities as justification for his own belief that New York's "Bolshevism" showed cities were "a menace to democracy." Advocacy of "mass action and the General Strike" as more efficacious than voting by the IWW and the Communist Party was adduced as evidence of their menace to American institutions. The industrial unions the left championed amounted to a new form of "despotism"; their spread would imperil the "personal liberty of [the] capitalist." "The Bolshevist platform," as parodied by Imperial Wizard Evans, was "produce as little as you can, beg or steal from those who do produce, and kill the producer for thinking that he is better than you." The "canker of bolshevism" meant the "overthrow of all authority."[19]

In interpreting the threat from below, Klansmen reminded their compatriots of Jefferson's fear of the unskilled wage-earners. "Jefferson was right," affirmed Simmons. "Unless" the cities were "reformed they will destroy both democracy and civilization." Simmons believed that "real Americans" simply could not survive factory discipline or urban life; farm life was an essential component of American manhood. "Factory work," he said, calling on Social Darwinism to brand the unskilled as biologically inferior, "progressively selects those who are more and more unfit to be Americans." Like the "idle rich," they were "physical weaklings," "fit only to be the subject of a more or less absolute monarch." Indeed, in the view of the Imperial Commander of the Women's Klan, workers were "those least fitted by blood and training to rule."[20]

Such open hostility toward workers was rare, however; more commonly it was packaged in racism. Klansmen blamed virtually all labor trouble on immigrants and "foreign agitators"—those its publications depicted as "the riff-raff and outcasts of Europe." Simmons maintained that already, in the cities of the North and East, ethnic lines had become class lines. The urban working class was split between skilled tradesmen from Britain, Germany and Scandinavia, whom he respected as the modern heirs of artisans, and unskilled workers from Southern and Eastern Europe, whom he detested. The new immigrants lacked the capacity to appreciate republican institutions. "Rebellion against tyrants to them," said Simmons, "means acceptance of Anarchism or Bolshevism, or at least German state Socialism." "Just as long as there is a tendency of foreign domination in any industrial section of this country," concluded Leroy Curry, "there will be war—eternal warfare."[21]

Klan leaders' barely hidden disdain for industrial laborers was indicative of yet another way they turned old political traditions to new ends. One of these was producerism, the nineteenth-century popular ideology that asserted the dignity of labor and scorned wealthy "parasites" who prospered off the honest toil of others. In the Klan's hands, producerist language was used to attack, not the idle rich, but men out of work, unions who challenged employers' prerogatives, and those who wanted to get "something for nothing" from the state. Local Klan chapters thus sometimes showered their towns with leaflets that proclaimed that no man "has a right to consume without producing" and warned the unemployed and vagrants to "find work." In 1922, the Athens Klan applauded the Mayor's "War on Idlers," in which police combed the city to arrest those without jobs. Among the supposed idlers were many whose only crime was their inability to find work in a depressed economy. Just as the Klan censured the unemployed, so it urged on workers a return to "the gospel of industry," as if sloth were the source of the economy's troubles.[22]

The Klan also called on time-honored American values to oppose public provision for social welfare. Evans asserted that the hallmark of Americanism was "individual liberty." This necessitated "individual responsibility," which "automatically puts us in opposition to paternalism, socialism, and all other schemes for removing such responsibility." When Athens Klansman *R. P. Yarborough* ran for county commissioner, the only concrete position on his platform was thus "the lowest possible tax rate." "When the government takes a too personal interest in the people," E. Y. Clarke explained, "the people lose their will, their sense of responsibility." Klan publications and speakers asserted that no citizen should be a "sponge on society." They bemoaned increasing reliance on state aid and the "marked tendency toward paternalism in government."[23]

Klansmen much preferred the older tradition of charity, in which only the "worthy" poor received aid, and the act of giving boosted the status of the providers. One Klan propagandist thus asserted that *"Charity"* was the greatest virtue of Klansmen's God. Klaverns around the country made private giving a staple element of their practice. To be "unselfish" was one of the Klan's principles, according to an Athens Klan lecturer, "not for self but for others." The members of his chapter implemented this dictum by going in their robes to do "charity work," among poor widows and children in particular. Klan chapters also collected alms in meetings for the "deserving poor." This practice became so institutionalized that many whites in need would ask the Klan for material aid.[24]

That charity was more than a public relations gambit was evident in the prior commitments of some Athens Klansmen and their wives. Local fraternal orders, which shared many ideals and members with the Klan, had long engaged in charity work. Some leading Klansmen, along with the wives of a number of others, for years had assisted a local Industrial Home that aimed to train the

poor woman "to earn an honest living by the 'sweat of her brow.' "
It found work for "the reliable poor," spoke of the need to "aid,
without pauperizing" the "worthy," and explicitly excluded any
"tenant with any objectionable feature." Such aid was thus hardly
innocent of politics: it emerged from and gave legitimacy to a form
of government based on private paternalism in an organic, hierarchical social order. In charitable giving, the Klan acted out its opposition to both reformist and radical approaches to the problem of
poverty and the question of entitlement.[25]

Yet, as their fears of changes in the federal government suggested, the threats Klansmen discerned came from above as well as
below. "Increasing economic inequalities," Evans warned,
"threaten the very stability of society." The Klan's involvement in
the Athens anti–chain store campaign drew on a critique of economic concentration the national organization had developed over
the decade. Just as Rivers charged chain stores with failure to con-
tribute to the churches like local businesses did, so another Klan
lecturer told his audience that "THE WALL ST[.] CRACKER
TRUST . . . [was] encorach[ing] on your city without paying
TAXES." The allegation that trusts lacked civic commitment
flowed from a more general condemnation of "materialism" or
"Mammon" worship in American culture, which the Klan adapted
from nineteenth-century popular protest movements. Klansmen
complained that contemporary society, as Imperial Wizard Simmons put it, valued "money above manhood." This "love of
money," asserted E. F. Stanton, was "the root of all evil." "Who
loves Mammon, hates God." "In the strenuous rush of big business," Klan leader Edward Young Clarke mourned, "we have for-
gotten the spirit from which came . . . this great nation."[26]

The order held the unbridled quest for wealth responsible for
the decay of communal ethos. Klan propagandist Leroy Curry thus
accused "materialism" of "poisoning the minds and shriveling the
hearts" of America's young men, who looked out only for themselves now with no thought for "the advancement of the common
weal." Such criticisms suggest Klansmen's nostalgia for the
nineteenth-century petty-producer ideal, in which communal obligations and a sense of fair practice tempered the voracious self-
seeking that private enterprise might otherwise promote. Looking
back to the turn of the century from the 1930s, the wife of a local
Klansman gave voice to that yearning. "Everybody used to be
neighborly," she reminisced, "helping them that couldn't help
themselves. Now unless you are organized or belong to some club,

nobody pays any attention to whether you're starving, half-clothed, or sick." "Times sure have changed something terrible," she concluded.[27] In short, in the scramble for progress, society had lost its humane features.

Klansmen sought to compensate for that loss by practicing what they called "vocational Klannishness." It entailed "trading, dealing with and patronizing Klansmen in preference to all others," even if that meant sacrifice of time, money, or former friendships. At least ninety-one local Klansmen had co-workers in the order, a number in businesses owned or supervised by Klansmen. Some perhaps hoped that membership would secure their employment or promotion. Combining commitment to the order with a bid for the trade of members, many Klan employers instituted—and advertised—"100% American" employment policies: they would only hire Klan members. Others made their sympathies clear with firm names such as Kwik Kar Wash, Kountry Kitchen, or Kars, Kars, Kars. Athens Klansmen, for their part, contracted work out to chapter members, backed fellow Klansmen for appointive jobs, employed Klan members, and urged residents to boycott "alien" capital.[28]

The Klan also gave voice to apprehension that middling folks would lose power in the emerging political order. On the local level, Klansmen often pitted themselves against the élite sponsors of municipal "reform." Klansmen saw in so-called Progressive proposals for appointed city managers and commission governments attempts to constrict popular control over the state so that it could better serve business interests. In Georgia, Klansmen in several cities butted heads with wealthy élites over such "reform" proposals. The Atlanta-based *Searchlight* denounced those in its city as "imperialistic" maneuvers for the benefit of "the big interests through their well-organized commercial clubs and autocratic Chamber of Commerce." To thwart the Columbus plan, Klansmen beat up the new city manager and bombed the mayor's home.[29]

In Athens, the commission issue provoked conflict well before the establishment of the Klan. Élite civic and business leaders, many future Klan opponents among them, continued to push municipal reform proposals in the face of a stiff popular resistance. Some future Klansmen, for their part, rallied opposition to the measures. Thus, while mayor in 1914, *Wiley Doolittle* spoke out against a plan for commission government. Klan alderman W. R. Tindall took the offensive in 1924, putting forward a proposal to make the Civil Service Commission elective rather than appointive so that the body would "belong to the people." *O. M. Martin,* an elected official and Klan leader, likewise condemned a city manager plan to the Athens League of Women Voters in 1926, knowing it would cost him their votes in his campaign for state representative.[30]

Since the *Banner Herald* curried favor with the élite and generally ignored other points of view, little record survives of the arguments made by opponents of Progressive reform, Klan or otherwise. Still, once in a while they got to air their views. One such occasion occurred in 1926, when the Clarke County Democratic Executive Committee ruled Klansman James Willie Arnold, a local lawyer, off the primary ticket as a candidate for a judgeship. In addition to having served as the Great Titan of the regional Klan in Georgia, Arnold belonged to the Elks and the Woodmen of the World, and had served as treasurer of the Century Club. He took part in the men's Sunday school at First Methodist Church, where his wife belonged to the women's Bible class and the Women's Society for Christian Service. Nevertheless, Arnold was not good enough for Democratic Party leaders; or so he interpreted their challenge to his candidacy.

In a populist appeal to local voters, Arnold traced his roots in the district back to its frontier days. "My people," he maintained, "are all plowmen," "sturdy farmers and therefore common people," God-fearing men who had tilled Georgia's soil, voted for its party, and enlisted in its wars. As a judge, he had hoped to protect the rights of such common people from "the special privileged apostles" who sought to deprive them of their sovereignty. Arnold implored his supporters to fight efforts to substitute appointment for election of public officials. Specifically, he condemned "a certain element here, who claims by some sort of divine right, to be the Crown Prince of all the Political power in this section." These élites, among whom he included Democratic Party leaders and the League of Women Voters, were "afraid of the voice of the people." They had "inherited" this "bigotry" from their forebears, who had crushed the Independent revolt of the 1870s and later popular dissent in the Democratic Party. "I believe with all my soul," Arnold concluded, that "sovereignty rests in the people"; he had "dedicated [his] life along those lines" and would not stop now.[31]

But the question of what ends "the people" should serve and who, exactly, "the people" were elicited different kinds of answers. The purpose of charges that the local élite operated undemocratically was to rally support behind an agenda that was itself anti-democratic. The Klan thus pitted popular sovereignty for WASPs against the right of Catholics and Jews—to say nothing of African Americans—to hold office. A few years before Arnold's missive, another local Klan campaign had thus criticized the town's "exclusive patrician aristocracy" and called upon "the good citizens of Athens" to overthrow the regnant "corrupt political ring" and "take over the reins of power that are justly theirs." Yet, thus used, the "ring" seemed to refer to the modicum of electoral influence wielded by Catholics and Jews. One of the main problems restoration of so-called popular control was designed to solve, moreover, was "the Semitic influence in Athens"—in particular, the "unfortunate" fact that three Jews sat on the Board of Education. One suspects, too, that these individuals were singled out for attack, not for their ethnicity and wealth alone, but also for their liberal records. One of the three, at any rate, Moses G. Michael, had led the fight to abolish corporal punishment in local schools—in so doing winning unmatched affection from the city's children—and had worked to promote tolerance and cooperation among residents of different faiths.[32]

A similar spirit of reactionary populist dissent infused the Klan's hostility to the League of Nations and the World Court. In Athens, as elsewhere in the United States, Klan chapters fought against these initiatives. Such efforts, in the view of the *Kourier*, aimed at establishing a "Super State," a "gigantic trust" that would "rule the world . . . in the interest of a few." As the self-appointed representatives of small, local business, Klansmen perceived that they would lose out if large capital in the United States cooperated with its counterparts in Europe and Japan for a less contested division of the world's spoils. Klansmen felt keenly their "inability to compete . . . with great corporations . . . who do not hold their allegiance to one flag and government." They believed the World Court and like efforts at international cooperation were plots by "the international bankers."[33]

As had the Populists, Klansmen thus charged finance capital and its political allies with responsibility for public policies inimical to the interests of petit-bourgeois Americans. Hence a Grand Klokard (lecturer) wrote to Tom Watson in 1922 to hone his own arguments about how "the Wall St. Bankers" and Federal Reserve policy had "brought hard times upon us." Watson's protégé, Klansman and state secretary of agriculture J. J. Brown, blamed the agricultural disaster in 1920 on the Federal Reserve's refusal to ease credit for farmers. Other Klan politicians earned accolades from their fellows for continuing Watson's attacks on President Harding and his alleged Wall Street paymasters for having produced the "wrecked farmers, banks, [and] small merchants of the South and West." More generally, the Klan accused "big financiers" of "robbing the people." The damage caused by the boll weevil was said to pale compared to that inflicted by "gamblers and speculators in manipulating the market."[34]

Klansmen inherited hatred of Wall Street from nineteenth-century petit-bourgeois radicals who located the sources of inequality not in the economic system itself, but in relations of exchange and unjust laws. The analysis they held in common maintained, as Watson put it in 1921, that "the money question . . . is the greatest of all economic questions." "In all ages," *The Searchlight*'s editor explained, "the financiers have been able to completely rule and ruin the nations." The source of the current troubles in the economy was that the United States had been "turned . . . over to the great financiers and the transportation companies." "And when the government loses control of those two things," the editor explained, "the citizens have but little to hope for."[35]

Like the resentments against Democratic Party élites, the critique of finance capital had deep roots in the South. Farmers ensnared by crop liens, mortgages, and monopoly control of the marketing and transportation of their crops had good cause to hate financiers and trusts. The economic crisis of the 1920s helped revive this antipathy as the banks' tight-fisted credit policies exacerbated the plight of hard-pressed residents. Local Klan leaders gave voice to the resulting popular hostility. Klansman George D. Bennett, for example, campaigned in 1926 on a platform that included "better banking laws." "I feel sure," Bennett intoned, that "the people of Georgia have suffered enough from high finance and rascality in high places."[36]

Klan leaders employed these complaints to their own ends. They made finance capital the scapegoat for a corporate order in which it was inseparable from industrial capital. Klansman E. D. Rivers thus implicitly exonerated industry when he told an Athens audience that unemployment and hard times resulted from mergers of big banking institutions. Over and over again, the Klan counterposed "the genuine Americanism of Henry Ford"—not coincidentally a virulent anti-Semite—to the alleged cupidity of John D. Rockefeller and his fellow "international money sharks."[37]

On the rare occasions when the Klan criticized not just monopolists or financiers but capitalists as a group, the charge was usually that they were insufficiently patriotic and racist. Capital's "love of money," the Klan alleged, had led it to import people of "inferior races" to the United States. A meeting of Grand Dragons denounced opposition to immigration restriction by "the big employers of pauper labor" reliant on this "European 'riff-raff' . . . the very scum of the earth." Employers used this imported "cheap labor," according to the Klan, to lower the living standards of white, Protestant Americans.[38] Klansmen's criticisms of large employers thus centered, not on their exploitation of workers, but their infidelity to their "race" and nation.

The communalist spirit of "vocational Klannishness" was similarly double-edged. While its advocates condemned the dominance of market values, they did so in a calculated effort to drive out Catholic, Jewish, and African-American entrepreneurs whom Klansmen otherwise had to weigh in against in an impersonal marketplace.[39] Klansmen's criticisms of Mammon had the same quality: insubstantial at best, reactionary at worst. Like "selfishness," "materialism" was a moral failing. Combatting it required, not systemic change in the economy, but rather a spiritual awakening.[40]

▲ ▲ ▲

For that change of heart, Klansmen looked to evangelical Protestantism in general, and to fundamentalism in particular. The Klan overlapped with, and helped feed the larger upsurge of, fundamentalism in the 'twenties; indeed, the very term "fundamentalist" only came into use in 1920. The 1920s "marked a crucial transition in American religious history," according to one of its leading interpreters. The evangelical Protestant establishment found itself on the defensive in the face of unprecedented competition from Judaism and Catholicism, as well as from advocates of the social gospel and higher biblical criticism. Belief in the inerrancy of the Bible, the truth of the creationist account of human origins, and an otherworldliness that justified the status quo on earth became defining features of the fundamentalist world view.[41] In it, Klan leaders found the spiritual anchor for their ideology. To combat the rival extremes of "democracy" and "monarchy," the Klan advocated fundamentalism as a parallel, all-encompassing explanation of and prescription for social order: a reactionary populist vision backed by the authority of the Almighty. In the Klan's hands, Protestantism was thus very much a political creed: it provided answers to the basic questions about who should wield power over whom, and how and why.[42]

Above all, "Christianity" was the natural "foe of Bolshevism." A revival of old-time Christianity, said the national leader of the Women of the Ku Klux Klan (WKKK), was "the only thing that will save our nation in these days of unrest and disturbance." "Radicalism can be combatted through religious education," the Klan press explained, for "it brings a contentment of mind that does not hearken easily to disruptive theories." As one internal Klan publication put it, "a revival of the old style religion" was "the most essential program throughout the Invisible Empire." "This will lead to victory in other aims."[43]

The last fortress of hierarchy and discipline, "old-time religion" was also the sanctuary of nationalism in a dangerously cosmopolitan world. The Klan considered Protestantism "the very foundation of our peerless civilization." As proof, the Klan press noted the Bible's importance to the nation's founders and to a string of Presidents, including Woodrow Wilson. Georgia's Grand Dragon described contemporary challenges to Protestantism as having created the "most dangerous situation . . . since the Revolutionary War." "*The Constitution of the United States is based upon the Holy Bible and the Christian religion*," he wrote, italicizing the words lest his readers fail to appreciate the connection, "*and an attack upon one is an attack upon the other.* If Christianity is destroyed in America," he asked, "how long would our Government endure?" The answer put forward by another Klan leader, himself a minister, was unequivocal: it "would in short order be a heap of ruins. In the present order of civilization, a nation without churches cannot exist."[44]

Yet the Klan's religion was not conservative in a conventional way. Rather, Klan fundamentalism had the same anti-élitist thrust as the reactionary populism it served to sanctify. Protestantism, explained Imperial Wizard Evans, was "more than religion." It was an expression of the "spirit of independence, self-reliance and freedom" of northern Europeans that promoted the rise of capitalism hundreds of years ago. America's triumph in the world of nations came from its founding faith, for "only men and women who dared to speak to their God face to face could have had the courage and self-reliance necessary to the mighty work of the pioneer." Evans thus interpreted Protestantism as, at root, a "protest against human authority over the souls, the bodies, the rights and the earnings of men."[45] In other words, Protestants' historic rejection of papal authority and belief in God's direct relationship with believers was of a piece with their rejection of feudalism.

Through Protestant fundamentalism, the Klan consolidated middling men against the impious classes beneath and above them. "The rabble" and "the mob," explained the Klan's Imperial chaplain, were to blame for the death of Christ. Just as the rabble did the work of the money changers in the Temple, so did the "riff-raff" and "the 'Upper Tenth'" share a taste for ridiculing men of the cloth. Indignant at the "burlesquing of the Protestant clergy," many Klan chapters led campaigns against offending movie theaters and literature. The Colonial Theatre in Athens thus cancelled a showing of *Simon, Called Peter* after protests that the film belittled the Protestant faithful.[46]

The promotion of Protestant fundamentalism also aided the Klan's mission in less direct ways. By mystifying relations of power and promoting uncritical awe of "higher authority," fundamentalism performed a vital service for the Klan. Klan leaders consciously employed the power of "ritual and formalism," in the conviction that "the people want and need faith and symbolism." "I wish to believe, and I do believe," was how one Klan tract described the order's rejection of critical readings of the Bible. The appeal of "the old-time, old-fashioned Gospel" to Klansmen, according to one, was precisely its otherworldliness, its "sermons . . . full of spiritual pathos and the power of the Holy Spirit."[47]

Central to the old-time religion the Klan was so enamored of was belief in the literal truth of the Bible. "The Klansman," explained one manual, "pins his faith to the Bible as the revealed will of God." "Human reason bows" before this "Impregnable Rock," "afraid to doubt." Such magical thinking helped Klan leaders subdue reason and elicit unquestioning loyalty to the status quo. Rational thought derived from philosophical materialism, in contrast, was deeply threatening. Darwin's theory of evolution was thus said to be part of the larger threat of "materialism" to "moral integrity." Indeed, Georgia legislators found the teaching of evolution to be not simply "improper" but "subversive."[48]

The Klan's opposition to religious liberalism was thus eminently sensible. Once the right to interpret the Bible was conceded, much less the right to enlist it in the cause of social reform, the whole structure of feeling that made fundamentalism such a formidable buttress of social order would collapse. Seen through Klan eyes, the social gospel acted as the entering wedge of communism. The "wider ramifications" of the social gospel, according to the Klan, included "pacifism, Bolshevism, attacks on the government, bitter arraignments of the churches, anti-capitalism, [and] sympathy with Russia." Klansmen's feelings of imminent disaster led them to renounce not only reason but tolerance as well. No "man or woman who opposes the government," proclaimed Klan luminaries, "has any right to protection." They "should be deported or placed behind prison bars." The national office instructed local Klans to fight "any kind of propaganda that belittles this government . . . or that calls into question constituted authority from any source." Critics "must either shut up or get out."[49]

To bring the Bible's missives to the public, Klansmen actively promoted religious revivals. Baptist and Methodist pastors of the Athens Klan worked together to organize one in 1925. Afterwards, V. D. Scroggins wrote on behalf of his church committee to thank his fellow Klan members for their "interest in the services and . . . liberal contribution," delivered in a robed appearance with the permission of the presiding minister. The "greatest Revival in many years," it won new members for the church and "brought back to the Master" many lapsed Christians as well. Local Klansmen also paid tribute to the individual apostles of "old time religion." They attended a service of the Young Harris Memorial Church en masse and in full regalia to make a donation in praise of the minister's work "in winning souls for Jesus Christ."[50]

Revivals were much more than religious events. They provided arenas in which believers worked out their social fears and aspirations through potent, visceral experiences. "Like political demagogues, the evangelists," as Lillian Smith once observed, "won allegiance by bruising and then healing a deep fear within men's minds." Just as the emotional character of revivalism served the Klan's purposes, so, too, did the interpretation of the Bible at these gatherings by fundamentalist preachers. Their jeremiads invested conservative political messages with divine sanction. In one of the revivals supported by the Athens Klan, for example, the evangelist "flay[ed] modern day youth." Girls, he said, spurned femininity, boys acted "sissy," and their nighttime joyrides were taking them down "the surest road to hell." For his part, leading contemporary evangelist and Klan fellow-traveler Billy Sunday claimed God's benediction for racial segregation, nativism, anti-communism, free enterprise, anti-Catholicism, anti-Semitism, Prohibition, and law and order.[51] By tapping participants' fears of damnation and hopes for salvation, clergymen like Sunday infused right-wing political commitments with an emotional force that made them impervious to rational disputation. There was thus little contradiction between the Klan's religious enthusiasm and its larger political program.

Like its religious enthusiasms, the Klan's religious antipathies were shot through with political content. In Klansmen's vociferous and oft-repeated complaints against Catholicism, the movement's dual animus of anti-radicalism and anti-élitism was much in evidence. And here, too, the order drew from an older republican tradition. That tradition had a long history of antipathy to the Catholic Church for its alleged monarchical ambitions; hence, it is not surprising that the Klan referred to Roman Catholicism as "the extreme of monarchy."[52] To a population schooled in the notion that individual conscience and reason were the proper basis for political action, Rome's authority over its parishioners appeared threatening indeed, particularly when wielded over hotly contested public issues such as Prohibition.

Yet the Klan did not simply reassert a static cultural tradition of anti-Catholicism. Klansmen adapted anti-Catholicism, as they did republicanism itself, to the concerns of their era.[53] Most commonly, the Klan charged Catholics with political offenses that echoed time-worn themes of anti-papal agitation. "The trouble with the Roman Catholic Church," explained an article in the Klan press by Warren Akin Candler, bishop of the Southern Methodist Episcopal Church, was "that it seeks to be both a church and a political party." An Athens Klan lecturer likewise characterized the Catholic Church as "a great political machiene," referring to its mobilization of Catholic voting power. Echoing the arguments made for disfranchising black men, Imperial Wizard Evans pointed out that eight million Catholic voters could now determine the outcome of elections in which Protestants divided.[54]

The Klan's diatribes centered on the implications of Catholic power for the future of the republic. The Georgia Klan thus prided itself on its fight "to maintain a free republican form of government against the subtle political encroachments of the self-arrogated, infallible, universal autocracy known as the Roman Catholic hierarchy." Such charges had a long history in America, dating back to the fear of Catholicism as a worldwide "conspiracy against liberty" in the age of the American Revolution, a history the Klan used to defend its positions. The *Kourier* thus quoted Thomas Jefferson to the effect that the United States would interpret an influx of priests "as a National aggression on our peace and faith."[55]

The crux of the problem, so Klansmen said, was the authority of the pope over the laity. Evans charged that Catholicism was imbued with "the monarchical idea of the individual as subject instead of citizen" in contrast with the republican idea, which "exalts the individual, clothing him and her with all the attributes of sovereignty." The pope's intervention in politics thus posed a choice for Americans, said George Estes: "the rule of the people or the rule of a monarch?" Klansmen believed that the church aspired to a theocratic state, and read in the activities of the church in America and the Catholic political parties of Europe evidence of this aim. Indeed, they charged the Knights of Columbus with being a "trained and equipped army" bent on conquest for Rome. Klaverns around the country, Athens included, used fraudulent "oaths" from Catholic lay organizations to "prove" a conspiracy to violently overthrow American institutions.[56]

Yet, if populist themes animated the Klan's case against the Catholic church, its core was preservationist. This urge makes sense of the otherwise ludicrous charges that the Catholic hierarchy fomented revolution, even that it collaborated with Communist parties. The logic of the accusations came from the republican conviction that tyranny bred rebellion. Klan writers thus blamed the Russian Orthodox Church (equivalent, in their minds, to the Roman Catholic Church) for the Russian revolution; they implied that the expansion of Catholic political power in the United States would ultimately produce a similar upheaval. Indeed, the combination of "commercialism" and Catholicism in America was paving the way for communism.[57]

In fact, the Klan's complaints against the church followed the pattern of its criticisms of élites more generally. Insofar as the Klan reproached the Catholic hierarchy, it was, at the end of the day, less for its conservatism than for the way its collaboration in injustice furthered tendencies toward socialism. The Klan's denunciations of the church hierarchy's parasitism on its parishioners should be read in this light. "My God," George Estes thus proclaimed, "does not build cathedrals and temples . . . from treasure wrung like blood . . . from the wearied frames of millions he has impoverished." Yet Klan empathy was selective. The hardships imposed on employees of miserly Protestants did not arouse it. More importantly, the Klan used the rhetoric of protest against Old World despotism to attack, not an élite, but an already beleaguered, subordinate people. When ethnic Catholics made up so large a part of the wage-earning population, attacks on Rome but thinly veiled anti–working class sentiments. The Klan thus assailed parochial schools for encouraging "class-consciousness, bitterness, and defiance."[58]

All the talk of rights and freedom notwithstanding, in fact, the Klan's populist assault on Catholicism was hardly democratic. Denunciations of Catholic "block-voting" implied that all cohesive voting, notably by trade unions and immigrant working-class communities, was wrong. Earnest backers of the Protestant minority in Ireland, Klansmen repudiated democracy when it meant Catholic self-rule. At the most basic level, then, anti-Catholicism offered a way to repudiate the consequences of democracy without appearing to reject democracy per se. The indispensable ideological load carried by anti-Catholicism helps explain its otherwise paradoxical prominence in a town such as Athens, which counted only two hundred Catholic residents in 1916, and a state such as Georgia, which had fewer than fourteen thousand—equivalent to less than two percent of its total population.[59]

Nevertheless, as the case of anti-Catholicism illustrates, the Klan rarely expressed its class politics directly. It could not. For to do so would have undermined the movement's attempt to defuse class conflict by denying the existence of class as a meaningful division in society, and offering in its stead religion, race, and sex as proper poles of identification. Its mission here would be aided by the way that, in the white petit-bourgeois world most Klan members came from, the very meaning of class was constituted through relations of race, sex, and generations—most commonly, through an obsession with respectability.

5

▲ ▲ ▲

"Cleaning Up" Morality:
The Politics of Sex and Age

As a site of debauchery, the Imperial Palace of the Klan rivaled some of the motion picture sets its representatives so habitually rebuked. The national chaplain of the Klan, Caleb Ridley, a well-known Atlanta minister and Prohibitionist, was arrested in 1923 for driving while intoxicated. Imperial Wizard Simmons, according to numerous people in a position to know, drank heavily, relished pornography, and regularly patronized prostitutes. His assistants E. Y. Clarke and Mary Elizabeth Tyler, the masterminds of the Klan's growth after 1920, were once arrested together, inebriated and in the flesh, during a tryst in a hotel. The list could go on. The unsavory conduct of the highest leadership helped cause its overthrow and replacement in 1922—and cost their movement much of its following over the decade.[1] The hypocrisy of the order's national officials is more than ironic, however. It underscores, in a circuitous way, their Machiavellian appreciation of the appeal of morality campaigns for the people they hoped to attract. And, indeed, rarely did local chapters suffer such scandals. To rank-and-file members, the Klan's professed commitment to purity meant a great deal.

The Klan in fact staked its bid for power on its value as a tool for restoring "traditional" values. Purity campaigns became the core of recruitment drives in localities around the country. A former Klan organizer reported in 1924 that officers from the Imperial Palace instructed local Klans, once organized, to "clean up their towns." One of the Klan's oaths of citizenship thus bound the member "to correct evils in my community, particularly vices tending to the destruction of the home, family, childhood and womanhood." Klaverns around the country boasted that their efforts had arrested the drift toward immorality: "whole communities that seemed travelling fast on the road to hell suddenly have turned toward Heaven." Clearly, national officials thought these clean-ups an excellent way to prove the value of the Klan to its constituency. They were right. Across the country, the Klan ingratiated itself with solid middle-class white citizens on the basis of its unrivaled commitment to community moral "clean-ups."[2]

The demand for purity served several ends. Its immediate object was to subdue internal threats to family discipline: from men who failed to act their parts, from children who repudiated the Spartan tastes of their parents, and from wives who sought to renegotiate the marital compact. Yet the movement's emphasis on moral purity was more than a strategy to boost enrollments and bolster parental authority and male dominance. Respectability was also a prime marker of status for the lower-middle-class men who joined the Klan in such numbers. Their moral standards were the visible sign of their difference from, and purported superiority to, both the working class and the élite.[3] Taking a stand for the old-time virtues was thus also a means of building internal cohesion and collective confidence among the white Protestant petite bourgeoisie the Klan hoped to mobilize.

▲ ▲ ▲

A postwar moral panic paved the way for the growth of the Klan in Clarke County. It was not the first time local residents had worried about alcohol, gambling, and prostitution; agitation over such issues had a long history in Athens as elsewhere in the nation. But when such affronts accompanied economic disaster and social strife, and when premarital sex joined the list of transgressions, the mixture proved volatile. Experimentation in marriage or morals, explained one Klan-recommended writer on the family, was particularly risky "in this age of unrest and discontent and instability." The editor of the *Athens Daily Banner* summarized the common diagnosis in 1921: "The tendency of the times toward disorder and crime and revolution and unrest" had a taproot: "disregard for authority both parental and governmental." "Bolshevism and socialism and all the radical isms rampant" were the harvest of lenient parenting. The return of the state's authority would "have to start in the home"; parents must "compel obedience and respect."[4]

What both writers, one a mainstream journalist and the other an ardent anti-democratic theorist, had in common was a conviction that ordered, hierarchical families undergirded ordered, hierarchical society, and a perception that the behavior of contemporary young people, women in particular, endangered both. Unnerved by postwar unrest, they looked to the family to provide a model of stability and authority for the rest of society. In "purifying" domestic life, they would rid the wider world of its evils.

Evangelical ministers first sounded the alarm in Athens. In a 1920 sermon, the Reverend W. F. Dick of the Athens Free Methodist Church denounced "the evils of our day" abundant in Clarke County. Dick flayed "the public dance hall . . . the pool room, the card table and other kinds of sin practiced in Athens." Many adults, locally and nationwide, shared his anxiety about the implications of the "dance mania" for the sexual mores of youth, particularly girls.[5] The coupling of dances with other alleged sins illustrated how hostility toward youthful sexuality merged with a more sweeping urge to control subordinates.

Pool halls, for example, affronted the very premises of the Protestant work ethic. "The pool room," explained a representative of the Georgia Anti-Poolroom League in 1922, "diverts man power" from economic development and family support. It encouraged general moral laxity, particularly in young men: "idleness . . . dissipation . . . intoxicants . . . profanity . . . lascivious stories . . . [and] the love of chance." Woodruff and his associates sought a ban on pool rooms, a topic the Georgia legislature debated throughout the 1920s. These efforts evinced the belief widespread among middle-class Southern whites that pool rooms were the hang-out of "trashy" mill people, and a disgrace to their communities.[6] They were also tangible symbols of their patrons' defiance of the mores, hence the authority, of town leaders.

The Klan first sallied into public view in Athens in 1921 as the war on vice gained momentum. In that year, bootleggers mocked the élitist pretensions of the local guardians of morality and in so doing, confirmed their fears about the correlation between alcohol and rebellion. When the mayor warned bootleggers to quit their trade or leave town, they responded with a "defiant letter" of their own. In it could be heard echoes of the labor revolt that had recently swept the nation. Banding together as "Bootlegger's Union Local No. 13," the writers defended the "small pint man." They served notice to the Athens press that "a bootlegger is making his money as honest as some of these nice honest-to-goodness people, who are always kicking about what the other fellow is doing." And they vowed to continue selling to anyone who wanted to buy their wares.[7]

Their humor exposed the class tensions over Prohibition rife in the Piedmont. Many poor farmers found that their produce brought in more money when converted to bootleg, and ready cash was a scarce commodity in the farm depression of the 1920s. These people and many of their mill-working counterparts also accepted spirits as a natural part of social life, at least as a personal choice. They had little enthusiasm for the stringent mores of the town middle class. Southern mill owners, after all, often built the churches and paid the preachers who promulgated these values, along with the notion that workers' poverty was the fruit of their own vices. Hence it is not surprising that the Georgia Federation of Labor, like the labor movement at large, distrusted the prohibition movement. Indeed, the GFL voted down every resolution for prohibition that came before it in the first few decades of the twentieth century.[8]

The insurrectionary spirit of the bootleggers was more than many upstanding Athens residents, no doubt conscious of these class tensions, could bear. Charges mounted through the next few months that the police refused to stop the open traffic in liquor and gambling. Finally the respectable lost their patience with equivocation. The Ministers' Association of Athens and Clarke County issued a statement in September proclaiming that "the fearful disregard of law now rampant," if allowed to continue, would "bring us a harvest of revolution and an orgy of anarchism." Hyperbole notwithstanding, these concerns were deep-rooted. Earlier that year the same organization had warned that "at a time when the world [was] seething in social unrest and disquietude, when anarchy overwhelms whole sections of the world," no law-breaking could be tolerated.[9]

The ministers' warning roused others to the assault on sin, which it seemed not only the defiant bootleggers but also the sons and daughters of upstanding families now indulged in. A few days later, the Athens press issued a call for the re-establishment of the "old family altar," a condemnation of youthful nighttime automobile riding, and a demand for mothers to keep a better watch on their daughters. Judge Blanton Fortson asked the grand jury convened the next month to go on record against "loose morals" in the county, in particular the "lax moral" state of young people.[10]

The jury complied. It, too, registered disapproval of the "excesses" of "the younger generation." The jurors amplified Fortson's charges, however, as to gender, social disorder, and the need for rigorous state intervention. Ignoring young men's part, they cautioned parents not to allow their daughters out for nighttime "joy-riding." They expressed concern that "a reckless abandon" had developed among youth in recent years. And they charged authorities with either "absolute inefficiency . . . or a wilful neglect of their duty" in regard to Prohibition enforcement. Their recommendations included replacement of these officials if necessary, chain-gang sentences for bootleggers, and the placing of all "places where people are liable to congregate . . . under the most rigid scrutiny." In effect, they declared a moral state of siege. A local revivalist then took up the campaign in his sermons. These combined efforts prompted city officials to a new crackdown on prostitution and Prohibition violators.[11]

Still, local purity advocates remained unsatisfied. Several established local men, including Klansman *Wiley Doolittle*, incorporated a new newspaper over the summer, the Athens *Daily News*. It began publishing sensational exposés of local vice and chastising public officials for inaction. Some candidates for local office charged the city administration with having allowed "immoral houses" to remain.[12] In the resulting furor, voters turned Mayor Andrew C. Erwin, a liberal and later outspoken Klan opponent, out of office and replaced him with purity advocate George C. Thomas.

As the community divided, local Klansmen moved into the breach. In March of 1922, they passed and published resolutions applauding the new mayor for "making our city a better, cleaner and safer place to live in" by upholding his pre-election pledges. They commended as well the Athens *Daily News* editorials "calling for better ideals and morals." The Klansmen explained that since their organization "consists of 100% Americans and stands for law and order and for a cleaner and better government," they would always be ready to assist the authorities "in matters of this nature." Indeed, a few took part in a liquor raid that resulted in a brawl as one of the targets defended himself against the intruders. Local evangelical ministers, for their part, kept up the squall. The Reverend Morgan a few weeks later mounted a "virile attack" on the "lust" he found "rampant" in Athens, stoking suspicions about what young people did at their dances after their chaperons left.[13]

Shortly thereafter, two special committees were set up to inquire into the extent of local vice. The grand jury committee, which included two prominent Klansmen, issued a stinging report indicting the moral caliber of the community. The report encapsulated the spirit of the era's anti-vice activity. It explicitly connected the decline in moral standards with the worldwide "vicious aftermath of the world war" that had taken its toll of the economy and other social relations as well. The jurors vented frustration at their inability to make their authority felt; their efforts to investigate had "been met by perjury, evasion and intimidation of witnesses." They complained of "the thorough organization of the criminal classes," who "are so banded together" as to make effective prosecution nearly impossible. They alluded resentfully to the complicity of members of the upper classes in vice and consumption of liquor. And they called for harsh penalties "to stem the tide of lawlessness."[14]

The jurors reserved their most animated comments, however, for their discovery of widespread illicit sex. Various wooded areas around the county were being used "as places of assignation" by local men and women, "many of them young and unmarried—from all sections of the city." In other words, not only mill youth and others of less concern to the establishment, but also middle- and upper-class white young people now engaged in forbidden sexual activity. "The evidence as to the objects of these visits are so conclusive and so appalling," the grand jury reported, that Athens parents would be horrified "beyond expression" if they investigated for themselves. In assigning blame, the committee specifically accused local mothers of "proving derelict in their duty toward their daughters." The report called for a series of draconian measures: hiring extra officers to conduct round-the-clock patrols of the woods, arresting those caught, and shaming them through the publication of their names. It also called for more rigid restrictions on dances, out of the conviction that the sexual suggestiveness of young people's dance styles "open[ed] the door toward excesses which we find it horrible to contemplate." That a newborn infant had been found dead recently in a sanitation truck hardly helped allay their fears.[15]

Yet the politics of morality in Athens was never separate from the racism, nativism, and class prejudice earlier campaigns had first infused it with. For the grand jurors, as for many of their peers in and out of the Klan, hostility to sexuality fused with concern about the security of the larger social order. The jury now insisted that since sexual control was "of such fundamental importance to the very integrity of our social structure," community leaders must overcome their reticence and promote "public discussion" of the danger. Their own recommendations to combat sexual indulgence among local youth were "concerned with the preservation of the social structure and racial purity."[16] These frank comments indicate the conviction among white civic leaders that the hierarchies of private life served public order, that regulation of women's sexuality buttressed white supremacy. Tolerance of sexual autonomy for women might erode the social order.

The Clarke County Klan organized a large rally, its first reported one, soon after the jury announced its findings. As significant as the timing was the way the rally bolstered the Klan's image as an ardent defender of public morality, with all the broad connotations this morality had. Shortly afterwards, a local nonmember announced publicly his admiration for the order's stand "for good citizenship." He maintained that, whatever might be said against the Klan elsewhere, in Athens, people viewed it as an asset to the moral health of their community.[17]

By this time, some members of the local governing class realized that all their talk of moral declension had helped create a monster. They now tried, in vain, to put the genie back in the bottle. "It is no exaggeration," bemoaned the *Banner* in late April, "to say that the Athens public is fast drifting into hostile camps." Its editor, who had earlier insisted that only the reassertion of parental authority could save ordered society, now accused purity advocates of inciting hysteria. "No town can exist," he warned, "if half the population is spying and reporting . . . against their fellow citizens." He urged residents not to allow "politicians to make political capital out of" local morals, but to leave them to the mothers, churches, and social institutions that had long controlled them. And he begged "some of the older, more conservative citizens" to step in and avert Armageddon.[18]

The fears of some prominent citizens about the campaign went deeper than they cared to detail in print. For, by this time, the Athens Klan had begun terrorizing area residents through vigilante forays, which the *Banner* alluded to in oblique terms. Joined by the police chief and the majority of the city council, the editor insisted that the moral decay had been "greatly exaggerated." In an effort to reason with the purity forces, they pointed out that Athens was "no worse than other cities." The whole nation, after all, was "morally suffering" since the war.[19]

Local purity ultras found scant comfort in the idea that Athens had company on the road to hell. They refused to relent. The rene-

gade *Athens News,* backed by the Klan, kept the controversy alive
with "screaming headlines . . . insisting that vice was rampant
. . . and that sinister influences were in control in the city," or so
the *Banner* charged. In the name of home protection, the Athens
League of Women Voters demanded to know the positions of alder-
manic candidates in the upcoming election on Prohibition enforcement; prominent individual women complained about public vul-
garity. The vigorous campaign of the new mayor and the courts
yielded results, though, and activity died down for a time.[20] After
a year-long hiatus, the uproar resurged in early 1924. This time,
the Klan emerged as *the* leader of the purity forces; its participation
would have explosive consequences for all concerned.

In late March of 1924, in the midst of a statewide recruitment
drive, the Klan renewed agitation over the "exceedingly deplorable" state of local morality. As evidence, it cited bootleggers who
sold their wares on the courthouse square and "a corps of [prosti-
tutes] who prey constantly on the young college students." In order
to put its own stamp on the issues, the Klan pointed in particular
to the operation of two pool rooms a stone's throw from the university. Patronized by "the vermin and dregs of society," they were
"harbors of vice and corruption." They also had Italian owners, as
the Klan was at pains to point out. The order's publicist and his
colleagues argued that in order to right the moral climate, "the
Semitic influence in Athens must be checked," and the leverage of
Catholic and Jewish voters—the alleged "ring" to which the Klan
attributed all ideas or values contrary to its own—"broken by an
organization just as powerful." In speeches and private discussions,
the anti-Semitism was apparently even more blatant; when a group
of university students lampooned the Klan in their yearbook, they
all added Eastern European suffixes to their names, associated
themselves with moonshine, gambling, and flappers, and changed
the school's name to "Solomon University."[21]

The grand jury soon picked up the ball, if not with the precise
spin the Klan had put on it. The grand jurors' April report mingled
grudging awareness of their inability to control the population, fear
of youthful sexuality, and anger at the boldness of lawbreakers.
The jurors realized that Prohibition was floundering for lack of ade-
quate "public sentiment" behind it. They reported that their dis-
coveries had convinced them of "the necessity of having police officers attend" the dances of local youth "to maintain order." And
finally, they expressed rage at "the audacity and effrontery" of the
Prohibition violators who had appeared before the jury "openly

flaunting their opinions."[22] The report encapsulated the plight of even moderate local purity advocates. They commanded the jury, the courts, and the press, yet they lacked real authority. A significant portion of their fellow citizens ignored them. Law-breakers mocked them. They needed police to monitor their own children. From their perspective, the signs hardly boded well for the future.

Mature female purity advocates, for their part, may well have felt shamed in their maternal roles. After all, the press and the jury had repeatedly blamed them for the going astray of the daughters they were charged with protecting. But they also had reasons of their own for interest in Prohibition, as a request for help to the state Women's Christian Temperance Union in that year from a North Georgia woman illustrated. The woman complained that in her community, whiskey production and fights stemming from it were rife, yet local authorities did nothing. "Many a poor womon and little childern . . . [were] all most starving on the acount of it." The "fighting and cussin" had gotten so bad that mothers felt "it isn't safe to let our childern out of the [yard]." The situation in Clarke County was less dramatic, but many women there shared the author's views on the dangers of intoxicants to their families' well-being and her conviction that not enough was being done. With representation from the allied League of Women Voters, the Women's Christian Temperance Union held a rally in May to devise strategies for Prohibition enforcement. They met at First Christian Church, whose two ministers led the Athens Klan and its anti-vice efforts.[23]

Meanwhile, the Klan continued to circulate the *Searchlight* article about the putrescent moral state of Athens. Promising future sermons on "the International Jew," the "Roman Catholic Hierarchy," the "Ku Klux Klan and some other things," the Reverend S. E. Wasson applauded the piece from his pulpit in the First Methodist Church. Wasson further charged that county officials failed to enforce the law because they were all beholden to "a certain power in a pool room on Broad Street." Over 150 members of his congregation shared his sentiments, Wasson said, a consensus not surprising since at least twelve Klansmen participated in his men's Sunday school class. The following week, the League of Women Voters chimed in, "bombard[ing]" the civil service commissioner with questions about why Prohibition violators were tolerated in Athens. Local League members had earlier resolved that the Eighteenth Amendment "can be enforced," but not by "men who are addicted to the habit of drink." The state office of the Klan, for its part, now trumpeted Prohibition enforcement as "the real issue" in the upcoming Democratic presidential primaries.[24]

In the midst of the uproar, the Athens Klan organized a massive outdoor recruitment rally on June 19; thousands of Klansmen from all over North Georgia attended to parade through town behind the Imperial Drum and Bugle Corps and then watch the initiation of "hundreds" of new members "in the glare of a giant fiery cross." These social purity efforts helped win the Athens Klan the state record for percentage increases in membership. Yet, if they won members for the Klan, they failed to discourage the practices they aimed to end. In October, "flagrant violation" still absorbed the grand jurors. They concluded, prophetically, with recommendations for "the employment of a special detective" and "renewed vigilance."[25]

Vigilance being its stock in trade, the Klan took matters into its own hands. Its Law Enforcement Committee concocted a plan over the summer to bring to town a private detective. With his help, Klansmen collected evidence that resulted in a host of indictments. They obtained it illegally, through deliberate entrapment of whiskey sellers, drinkers, and gamblers. Public exposure of these underhanded methods led ultimately to a scandal, a split among purity advocates, the abandoning of the indictments, two threatened lawsuits, and the temporary abrogation of the local Klan's charter by the national office. Yet, in the final analysis, these side-effects did not really matter. Through the entrapment scheme *and* the exposure, the Klan had polarized the community. It had proved itself the most determined anti-vice force in the county, the premier voice for local people who feared changing family relations and sought to compel obedience to older norms. That fall, the Klan's candidates swept the local elections.[26]

The incident itself offers an unusual window on the inner workings of the Klan. In the first place, the members of the Klan's Law Enforcement Committee were not ordinary rank-and-filers, but established civic leaders. The Reverend M. B. Miller, in addition to serving as the Klan's Exalted Cyclops, was also the minister of the First Christian Church. He was described by the police commissioner as "the highest type of Christian manhood." Miller's assistant in his church and in the Klan was the Reverend Jerry Johnson. While acting as Kligrapp (secretary) for the Athens Klan, Johnson served as director of Young People's Work at the First Christian Church and field secretary of the Christian Endeavor Union. He spent his days, by his own account, "fighting for victory in bringing our young men and women to the feet of Jesus Christ."[27] Dr. Henry Birdsong, a surgeon, was a charter member and future president of the Athens Lions Club, a leader in the Chamber of Commerce, and a member of the Prince Avenue Baptist Church. His wife taught in the church's adult Sunday school and presided over a circle of its Ladies' Missionary Society. Bela Dunaway, a leading local businessman, was a longtime prohibitionist, a Grand Noble of the Odd Fellows, a member of the men's Sunday school class at the First Methodist Church and of the Woodmen of the World, a founder of the local Law Enforcement League, and father of two veterans of the Great War.

Klansman *Wiley Doolittle,* who served on the grand jury that issued the inflammatory 1922 report and founded the newspaper that broadcast the moral crusade, had a long and distinguished record of civic activism. Described by Klan opponent Lamar Rucker, an Athens attorney, as "the top-most bird on the top-most twig" in Athens, *Doolittle* was a local economic mogul who had served three terms as mayor of Athens in past years. One of the city's "most influential citizens," in the view of the *Banner,* he was a past president of the Chamber of Commerce, an organizer and president of the Athens Kiwanis Club, chairman of the Clarke County Democratic Party Executive Committee, a leading Mason, an avid churchgoer, and a man known as being among the city's most "benevolent" in giving to the poor.[28]

The fact that the Law Enforcement Committee included *none* of the Klan's poorer, less prominent members is significant. The order seems to have purposely staffed the committee with men whose claim to authority would be difficult for their non-Klan peers to question. In so doing, Klan leaders most likely aspired to attract others of the same class. Whether intentional or not, the composition suggests a clear hierarchy within the Klan. It replicated internally the chain of command it prescribed for the rest of society. Indeed, internal Klan documents indicated that only better-off members were trusted to plan activities and speak publicly. These practices illustrated how intimately the quest for respectability was tied to class differentiation. The episode also exposed the roots the Klan had sunk among leading citizens. The most telling revelation of these roots occurred when the grand jury, a body said to be "as a rule largely composed of representative business men and merchants," ultimately had to throw out all the indictments stemming from the Klan's investigation. Repeated efforts had proved it impossible to assemble a jury without Klan members or those tied to them by kinship.[29]

The subsequent confessions of participants in the entrapment scandal further exposed the usually concealed operations of the Klan in such "moral clean-ups." When censured for using illegally obtained evidence, Miller claimed legitimacy for the plan on the grounds that Governor Clifford Walker had worked with the Klan's state office to devise the scheme at the outset. The Grand Dragon and the governor both denied the allegations, of course. More disinterested witnesses confirmed Miller's account. The Realm office even suspended the local Klan leaders involved in order to substantiate its claim that the order did not condone their methods. This dissembling was strictly for public consumption: to avoid harmful publicity and ward off investigation or prosecution of the Klan. The latter seemed imminent since the lawyer of one of the alleged whiskey sellers, Mrs. Mae Eppes, the wife of a local contractor, had obtained a list of Klan members and threatened to sue them for defamation of character. The Atlanta office's fear of exposure was in fact so great that it attempted a burglary to remove local office records.[30]

More interesting than the duplicity of Klan officials were the reactions of local Klansmen to the hypocritical censure. These made clear the rank-and-filers' unshaken faith in the legitimacy of their efforts. Miller, who had lectured for the Klan in several neighboring towns, was shocked by the Klan officials' denial of their complicity. Breaking his vow of loyalty, he declared his outrage publicly. Miller's youthful assistant, the Reverend Jerry Johnson, went further. He issued a statement repudiating Klan officialdom to the "young people of Athens" whom he served in his ministry. No one, Johnson intoned, should be allowed to question the word of a "Man of God" like Dr. Miller the way Georgia's Grand Dragon had. Johnson said that he still believed in the Klan "heart and soul," but its deceit and its shameful abandonment of Dr. Miller had convinced him he should resign.[31]

Then, when state officers continued stonewalling, Miller lost his patience and spilled everything. Accusing them of having "betrayed" local members, he explained that, not only had Governor Walker known of and the Atlanta office designed and paid for the investigation, but Mayor *R. D. Moore* (also a Klansman), the chairman of the Athens Civil Service Commission, a member of the police commission; police captain *Vance Anderson* (a Klansman);

and several local policemen had all assented to the methods employed in the investigation. In Miller's view, moreover, the Klan
had merely implemented the grand jury's own October recommen-
dation for use of a private detective, after the police had claimed
their hands were bound by the need to respect the rights of suspects.[32] He was, in short, wholly unapologetic.

In his conviction that the pious end fully justified the illegal
means, Miller gave voice to a staple of Klan thought. Indeed, he
impugned the honor of the leaders who now shrank from public
affirmation of this principle. Miller charged that in the scramble
to protect their organization, they had "dodge[d] the real issue."
Gambling and bootlegging *were* undermining local morals, and the
Klan had taken action to remedy the situation when no other force
would. Miller's indignation made it clear how potent these issues
were for ordinary Klansmen. He defended the project as "a sincere
effort to improve moral conditions . . . for the benefit of the young
people being educated in our town," a concern other Klansmen
echoed. Above all, he invoked the appeals of women to vindicate
the Klan and silence its critics, appeals that the association of the
WCTU and the League of Women Voters with his church make
plausible. "Sisters have called me," Miller chastised the faint-
hearted, "with tears in their eyes because of brothers who have
been led into lives of dis[s]ipation by Athens associations, mothers
weep over wayward sons and daughters, while 'Peaceful Athens'
sleeps on." The mayor shared Miller's vexation with the critics,
although he distanced himself from the investigation's methods
and affirmed his willingness to retire if "the better class of citizens" thought he was wrong. But *Moore* agreed with Miller on the
key point: "there can be no middle ground." "If you stand for morals," he chided, "put your shoulder to the wheel and help enforce
the laws."[33]

Local rank-and-filers, for their part, had no desire to disassoci-
ate themselves from the methods of their organization. On the contrary, they backed the deposed officers of their chapter against the
chicanery of their superiors. Guileless, they called a mass meeting
in late January to decide what to do. After voting confidence in
their local leaders, they severed ties to the national order and returned their regalia to state headquarters. Confused and dispirited
by the whole affair, they passed a resolution stating that they
would forever remain faithful to "the spirit of Klancraft," though
they groped for other organizational means of expressing their devotion. They briefly courted dethroned Imperial Wizard William J.

Simmons, for example. When ex-Klansmen brought him to speak about his new organization, the respectable were again prominent: two ministers and a city alderman sponsored the lecture, while another reverend shared the stage with Simmons.[34]

In the end, it took only a few months for the wounded pride of these disaffected members to heal. Just as earlier they had viewed illegal entrapment as a legitimate way to fight immorality, so now they came to accept the deceit of Klan officials and their own public humiliation as vindicated by the ends it had served. A majority of the old members joined to petition for the readmission of their chapter. Having had time to reflect, they now pledged full confidence in all levels of the Ku Klux Klan's leadership. Indeed, they commended Georgia's Grand Dragon for "preserving the integrity of the members" through the lies with which he had averted investigation, and celebrated the fact that they were now stronger in Athens than ever before.[35]

All was forgiven. By December, even the Reverend Miller, since transferred to an Indiana parish, sought to mend bridges. Sending Christmas greetings to his old associates, he notified them that "I still wish to hold my membership in Athens." Miller having admitted his "mistake" and promised to "prove his loyalty" if readmitted, the Grand Dragon believed that the "lesson" had sunk in and "we will never have any further trouble with him." He advised the Athens Klavern to take Miller back, with the stipulation that he "never again be allowed to be elected to Klan office." With his mentor now back in the order, the Reverend Johnson, for his part, quietly dropped the suit he had initiated against the Klan to collect damages for his tarnished reputation.[36]

▲ ▲ ▲

The Athens purity crusade may have been unique in its particulars, yet it emerged from ideas and fears that local members shared with their counterparts elsewhere in the United States. The Klan promoted moral "clean ups" to exorcise dangers to the ordered homes it believed a stable and powerful nation depended on. Unless the country realized that its "life depends upon a moral 'clean up,' " as one official put it, it was "doomed to inevitable destruction."[37]

The rationale for these endeavors came from the conviction, common to both evangelical Protestantism and republicanism, that immorality would erode family commitments, demoralize the citizenry, and undermine the state. Imperial Wizard Evans associated the idea of freedom as license with the Old World: such libertinism

endangered not only its practitioners "but the social structure."
Since the United States government was of, by, and for the people,
he reasoned, the state had the right—indeed, the obligation—to patrol the "moral standards" of the populace. The reason it had to do
so was straightforward. "Our young," Samuel Saloman explained,
"are apt pupils." "Moral teachings they may receive in well-
regulated homes will avail but little if they see such teachings constantly set at naught" in the outside world.[38]

Klansmen feared the waning of parental power. Modern "children rule the roost," deplored Klan author E. F. Stanton. "Lack of
control of parents over their children," another writer alleged, had
led to the "most deplorable results," prostitution among them.
Some Klansmen worried about how the unprecedented leisure enjoyed by young people would affect their characters. Others found
the values of youth scandalous. "Pleasure has become the god of
the young people of America," remarked one Georgia Klansman,
"and a very unwholesome and lascivious pleasure it is." Still others found the peer culture of youth impertinent. *Kate Yarborough*,
the wife of Athens Klansman *Scott Yarborough*, later complained
that "my children calls me old-fashioned, cause I don't try to dress
like they do, and talk proper."[39]

Young people's default on time-honored obligations to parents
made Klansmen angry. One Athens member complained that his
nephew "don't give his mammy" but a third of his wages; "the
rest he blows on himself." Another griped that although he needed
good help in his barber shop, and his son was "as good a barber as
ever stepped in shoe leather," the city-bred boy refused to work for
him. Klansman *A. J. Boyd*, for his part, criticized emergent dating
practices. "Now, a boy will drive up in his car, blow his horn and
the girl will come running out . . . if her parents . . . ask where
she's going, this is what they get, 'None of your business,' and
away they go." That *Boyd* did not even *have* children of his own
indicates how such complaints resonated as symbols of the more
general abatement of adult authority. Indeed, across the country,
others shared the complaint: the Chippewa Falls, Wisconsin,
Women of the Ku Klux Klan, for example, scheduled discussions
on "obedience of children."[40]

The Klan proposed to elicit filial obedience in good part
through simple compulsion. While the professional middle class
was turning toward consensual models of child-rearing, Klansmen
defended physical punishment. "The reason the old-fashioned boy
worked more than the modern boy," the *Searchlight* quipped, "was

because of the hickory stick that lay in state above the kitchen door." In Athens, when liberal school board member M. G. Michael proposed to abolish corporal punishment as a "barbarous . . . relic of the dark ages," Klansman *I. A. Hogg,* a fellow board member, protested. "King Solomon says beat the devil out of them," *Hogg* argued, "and save their souls from hell." *Hogg* maintained that only the threat of force could stop children from taking "advantage." His argument carried the vote.[41]

While young people's attitudes and work habits caused concern, their sexuality appeared by far the greatest threat. Imperial Wizard Evans adduced as one of the Klan's primary motives "the moral breakdown that has been going on for two decades." It had gone so far, he said, that "those who maintained the old standards did so only in the face of constant ridicule." The order excoriated a range of expressions of youthful sexuality: suggestive fashions for girls, "petting parties," and "parking" in cars. More generally, Klansmen declared that they were "implacably opposed . . . to all the amatory and erotic tendencies of modern degeneracy." The Klan pitted itself against those who would follow "the fleshly propensities which are all selfish and sensual and which produce anarchy."[42]

Klansmen found particularly insidious the commercial entertainments contemporary youth flocked to. "Degrading, depraving or disgusting" movies, the *Kourier* charged, were "undermining morals." Not only films, Imperial Wizard Simmons maintained, but also jazz music and "filthy fiction" were "polluting" society. Contemporary young people were being "submerge[d]," according to another leader, "in a sea of sensuality and sewage." The profusion of such pollution metaphors was an index of the intensity of Klansmen's desire to buttress the older chain of command in relations between parents and children.[43]

Indeed, the Klan expected family members, especially women and children, to subordinate their individual aspirations to the needs of the corporate unit. The hierarchy of power in the home was offered as a model for other institutions in society. One Klan-recommended author used the untrammeled sway of the head of the family as an example business firms should replicate. Another explained that the merit of the ancient Roman citizen was that "in obeying his father he learned to obey the state."[44]

The order's discussions of divorce made evident some of the threats to such obedience. The *Searchlight* attributed the soaring divorce rate to the selfishness of modern partners. It condemned those who approached marriage not with a sense of the "duty they owe to their children and society," but instead with an eye to their own fulfillment. Such individualism threatened the foundations of the home and with it "the social and economic structure of the nation." In a 1930 speech in Athens, Klansman E. D. Rivers blamed the popularity of divorce on the emerging notion of "companionate marriage." Presumably, it imperiled the traditional family by making partners' loyalty conditional upon the meeting of both of their emotional and sexual needs.[45] To combat this dangerous individualism, the Klan prescribed clearly demarcated roles that would insure the mutual dependence of men and women.

Klan tracts and speakers dwelt far less on men's behavior than on women's. This was in part because male roles were changing less than female roles, and in part because Klansmen were more interested in controlling others than in self-scrutiny. Nevertheless, they expounded a particular model of masculinity. Klansmen expected men to marry, to provide for their families, and to exercise control over their wives and children. "God intended," affirmed one Klan minister, "that every man should possess insofar as possible, his own home and rule his own household."[46]

Rule over one's women was mandated by another staple of the Klan's conception of masculinity: "honor"; or, as it was sometimes called, "chivalry." Honor dictated a commitment to protect the virtue of "American" women. Historically, honor in fact rested on a man's ability to control the sexuality of his female relations. Their "purity" was the complement of his "honor"; hence Klansmen's insistence on "the chastity of woman." Yet it seemed that young women no longer shared these "high ideals." The job of Klansmen, then, said Imperial Kludd (chaplain) Caleb Ridley of Atlanta, was "to make it easier for woman to be right and do right." Similarly, an Athens Klan lecturer spoke of "fathers' duties to their daughters" as an essential aspect of the order's commitment to "protection of the home and the chastity of our womanhood."[47]

Klansmen's patriarchal insistence on the need to "protect" young women gained credibility from the reality that they *were* vulnerable, as those who knew unmarried or deserted mothers could see. Throughout these years, the Athens press reported cases of abandoned children and infanticide. Whether buried in shallow graves or left in satchels on the roadside, they offered grim testimony to the dire straits of their mothers. Athens Klansman *P. F. Vick's* daughter was one such victim of male inconstancy. When a poor young man impregnated her in 1912, only a court order persuaded him to live with and provide for her as the law required. Other local young women lost their lives from botched abortions. Similarly, hard-drinking husbands and fathers often mistreated the women and children in their lives, a fact Klan propaganda regularly emphasized. Citing pitiful examples to support its case, the *Kourier* thus argued for the retention of the Eighteenth Amendment on the grounds of "the wife's right to the home and protection that was promised her and the little children."[48]

Yet, women's troubles notwithstanding, the Klan's rhetoric of chivalry scarcely concealed the urge for command. Indeed, on occasion it became clear that young women's defiance of their parents' mores agitated Klansmen at least as much as their vulnerability. "Girls have lost their timidity," E. F. Stanton asserted, "and are [now] more brazen than boys." Girls were cautioned that they should return to chaperons. "Any privilege given invites further encroachment," the writer intoned; "things that are easy to get are seldom much valued." The *Searchlight* urged modern girls to confide in their mothers as their forebears had; if they did, "fewer would go wrong."[49]

The focus on young women illustrated how parents' authority over children was of a piece with that of men over women in the world the Klan sought to defend. Many of Klansmen's less self-conscious pronouncements expressed resentment of female self-assertion. "We pity the man," taunted the Klan press, "who permits the loss of manhood through fear of wife." Similarly, the author of *Christ and Other Klansmen* censured "women [who] blaspheme God by disobeying their husbands." Other writers upbraided "the gadabout mother" and the wife "who care[s] more for the unhealthy activities of social duties" than for her obligations to her husband. As the reference to health suggests, Klansmen tried to naturalize gender hierarchy. American women already *were* emancipated; change should go no further. Feminists should cease complaining about such "pretended wrongs" as the burdens of housework. "God or nature" dictated women's roles, not men and society, so nothing could change them.[50]

Having options outside the household might lead women to deviate from their cardinal function. That, according to the Klan, was childbearing and childrearing. The entry of women into the labor force and politics, Klan leader E. Y. Clarke stated, could thus pose "a real danger unless there is some strong organization constantly preaching that women's place is in the home." "Citizenship for our young American women," explained Imperial Wizard

Simmons, "includes the essential duty of motherhood," just as for men it included breadwinning. Only one or two children would not do, moreover. The Klan felt it necessary, said the Imperial Wizard, to "insist upon . . . 2, 3, 4, and sometimes . . . 5 or 6 children." Women who rejected childbearing "as a burden" needed a "socializing education." "If society is to live," Saloman explained, women must "cheerfully" accept their familial duties. The Klan was therefore, as one speaker put it, "violently opposed to birth control."[51]

Although hostile to sexual emancipation, the Klan was not an outright foe of all women's equality. The order's commitment to moral uplift in fact led it to support rights for white Protestant women. "Subjugated women," the *Searchlight* explained, "means subjugated morals." Hence "the Klan believes," as one of its statements of principle read, "in the EQUALITY of the sexes without hesitation." Klan members recognized that women could no longer be confined to the household; rather, they must work with their menfolk to achieve common political and social goals. The order even asserted on occasion that some of the world's problems resulted from the exclusion of women from power.[52] In short, Klansmen were not hidebound conservatives; grudgingly or not, they accommodated the desires of their female counterparts for expanded roles.

Indeed, the Klan championed suffrage for Protestant white women. As did many others, Klansmen viewed female suffrage as the best defense for Prohibition. Two Klan-associated politicians in Georgia, W. D. Upshaw and W. J. Harris, the former a past president of the Anti-Saloon League, became the first men in that state to vote for the Nineteenth Amendment.[53] An Athens Klan leader and city alderman, W. R. Tindall, beat even the League of Women Voters to the fray when he attacked a 1924 plan to levy a special tax on women voters as "unfair, unjust and contrary to our principles of government." The Klan also counted on women's votes to thwart Al Smith's bid for the presidency in 1928; after his defeat, the order paid tribute to "the Protestant Women of America."[54]

Just as it endorsed women's votes, so the Klan respected the activism of like-minded women. Throughout the United States, it frequently praised the work of the WCTU and helped its chapters with particular projects.[55] The Klan even adopted a take-off on the female temperance motto "For God, Home, and Native Land." In turn, the WCTU in many places backed the Klan's moral reform work.[56] In Clarke County and Georgia generally, the efforts of the two organizations were mutually reinforcing. In some cases, local affiliates of the General Federation of Women's Clubs also supported the Klan's efforts, with which the Georgia division at least had an ideological affinity.[57]

Nonetheless, recognition of women by Klansmen was always shot through with ambivalence. Klansmen's ideal, after all, was the nineteenth-century petty proprietor—whether farmer, artisan, or merchant. His vaunted independence as a citizen presumed his control over the labor and behavior of the dependents in his household.[58] However much Klansmen might try to cooperate with women who shared their social goals, female initiative set them on edge; the undertow of patriarchal prerogative impeded full solidarity. This undertow was most obvious in Klansmen's relations with the Women of the Ku Klux Klan (WKKK).

Established by the Imperial Palace in June of 1923, the WKKK was announced as "a Protestant Women's Organization which is *for, by,* and *of* women." Yet the reality was murkier. Leaders of the men's Klan maintained ultimate control over the women's Klan, a control symbolized by their ability to appoint their wives to office in the WKKK. Male Klan propagandists also tended to describe Klanswomen as the "mothers, wives and daughters" of Klansmen, despite the fact that many were not related to male Klansmen, and to portray the WKKK as "auxiliary" despite its claims to independence. These ambiguities hint at the tensions in gender relations that the Klan hoped to submerge: both the women's desire for control over their own affairs and the men's reluctance to surrender any real power.[59]

Similar troubles surfaced when Klansmen attacked other groups in women's names. As in local practice, so in ideology: the Klan used respectability as a weapon, a tool to differentiate its own constituents from those it would direct their passions against. Sexual politics figured prominently in the Klan's efforts to stir anti-communism and anti-Catholicism, for example. Klansmen's recitations of the sexual subversion practiced by Communists and Catholics aimed to convince their audiences of the essential "otherness" of the groups at issue, to dehumanize them so as to loosen inhibitions against aggression. Yet these disquisitions revealed more of the speakers' own attitudes toward women than they realized.

Communists came in for the lion's share of attack. Klansmen warned that communism would destroy the family as their constituency understood it. The argument was most amply developed in Samuel Saloman's *Red War on the Family*. Saloman's work was typical of the Klan's anti-communism: rife with exaggerated and absurd charges, it nonetheless had a rational core. That core was an assumption the Klan wholeheartedly agreed with: that the hierarchical family was the basis and guarantor of ordered society. Saloman's central charge against socialism flowed from this. In recasting marriage and morality, it would destroy civilization, as had similar experiments in ancient Greece and Rome. "Monogamic marriage" without easy recourse to divorce, Saloman insisted, was "the sheet anchor of our civilization"; whoever opposed it "must be regarded as a foe." If conventional marriage and its safeguards weakened, gender roles would also erode as "men became more effeminate and women more masculine." The blurring of these boundaries would result in chaos and social decay.[60]

Like the Klan, Saloman couched his argument in the chivalric tradition. He reminded men of their obligation to shelter women from danger. Yet, behind the pretense of care, lurked a sense of personal loss for the men who posed as women's Galahads. To shock male readers resentful at the loss of their gender privileges, Saloman quoted a tract asserting that, in socialist society, women "will be entirely independent" economically, socially, and in intimate relationships. Horrified, he cited communists' commitment to making housework and child-care no longer the responsibility of individual women but of the whole society. This theme of the loss of women's private services was echoed in the Klan press; in Russia, complained another writer, "the young and old are cared for by all the people impersonally."[61]

The Klan's anti-radicalism, in fact, exposed how the mask of chivalry concealed an unwillingness to surrender proprietary rights to women. Saloman's primary charge against socialists was that they advocated the "muck of free love." One clue to the nature of the threat is that virtually all such passages focussed on the liberation of *women's* sexuality. Indeed, Saloman complained that the philosophy of free love was "running wild among the enlightened and emancipated and sex-conscious women of America and Europe." He could see no goal in the emancipation of woman but allowing her "to devote herself to free and unrestrained love." Saloman simply could not imagine women as autonomous individuals making their own choices. Rather, their release from being the "property" of one man could only result in their becoming the property of all. Unions based on mutual attraction and commitment, free from state sponsorship, could mean only "the morality of the brothel." The liberation of woman would thus reduce her to a "harlot," a "clandestine prostitute." Why? The "community of property necessarily and logically involves the community of women."[62]

This equation exhibited starkly the way gender infused class and class, gender in Klan thought. As Marx and Engels once pointed out, the specter of the "community of women" could only make sense to men who saw women as property. In other words, it only had meaning for those who understood their control over their wives and daughters to be integral to their own class status, as it was in the republican tradition adapted by the Klan. Denying men of other classes access to Klansmen's "own" women was therefore necessary to police class borders. Klansmen thus interpreted workers' revolution as a challenge to their dominion over the women of their group. "Under dogs . . . [would now] satisfy their cravings of centuries," Saloman warned, in their effort to break the bourgeois male's monopoly on "luscious womanhood."[63]

If communism held the worst dangers to domestic order, Catholicism came in a close second. Here, too, the purported deviants were charged with violating the natural order of relations between the sexes. But here the immediate targets were priests and nuns. The core of the Klan's case against them was that their abstention from heterosexual marriage and reproduction necessarily led to unnatural, antisocial perversions. Tom Watson encapsulated the logic in the title of a 1917 anti-Catholic tract: *The Inevitable Crimes of Celibacy: The Vices of Convents and Monasteries, Priests and Nuns.*[64]

Whereas Klansmen criticized socialist society for giving men too little authority over women, they accused priests of having too much. Priests usurped powers that rightfully belonged to male household heads. Associated as it was with folk memories of aristocratic men sexually exploiting the female relatives of their male subordinates, this allegation resonated with the general case made against the Catholic church for monarchical pretensions. Evans abhorred the "galling subjection" of the Catholic feudal societies of Europe wherein "no man save the king truly owned anything; no property nor children, nor wife, nor life that could not be taken from him at the whim of every superior." A wife and children were thus among the pieces of property a man had a right to control; the horror of Catholicism was its alleged interference with this control. Klansmen's principal complaints centered on two phenomena: in the shorthand of an Athens Klan lecturer, "the confesional" and "the coruption of the priesthood."[65]

Attacks on the practice of confession were virtually transparent defenses of male dominance in domestic life. The Klan's case had been laid out clearly by Tom Watson in the 1910s. He charged that confession allowed a priest to act as the "confidante of another man's wife," to whom she divulged the couple's "inmost secrets . . . all that is sacred between her husband and herself." The information priests would obtain included "sexual procedures and techniques with her husband, extra-marital activities, masturbation, homosexuality, and unnatural fornication." This violation of the husband's privacy, Watson warned, would "rot out the heart" of the nation. That it was the husband's privacy, not the couple's, that was at issue was evident in the subtitle borne by another anti-confessional tract in the Klan's arsenal: *An Eye-Opener for Husbands, Fathers, and Brothers.*[66]

The sexual power attributed implicitly to priests in the critique of the confessional was made explicit in discussions of nuns. In *What Goes on in Nunneries*, Tom Watson had defined convents as places where "bachelor priests keep unmarried women under lock and key, and whose children are killed." Elsewhere Watson described the nun's obligations as those of "the temple girl." Portraying the convent as "a securely locked seraglio," he invoked the sexual exploitation of lower-class women in the harems of the East. The Klan took up this theme. Georgia Klan leader and *Searchlight* editor J. O. Wood won election to the General Assembly on a platform calling for more rigid inspection of convents.[67]

Klansmen were more ambivalent about nuns than about priests. They generally imputed to nuns a perverse hypersexuality. Yet, as in the complaints about the confessional, the charges assumed nuns to be the passive victims of priests, who exploited their social power to gain sexual access. Alma White, for example, condemned priests for exercising "tyranny over helpless victims behind convent walls." Houses of the Good Shepherd, according to the *Searchlight*, were no more than "slave pens" that promoted "the debauchery of Southern orphan girls." The common denominator in such allegations was an inability to come to terms with women who removed themselves from the institution of marriage. Klansmen simply refused to believe that women might choose to be free from husbands and children; hence their insistence that women in Catholic institutions were held against their will.[68]

▲ ▲ ▲

This bewilderment in the face of women's efforts to achieve more independence was not confined to Klan ideology. On the contrary,

Klan propaganda persuaded because, in however serpentine a manner, it connected with feelings that emerged from troubling encounters between men and women in everyday life. In prescribing the roles it did for women, the Klan sought to buttress a paternalism being threatened on many fronts. One of the most intimately challenging was women's agency in redefining the norms of relations between the sexes.

The wives of several local members, after all, filed for divorce in these years. The resulting proceedings made clear some of the conflicts between their own ideas about fairness and those of their estranged husbands. *Ilene Anderson* kept her husband from fleeing the state after their breakup. *Mollie Braxton* got a restraining order to deter hers from "visiting or interfering" with her; as did *Nell Henson*, who also prevented her husband from trying to sell their house or cause trouble with the children about her custody. Other Klan husbands found themselves having to explain to judges and juries why they should not pay alimony and court fees. No wonder the Athens Klan included divorce courts among the evils their movement would combat.[69]

A more intriguing illustration of the erosion of paternalism was the way white women in distress began to appeal to the Klan for aid against their male relatives or neighbors. Like Athens Klansmen, they saw value in the methods the Imperial Palace feigned to disavow and sought to deploy them to their own ends. In verbal and written requests, they asked the Klan for help. In effect, they drew it in to offset the power imbalance placing them at a disadvantage vis-à-vis the men in their lives. Clarke County women were not alone in this effort; their peers in other parts of the country had the same idea. The Georgia Realm office, for its part, reported receiving an average of twenty letters *each week* from women inviting the order to threaten or use violence against people whose conduct they disapproved of. One former Klansman who worked in the Atlanta headquarters of the order maintained that the Klan "gets hundreds of letters from women asking that some man they don't like be whipped."[70]

One of the best surviving examples of the genre is the appeal written by *Betty Thomas Perduto* in July of 1928. The story she told was a saga of psychological torture and physical abuse at the hands of her gambling, alcoholic husband. When she left him for the second time, taking their two children with her, he challenged her petition for divorce with malicious allegations of adultery. The result was a mistrial. With no divorce, no alimony, and no funds to fight the suit further, *Perduto* now faced the prospect of losing her "two poor innocent children" because of his widely broadcast slanders. She implored the Klan to force her husband to withdraw his case and give her a divorce on the grounds she stipulated.[71]

Perduto may have had a weak hand, but she played it well. With a curious mixture of cunning and sincerity, she connected her struggle for an equitable divorce to the larger issues in the Klan's program. Not only did she emphasize her history of devotion as a mother and perseverance as a wife, but she also drew attention to other strikes her husband had against him. If they had failed to sway the jury, they might still convince the Klan. While she was "an American woman," as she pointed out over and over again, her estranged husband was an Italian, and a "devote Roman Catholic" at that. Indeed, she expressed stupefaction that the jury would "deal so hard with an American woman and be so lenient [with] a devote Catholic." Further hoping to strengthen her case, she reported that Mr. *Perduto* had once gotten "very intimate with a colored nurse" hired for the children. "It [was] impossible," she said, "to keep any help on account of his conduct." Her conclusion made it very hard for the order to refuse her request. "Being an American woman," she implored, "I feel that I ought to have Justice. If I can't have it from the Courts I know of no other one to go to except your organization, who has it in your power to deal with such cases."[72] In effect, she issued Klansmen an ultimatum: do her bidding or forfeit their honor, the *sine qua non* of Klan manhood.

Like Mrs. *Perduto*, other female petitioners sought to deploy the Klan's force in order to enhance their bargaining power in domestic disputes or press their notions of propriety upon their communities. If established channels failed to satisfy their notions of right and duty, they felt justified in going outside the law to achieve their ends. They felt their cases further strengthened when those whom they accused were also the targets of Klan hostility. Hence *Perduto's* emphasis on her estranged husband's Italian birth and Catholic faith. These white women goaded their self-styled protectors into vigilantism by manipulating the contradictions of Klan ideology to compel its adherents to help them achieve their ends. "If you all are for the protection of a community," one rural woman thus admonished the Klan, it should drive out two prostitutes patronized by her husband, sisters whose parents tolerated their trade "because they are bringing in plenty of money, money that poor wives and children really need."[73]

Her use of a conditional sentence structure ("if . . . then") illustrated how women used the Klan's stated goals as leverage for their own demands. In light of Klansmen's commitment to shoring up male authority in the home, it seems unlikely that they would have taken the initiative against wife-beaters. More than a few Klan members, after all, harassed their own estranged wives, and the record fails to reveal any complaints to the Klan from men about wife-beating.[74] It appears likely, then, that it was the female victims themselves, pressing from behind the scenes, who charted for the Klan the dividing line between legitimate authority and unacceptable abuse—and compelled its members to patrol that border or lose face. The police even alleged that women who lived in the area of the Southern Mill, an area in which the Athens Klan enrolled many people, had "formed a league designed to oust undesirables from the neighborhood." Police officials maintained that these women had instigated at least one flogging attributed to the Klan, of a mill worker who spent his wages on another man's wife and beat his sister when she complained.[75]

Whether or not such a "league" existed, the women who called on the Klan were hardly the passive creatures of chivalrous mythology. On the contrary, their appeals constitute tangible proof of their attempts to contest their spouses' unilateral control. By claiming rights for themselves and duties incumbent on their partners, these women sought to modify their subordination. In their requests, they articulated a "moral economy" of domestic life, using the prescriptive ideology of male supremacy to achieve their ends.[76]

If they had performed their marital obligations, then they insisted that they deserved protection when their mates failed to honor their part of the marital bargain. The women thus made evident their conviction that they had a right to relief and to what they perceived as justice. They expressed a sense of outrage and violation: a feeling that they, or the women on whose behalf they wrote, had fulfilled their obligations only to be betrayed by their partners. *Perduto* thus expressed anger at having "tried very hard to make the best of a bad bargain" in her marriage—to no avail. If the Klan wanted a system of private patriarchy, these women implied, then its "Knights" owed protection to women wronged by their spouses. Indeed, another woman called on the Klan "in God's name for help." "If ever a woman needed your protection"—the protection Klansmen's creed after all promised—it was her neighbor, starved and beaten mercilessly by an unrepentant husband. The writer wanted the Klan to administer to this man "the last resort in *full*." Female petitioners thus manipulated the Klan's ide-

ology of chivalry to gain the redress that they could no longer hope
for in civil society and could not yet expect from the state. Their
entreaties no doubt flattered male Klan members, who enjoyed not
only the gratitude but also the legitimacy thus conferred upon
them.[77]

At the same time, women's goading must have caused Klansmen no little discomfort. For, while these women looked to the
Klan for help, their appeals probably added to its members' alarm
over flagging domestic discipline. Klansmen, after all, were wont
to portray the menaces to family well-being as external. Yet many
such white women located the primary danger within—in their
own husbands. Grasping the ambiguity of the Klan's rhetoric of
"home defense," these women pulled the tail to wag the dog. For
their part, Klan leaders worked to channel the anxiety thus excited
into directions that might aid their overall project. One such
spokesman thus described deteriorating moral standards as "secondary" symptoms of a "deeper" malady: the "strangle hold" of
Jews and Catholics on the nation's property and industry. Their
self-aggrandizement undermined households of small property, and
thereby the male authority that previous moral standards depended
on.[78] Race, it seemed, would serve as the lightning rod for the
charged relations of class and gender.

6

▲ ▲ ▲

The Approaching Apocalypse: The Politics of Race

The year 1921 was a pivotal one in Georgia. The trough of the postwar depression, it was the year the Klan enjoyed its most spectacular gains. Not coincidentally, it also saw an unprecedented cleavage among the forces committed to white supremacy. In late April of that year, outgoing governor Hugh Manson Dorsey issued a public statement on the plight of black Georgians. The fruit of much prodding from African-American activists and white liberals, Dorsey's booklet detailed 135 instances of documented violence against the state's black residents in the years 1919 to 1921 alone, including cases of forced flight, peonage, individual acts of cruelty, and lynching. The statement appeared in an already charged climate. The month before, pushed by the Commission on Interracial Cooperation (CIC), Dorsey had publicly refuted inflammatory lies told by Klan lecturers. In speeches in North and South Georgia, they claimed that Atlanta blacks had threatened a riot unless their demands were met; the city and state governments being unable to meet the challenge, only the Klan had proved able to suppress the incipient rebellion. Now, the CIC broadcast Dorsey's statement widely, mailing copies to every minister in the state to try to enlist support for reform.[1]

By today's standards, Dorsey's statement seems timid. If anything, his catalogue understated the violence visited on black Georgians. Not "a tithe of the terrors" had reached print, according to M. Ashby Jones, the Atlanta minister who served as chair of the

CIC. "In many of the rural sections of Georgia," Jones said, a black person led a "a life of constant fear." "To him, the beat of horses' hoofs on the road at night, the rushing sound of a motor car, or the sudden call of a human voice, may be his death summons." Set against this inferno, the governor's call for such measures as "publicity," repeal of the laws sustaining peonage, and changes in the state court system to make lynching easier to prosecute, seems woefully mild. Like the CIC, in fact, Dorsey remained committed to segregation, racially restricted suffrage, and "racial purity."[2] But, limited as the proposed reforms were, never before had a chief executive in Georgia so openly sympathized with African Americans and affirmed that they had rights deserving of protection. In the view of many white citizens, this stand was little short of treason.

Hard-core white supremacists, Klansmen among them, struck back fast and hard. Georgia politicians, from governor-elect Thomas Hardwick to the president of the state senate, charged Dorsey with "infamous slander" against his state. In short order, Dorsey opponents assembled a coalition called the Dixie Defense Committee. It mounted a propaganda blitz against the governor's statement and called for his impeachment. Among the key organizers of the mobilization were several up-and-coming Klan luminaries. They included the Atlanta Baptist minister Caleb Ridley (national chaplain of the Klan); Imperial Klaliff Edward Young Clarke; the Guardians of Liberty (a Macon-based anti-Catholic group whose head would become the whipping boss of the Klan there); and the Patriotic Societies of Macon (whose secretary was also a Klan sympathizer). Indeed, the Imperial Palace of the Klan acted as the nerve center of the Dixie Defense Committee. Clarke set up its public mass meetings, coordinated the attendance of area Klaverns, and gathered support from politicians. Ridley wrote the Committee's leading tract and served as its featured speaker, a platform he could also use to promote the Klan.[3]

Many Klan issues congealed in the campaign to ward off any reform of race relations in Georgia: racism, right-wing populism, opposition to labor, sexual conservatism, antipathy to the social gospel, even anti-Catholicism. "The Damnable Dorsey Pamphlet," campaign activist and Klan supporter Miss A. Benton alleged, was an "insidious jesuit scheme to destroy Anglo-Saxon Morale" and enlist the nation in "the Cause of the Negro." "Our pulpits," she claimed, in reference to ministerial leadership of the CIC, "are turned into 'converting' agencies; our Women's Missionary Societies have become Inter-Racial Committees." Worse, it was having some effect. Benton could see but one outcome if these activities were not stopped, and quickly. "Our PRICELESS white girls," she warned, would in no time be bearing the children of black men.[4]

Benton's tirade illustrates how race was, at one and the same time, at the core of the Klan's politics and a medium through which to fight other causes. In the immediate term, Klansmen wanted to put down challenges to white supremacy and WASP dominance, whether they came from below or above, using whatever means it took. Indeed, the Klan's approach to race relations was nothing short of apocalyptic. The order held that even the slightest concession would embolden African Americans to make further demands, which would in time undermine the whole apparatus of racial hierarchy. Access to better education, for example, would only increase the yearning blacks felt for racial equity; worse, it would equip them with some of the tools needed to win it. And without resort to extralegal terror, how could whites inspire the fear needed to keep blacks from fighting for things they believed to be rightfully theirs? As Jo Ann Gibson Robinson, a leader of the Montgomery bus boycott, observed of the thinking that led a later generation to obdurate resistance: "They feared that anything they gave would probably be viewed by us as just a start." "And you know, they were probably right."[5] Viewing gains for blacks as losses for themselves, Klansmen readied themselves for Armageddon.

Yet "race" worked overtime for the Klan. In addition to its use in stifling competition from African Americans, it performed other services. Beleaguered by conflicts of class and gender that their sensibilities left them ill-equipped to explain, Klansmen displaced these conflicts onto imagined racial Others—whether African Americans, Jews, or immigrants, all of whom were conceived as biologically distinct from and inferior to "real" Americans, members of the "Anglo-Saxon race." Individual psychological conflicts undoubtedly contributed to this displacement—one simply cannot read Klan propaganda without sensing this. But the critical point here is that this displacement was also socially and ideologically necessary for the success of the Klan's overall project. Race was the Klan's answer to the class division its members so feared; race moved to center stage to push class into the wings as a way of understanding and organizing social life.[6]

This double mission was much in evidence in the agitation against Dorsey. On one hand, Klan leaders charged the governor with giving aid and comfort to black radicals at the "worst" time, just when the farm crisis permitted no leeway in dealing with hired labor. Dorsey and the CIC ministers might say that they opposed racial amalgamation, Ridley charged, but in practice they served the NAACP and gave "courage" to ordinary blacks in ways that could prove perilous. Ridley even suggested that Dorsey might be "bidding for the colored vote," which, he warned, "the National Government is about to again make" a factor in state politics. The result, he implied, could only be a race war.[7] Clearly, holding down African Americans was Ridley's foremost intent.

Yet his discussion of the nature of the danger intimated other preoccupations. Thus, he summoned up images of "white women ravished" by black men of "uncontrollable passion." He hinted at the influence of New York Jews, and he drew attention to the backing Dorsey received from "ministers, lawyers, and capitalists," who failed to see the danger their "misguided schemes" held for "the great majority of Georgians." In short, the clash over the Dorsey statement replicated the kinds of frictions among élite and petit-bourgeois whites that came into play over chain stores, municipal reform, and the World Court.[8] Similarly, the evocation of interracial sex indicated that anxieties about changing relations between men and women and even parents and children helped drive Klansmen and their sympathizers to action.

In many ways, the conflict over Dorsey's statement was distinctively Southern, not least in the pervasive extralegal coercion and violence the governor raised his voice against. Openly or tacitly, the region's political economy sanctioned this state of affairs. Dorsey's criticism was noteworthy because it was so singular; in the North—at least in districts where politicians had to face black voters—he would not have seemed such a renegade. Claiming that Dorsey had exposed Georgians "to the contempt and scorn of the outside world," his opponents also exhibited—and played to—white Southerners' acute sensitivity to Northern criticism. By calling themselves the Dixie Defense Committee and the Guardians of Liberty, Dorsey opponents laid claim to a tradition of states' rights advocacy in defense of white supremacy going back a hundred years. Similarly, the insurrection panic the Ridley manifesto expressed had a history as old as Southern slavery. Even the rhetorical mode had a regional flavor, with its emphasis on threats to the purity of white womanhood and offenses to the honor of "red-blooded" white men.[9]

▲ ▲ ▲

And yet, for all that, the basic concerns and commitments animating the campaign against Dorsey were hardly unique to the South.

If not so obsessed with these issues as many of their peers in the South were, whites in the North were not oblivious, either. Shared fears and common values made it possible for Southern-bred Klan officials to round up defenders of white supremacy on both sides of the Mason-Dixon line. Around the country, in fact, Klan members and leaders complained of black ambition. The new confidence African Americans showed in resisting white racism in the wartime and immediate postwar years appeared to Klansmen as perhaps the greatest challenge of the day. "The real arraignment of the Negro" by the Klan, W. E. B. Du Bois perceived, "is that white America with its present machinery is not going to be able to keep black folk down." That is, while the pervasiveness of racism made a movement like the Klan possible, it was the resilience of blacks in the face of this antagonism that led die-hard racists to believe the Klan necessary. "It is a new Negro who inhabits the South today," the NAACP explained in a discussion of the Klan's purposes and prospects: above all, "a new Negro youth . . . that will not be cowed by silly superstition or fear."[10]

Dumbstruck, Klansmen shared in the assessment if not the applause. One internal publication thus complained of "the haughty ambitions and arrogant aggressiveness" of contemporary African Americans. Others used words such as "brazen" and "flaunt[ing]" to describe the attitudes of black activists. This language showed how the confidence of the "New Negro" took Klan members aback. Indeed, the *Kourier* complained that "black and pro-black agitators [sought] to make capital out of every possible incident." In these circumstances, Evans once informed members, maintaining white unity and supremacy at all costs was "our most important work." Georgia's Grand Dragon elaborated. The black man, he warned, "must be brought *again* to realize that he is of an inferior race and of a lower standard."[11]

As Klan leaders were painfully aware, the explosion of discontent among oppressed "races" was not confined to the United States. Following Lothrop Stoddard, contemporary racialism's guiding light, Simmons maintained that the central fact of the contemporary world was "the 'rising tide' of the colored peoples and the backward white peoples, [and] their ultimate domination of the human process." In other words, whites of Northern European descent made up a small minority of the world's population; the days of their supremacy appeared numbered. The phenomenon Simmons referred to was the surge of nationalist feeling among colonized peoples in the wake of the war and the Russian revolution.

"White Men, Beware," intoned an editorial in the Klan press: "The prestige and security of the White Race received a terrific setback in the World War." Recently, affirmed one Klan leader in 1922, "white men" had suffered the "most humiliating" experiences, such as near-expulsion from India and being "bullied by a far inferior people" in Turkey. Later in the decade, after the postwar rebellions had subsided, the *Kourier* still saw reasons to fear. It warned, for example, of "the menace of Moslem unity" to America and its European allies.[12]

Members shared their leaders' anxiety and their resolve. When Simmons spoke at the Decatur, Georgia, court house in 1921, the audience interrupted him with vigorous applause when he proclaimed that the Klan "makes niggers get in their place and stay in place!" Indeed, the Klan became the foremost advocate of resistance to any modification in race relations. "The Klan believes in white supremacy," as one West Virginia Klan leader put it, "and will not compromise on this issue." Throughout the decade, the Klan used instances where blacks claimed equal rights—whether through voting, NAACP activity, trade union organization, membership in the Communist Party, or intermarriage with whites—as reasons for whites to join the Klan, and for delinquent members to renew their commitment. As the attack on Dorsey and the CIC showed, whites who aided blacks especially enraged Klansmen, for their cooperation destroyed the fiction of cross-class racial loyalty. With "nigger lover" as their battle cry, Klan-led forces more than once routed reform-minded politicians.[13]

The Klan made it clear from the first that fending off challenges to white supremacy—whether they came from blacks or whites—was central to its mission. The most potent symbol of such thoroughgoing absolutism was the Reconstruction-era Ku Klux Klan; Klansmen in the 1920s appropriated its mantle and romanticized its methods. They consistently exalted their forebears for having "saved" white civilization. Indeed, the men of the first Klan were praised as "the greatest heroes of all history." Simmons maintained that his order was "the reincarnation among the sons of the spirit of the fathers" who took part in the original Klan. To demonstrate this filial devotion, the Imperial Palace absolved members of the "original" order of admission fees and dues obligations. Chapter secretaries kept tallies of the number who enrolled—presumably for advertising purposes. The Athens Klan boasted two such veteran white supremacists.[14]

Many Klan chapters in the South also adopted the names of

leaders of the original Ku Klux Klan, calling themselves, for example, the John B. Gordon Klan. The grandson and namesake of the Reconstruction-era Klan leader Nathan Bedford Forrest was appointed Grand Dragon of Georgia by shrewd promoters who knew the drawing power of his name for their intended audience. Athens Klansman *O. M. Martin*, for his part, defended his own organization from critics by portraying the Reconstruction Klan as the savior of Southern society.[15] Such practices expressed an identification with Southern traditions that the Klan ordinarily downplayed in the name of common white, Protestant Americanism. Yet, with the Dunning school of Reconstruction history at the height of its popularity in the United States, sympathy with Confederates was not as divisive among whites as it once was or would become.[16]

So panicked were Klansmen over adjustments in race relations that they repudiated reason as a method for understanding and improving human life. The Klan, Evans thus boasted, was "emotional and instinctive." Primal drives antedated reason in human history, he said; they still moved the common people, if not intellectuals who had lost touch. In place of rationality, Evans advocated "race pride and loyalty" as "fundamental instinct[s]" whites should rely on in these troubled times. Local Klan spokespeople applied Evans' abstract injunctions in visceral racist agitation. Thus, to discredit Andrew Erwin, a liberal Al Smith supporter, Athens Klansman J. H. Wilkins urged his listeners to "imagine Andrew Erwin's living room." "On his right sits a chinaman. On his left sits a wop and a nigger. Think of all those smells together. That's the Smith crowd."[17]

The overthrow of reason was less an end in itself than a means of short-circuiting related claims of fundamental human equality. The Enlightenment tradition of rational inquiry, it seems, was too intimately tied to liberalism and the left. Evans explained that in order to put their own children and race above "aliens," WASP Americans "had to reject completely the whole body of 'Liberal' ideas." To combat the subversive implications of liberal thought, Evans ridiculed the idea of fundamental human equality. By removing it from serious discussion, he would naturalize hierarchy. It was unnecessary even to argue about racial equality, he said, as "the average [white] American does not believe it." In any event, he added, prescribing as much as predicting, the question would never be settled by logic, but rather "by race instinct, personal prejudices, and sentiment."[18]

In place of the dangerously leveling ideas of liberalism, Klan leaders put forward a racialist theory of human society and history. Having rejected Darwin's theory of evolution in biology, the Klan embraced the crudest Social Darwinism to explain their world. "We believe," explained Imperial Wizard Evans, "that the races of men are as distinct as breeds of animals." Klan leaders posited racial difference as the unrelenting determinant of human affairs. Not environment, not economics, not intergroup relations, nor even culture shaped the evolution of society. Rather, as the *Search-light* summarized the Klan's position, "race forms the basis for all human actions and reactions." "The unchangeable differentiation of race," distinction that presumed immutable hierarchy, was thus the critical variable in social development. Since it was "decreed by the Creator," it could not be undone; to tamper with divine law was to invite disaster.[19]

Klansmen employed this racial theory to explain their world. They explicitly denied, for example, that past enslavement and ongoing discrimination accounted for the poverty that plagued so many African Americans; the cause was, rather, "racial degeneration." Klan propagandists likewise charged that the decay of particular societies resulted from their members' having disobeyed an alleged law of nature forbidding miscegenation. "Hybrid" breeding led to "race suicide." According to the Klan and like-minded thinkers, such mixing accounted for the ruin of the civilizations of classical antiquity; "a mongrel civilization" could not survive. Here was a lesson that Anglo-Saxon Americans ignored at their peril. Similarly, Klansmen attributed different forms of government to divergent racial proclivities. According to Simmons and his co-thinkers, for example, the failure of parliamentary democracy in Russia resulted from the "inferior" stock of its inhabitants. Evans, for his part, claimed that an inborn submissive "mental nature" in Celtic, Southern European, and South American peoples accounted for both their Catholicism and their poverty.[20]

Of course, a tendency toward racial exclusivity had long been part of the republican tradition the Klan drew from to oppose threatening developments. In the Klan's hands, however, republicanism's implicit exclusion and suspicion of African Americans as slaves or propertyless free people became a fully worked out ideology in which "race" was used to indicate "otherness" and with it inevitable deviance and danger. "We know," one Klan-endorsed tract reminded readers, "that a republic is possible only to men of homogenous race." Simmons, for his part, recalled for his readers a longstanding consensus among propertied Anglo-Saxons that, in order to function properly, their brand of democracy required great commonality among the enfranchised. Racial, ethnic, and religious differences, like class cleavages, impeded such unity of purpose. Disunity, Evans explained, made nations prey to stagnation or conquest. "Alien ideas and excessive liberalism toward them," he asserted, had put the United States in a vulnerable position.[21]

To demonstrate their fidelity to American traditions, Klansmen pointed out a history of chauvinism in American leaders. Indeed, the Klan claimed for its positions the authority of the founding fathers and the widespread belief that American expansion was divinely ordained, a "manifest destiny." Evans insisted that the authors of the Declaration of Independence claimed equality only for "white men." Their unrepentant ownership of slaves and denial to Native Americans of "all political rights," he asserted, made this clear. Klansmen also repeatedly noted that Abraham Lincoln believed African Americans inferior and opposed granting equal rights to them. Similarly, the *Kourier* quoted Henry Grady, premier spokesperson of the New South, on the innate superiority and legitimate dominance of "Anglo-Saxon blood." Klansmen also looked to political leaders of their own day for ideological support. Among these were President Calvin Coolidge, who asserted that "Nordics deteriorate when mixed with other races," and his successor, Herbert Hoover, who warned that immigrants "would be tolerated only if they behaved."[22]

In articulating its vehement racism, the Klan also drew strength from recent cultural developments. The first three decades of the twentieth century marked, in the words of historian I. A. Newby, the "zenith" of racialist thought in history, science, social science, and popular culture: "the years which produced the greatest proliferation of anti-Negro literature, and the years in which that literature enjoyed its broadest appeal." Eugenic thought also flourished in this period; it drew support not only from conservatives, but also from liberals, and many socialists and feminists. Finally, the 'twenties saw the emergence on a mass scale of political anti-Semitism. In widely published and repeatedly reissued works, old-stock American, Ivy League–educated authors like Madison Grant and Lothrop Stoddard informed millions of contemporaries that race was the primary force in history and the taproot of America's troubles. The Klan's gratitude to such respectable racist ideologues was great. Drawing confidence from their arguments, Evans maintained that it was "hardly necessary" to argue for inborn disparities since the notion of basic human equality had "been abandoned by all thoughtful men."[23]

Carrying common strains of racist thought to extreme conclusions, Klan leaders made up in fervor what they lacked in originality. In order to release inhibitions against aggression, they first had to dehumanize their targets. Internal Klan documents thus condemned "trash immigration," while a Klan-recommended writer referred to African Americans as "ten million malignant cancers gnaw[ing] on the vitals of our body politic." The first degree in the order's ladder of fraternal rituals, a paean to the Reconstruction-era Klan, for its part, alluded to black men as "lust-crazed beasts in human form." Having made their intended victims appear both subhuman and life-threatening, Klan leaders could then whip up enthusiasm in their own followers for combat in "self-defense." Indeed, one author warned readers that "the day of reckoning is upon us." Klan members readied themselves for this reckoning. They pledged to "preserve unto death the peculiar distinctiveness of the white race from the foul touch of a lower stock." In a ritual enactment of this commitment, Simmons used to open the Atlanta Klan's internal meetings by thrusting two guns on the table and calling out, "Bring on your niggers." "The inevitable conflict towards which all events tend," he explained elsewhere, "is the white man against the colored man," fought out on a world scale.[24]

Such bellicosity helped distinguish the Klan's brand of white supremacy. Members saw themselves as an army in training for a war between races, should that prove necessary to perpetuate the United States as "a white man's nation." Klansmen boasted that they had bonded into "an invisible phalanx . . . to stand as impregnable as a tower against every encroachment upon the white man's liberty, the white man's institutions, the white man's ideals, in the white man's country, under the white man's flag." If other white Americans failed to heed the call to suppress African Americans and exclude immigrants, Simmons predicted a United States "blood-soaked by one of the most desperate of interracial wars—a war at once civil and international." Georgia's Grand Dragon concurred with the prognosis. Evans also agreed that different races could never share the earth in peace. He insisted that world history was nothing more than "race conflicts, wars, subjugation or extinction." The law of nature dictated that "each race must fight for its life, must conquer, accept slavery or die. The Klansman believes,"

Evans concluded, "that the whites will not become slaves, and he does not intend to die before his time."[25]

These violent intimations extended to Jews as well. Indeed, the inclusion of Jews among the ranks of racial enemies helped differentiate Klansmen from more mainstream white supremacists. The malignancy of Klan leaders' anti-Semitism was illustrated by their filial regard for its German masters. In 1925, for example, the *Kourier* proudly reported that the German *Hammer Magazine* had welcomed the Klan as a fraternal ally, noting their common goal of "shatter[ing] the bonds in which the Jewish offender has smitten all honorable nations." Later that year, the *Kourier* reprinted an article from the press of the extreme right in Germany about the "terrible misdeeds" of "the Jew." "The people's Germany," the author ended, "knows only one task—the warding off and the annihilation of the blood-enemy of the Aryan peoples—the Jew." The murderous impulse of this author's disciples in the Klan was further illustrated by their references to Jews in such terms as "one of the greatest menaces" to society, as a treacherous group plotting "world dominion"—or simply, as "vermin" and "scum."[26]

To the extent that it makes any at all, this fury only makes sense in its larger setting. The upsurge of anti-Semitism in the 1920s—reflected in the mass readership of Henry Ford's rabid *Dearborn Independent* as well as in the growth of the Klan—was no mere quirk. Outbursts of mass Jew-hatred are almost always symptomatic of profound social crisis or malaise.[27] In modern times, this hatred has typically accompanied anti-radical ideologies, which use Jews as scapegoats for the evils of capitalism. "As a *political* ideology," commented a leading scholar of American anti-Jewish organizations, "anti-semitism without an anti-revolutionary aspect is so rare as to be almost unknown." Or, as Jean-Paul Sartre put it: "anti-Semitism is a passionate effort to realize a national union *against* the division of society into classes. It is an attempt to suppress the fragmentation of the community into groups hostile to one another by carrying common passions to such a temperature that they cause barriers to dissolve."[28] Racial exclusion and victimization thus became the strategy for a class revanchism.

▲ ▲ ▲

The Klan's harangues against Jews in fact revealed the class drives of its racialism more clearly than those against any other group. Although the more avid evangelicals in the Klan's ranks blamed Jews for the crucifixion of Christ and condemned them for blocking such fundamentalist initiatives as mandatory school prayer, most Klansmen said little at all about religion in their charges against Jews. More commonly, Klansmen, like Henry Ford and other leading anti-Semitic contemporaries, dwelt on economic and social themes. They accused Jews, first and foremost, of dominating international capitalism, particularly the world's finances. "The Jew has a monopoly on the monetary system of the commercial world," as Leroy Curry put the Klan's principal allegation. In his view, Jews were the cause of the "fact that during the last one hundred years we have witnessed the building of a gigantic money monopoly that is without parallel in a thousand years of human history." Curry, like other Klan writers, located the source of this alleged power in Jews' supposed lack of business scruples.[29]

In such allegations, Jews stood surrogate for capitalist behavior that disturbed Klansmen. Against a backdrop of revelations of endemic profiteering by American corporations during the war, Curry asserted that Jews' economic behavior was guided by the "degrading" and un-American "doctrine that money is more powerful than the character of the nation." Similarly, in a decade that saw spiraling inequalities in the distribution of wealth, Evans accused Jewish creditors of benefitting from the miseries of "the poor," as if this were unusual. Or, as an Athens Klan lecturer put it more agitationally, Jews "get rich and prosper off of what they make by cheating and swindling the Americans." Finally, at a time when the employment of young women alarmed many adults about the future of family life, another Klan propagandist accused Jewish bosses of having "procured" gentile female employees, as if hiring women were itself a sin.[30]

In this way, Jews served as the symbol of "bad," or big, capital, which Klansmen distrusted. By portraying exploitation, destructive competition, and economic concentration as unnatural anomalies caused by the perfidy of a small minority, the Klan's anti-Semitism implicitly defended the "good" or small-scale capitalism of the local commonwealth vision. Not the organic development of industrial capitalism from the world of small property-holders, but the malicious designs of racial "others," explained why the nineteenth-century dream had failed.[31]

The Klan gained confidence to attack Jews from longstanding American political traditions. Thus, when maligning Jewish immigrants (overwhelmingly urban), Klansmen sometimes invoked the hostility to cities and their labor forces characteristic of Jeffersonian republicanism. Similarly, anti-Semitism fed off the legacy of antagonism between direct producers, whether farmers or artisans, and merchants and creditors enshrined in the critique of "Mammon" developed by nineteenth-century American protest movements. The Klan's central allegation against Jews—that they had a "stranglehold" on finance and thereby the whole economy[32]—harked back to the People's Party in particular. Committed themselves to private property and production for profit, the Populists had criticized subsidiary features of capitalism; namely, the machinations of financiers and the monetary policies of the federal government.

Turning that latent potential into overt, malicious anti-Semitism, the Klan fused raw material from these popular protest traditions with the longstanding stereotype of "Shylock." The economic parasitism ascribed by the Populists of the 1890s to finance and monopoly capital in general became, in the Klan's hands, a characteristic of the "Jewish race." Thus, Samuel Campbell, Grand Klokard (lecturer) of the Klan, asked Populist stalwart Tom Watson in 1922 for more information about how "the Wall St. Bankers—mostly Jews" had "brought hard times on us" after the Great War. The following year, Campbell wrote an article insisting that Jews had become "a national danger." As evidence, he cited the efforts of "the International Banker" to dominate national governments and international relations.[33]

Yet, paradoxically, the Klan, along with other anti-Semites in the postwar era, also maintained that Jews dominated the paramount force opposed to international capital: communism. This notion found support in the so-called *Protocols of the Elders of Zion*, a forged document put into mass circulation by foes of the October Revolution to "prove" that the Bolsheviks' victory resulted from a Jewish conspiracy. "Jew" and "Bolshevik" in fact frequently served as synonyms in Klan vocabulary. "Bolshevism," as the *Imperial Night Hawk* expressed the oft-made allegation, "is a Jewish-controlled and Jewish-financed movement in its entirety." The paper went on to declare that Lenin, Trotsky, and all the other Bolshevik leaders were Jews. Internal Georgia Klan circulars alerted members to the danger America faced from "the radical Bolshevic forces, aided and led by the International Jew."[34]

The apparent inconsistency of the allegations made against Jews, their "magical" character, makes more sense when considered in light of the predominance of the petite bourgeoisie in the Klan and other anti-Semitic organizations. As petty proprietors caught between capital and labor in what appeared to them to be a zero-sum game, Klansmen harbored grievances against both sides, a Janus-like position conducive to the dual antipathies that characterize political anti-Semitism.[35] The economic place of the petite bourgeoisie, moreover, made it, at least potentially, the most nationalistic class. Whereas both large capital and labor had some incentives for cooperation with their counterparts abroad, petty capitalists could anticipate no such rewards.

Their rootedness in, and dependence on, local economies and tangible property helps explain the charge of "cosmopolitanism" levied with regularity against Jews and at the root of the oft-used term "the International Jew." One author thus complained, in the same breath, of "international unionism" and "international banking." The Klan accused Jews of being the secret force behind such proposals as the League of Nations, and explained their alleged motives thus: "having no national government of their own they seek to attain them all." Georgia's Grand Dragon, for his part, insisted on the need for education in "American Nationalism" to combat the "Cosmopolitanism advocated by International Jewry."[36]

The second of the principal charges made against Jews—their alleged radicalism—was also levied by the Klan against immigrants more generally.[37] Simmons thus warned his followers that they were being crowded out by a "mongrel population . . . organized into Ghettos and Communistic groups . . . and uplifting a red flag as their insignia of war." Indeed, the new immigrants "threatened to smother our working people with the noxious poison of Bolshevism." The numbers of native-born whites among the communists were "so few . . . that it is comparatively easy to eliminate them as a revolutionary strength," explained the *Searchlight*, "but the working masses of aliens or foreign-born and negroes are a strength to be reckoned with."[38]

Klansmen identified immigrants, like Jews and African Americans, as racial groups distinct from themselves. Tapping racism to subdue labor struggle, the *Searchlight* thus described the unrest among foreign-born industrial workers as "the present racial menace." Similarly, the Klan frequently proclaimed as one of the key points in its program "the prevention of unwarranted strikes by foreign agitators." In light of the Klan's antipathy to labor militancy, its complaints that immigrants undercut the living standards of "American" workers appear self-serving. But the obviousness of the contradiction suggests that more was at work. Agitation against "cheap labor" from abroad served the ideological end of bonding native-born white workers and their employers in the kind of intraracial unity the Klan counterposed to class struggle. The cheap-labor charge also provided immediate benefits to the Klan, which otherwise had no economic program to attract native-born white workers disturbed about their wages and job security.[39] As it was, by setting the Klan apart from élite conservatives who shared its other complaints about immigrants, this populist theme gave Klan nativism a mass appeal.

Class and gender themes were entangled at the roots of a third common allegation the Klan levied against immigrants, that they lowered the nation's moral standards. Republican thinkers had feared the Old World proletariat in part for the deviant family structures, gender roles, and sexual conduct assumed to flow from its propertylessness. Klansmen brought these assumptions to their discussions of immigrants, hence their use of such phrases as "European riff-raff" and "slaves of ignorance and vice" to describe them. This framework helped make sense of the new forms of commercial, urban-based mass culture, which Evans complained sought "to degrade us to the level of European morality . . . to the slums, ghettos and cesspools from which most recent immigration has come." Similarly, the Klan often asserted that the opponents of Prohibition were "mainly alien," and targeted the foreign-born in its raids on bootleggers.[40]

The battery of allegations Klansmen made against African Americans contrasted with the limited range levied against Jews and immigrants. Thus, the Klan charged that blacks were biologically inferior, unfit for democratic participation, criminal and immoral, lazy, oversexed, and on and on. In contrast, complaints about competition for trade, work, and housing were less frequent.[41] From this imbalance in coverage, one might deduce that economic motives were not a factor. But that would be a mistake. While class perspectives and goals influenced the Klan here as elsewhere, to have acknowledged them directly would have undermined the order's cardinal contention: that "natural" divisions like race and sex should take primacy over "artificial" distinctions like class. As a result, the order generally expressed class concerns in an opaque racial idiom whose drives only become clear when analyzed in context.

The subordination of African Americans, after all, undergirded the entire Southern economy: "the negro," as a representative of the Athens Chamber of Commerce put it, "is our only and best form of domestic and general labor." While Klansmen rarely spoke of this reliance openly—to have done so would have been to recognize black contributions to America—in practice they sought in numerous ways to ensure that blacks would remain a cheap, unorganized labor supply. Like its Reconstruction predecessor, the Klan of the 1920s may thus have posed its mission more often as the defense of white culture than as the restraint of black labor. But those familiar with the South understood that the distinction was academic: the purpose of the former was to safeguard the latter. "The races in the South may be divided into two classes," as an Athens mayor observed hyperbolically in 1923: "the Employer and the Employee." And, in fact, the second Klan, like the first, aspired to control black workers. When Simmons asserted of his predecessors after Emancipation, "it's all rot about the K. K. swinging [lynching] niggers—niggers were loafing and K. K. made 'em go to work," the potential uses of his own organization could not have been lost on his listeners.[42]

The anti-labor animus of the Klan's commitment to white supremacy appeared most clearly in other forums, however: namely, in the racial themes that pervaded the order's anti-communism and anti-Catholicism. Indeed, one key aspect of the threat Klansmen saw in communism was the unprecedented commitment of its white followers to black rights. The Klan thus denounced "Bolshevist agitation" for interracial trade unionism and pointed with alarm to the Communist Party's overtures to African Americans. Klansmen believed that communism would mean the end of racial hierarchy. The "worst" offense of Communist union organizers in North Carolina, in the view of the *Kourier*, was their advocacy of "negro equality."[43]

Similarly, the Klan denounced the Catholic church for furthering "social equality." Klansmen attacked the church, in particular, for its recruitment and training of black priests to serve interracial congregations. The doctrine of the equality of believers and the practice of integrated congregations had the potential, in Klan eyes, to upend the social order. A Mississippi female Klan supporter thus summoned the specter of slave rebellions to depict the dangers of blacks' converting to Catholicism. The result would be the "horrors of Hayti and San Domingo": black men "yearning for the fertile fields and fair women of their masters." More generally, the Klan charged the Catholic church with being "after the negro as one of its major steps in dominating the American republic," an appeal to time-honored republican fears of those with no stake in society being used as an entering wedge for despots.[44]

The Klan's varied attacks on African Americans, Jews, and immigrants in fact converged on a common core goal: securing the power of the white petite bourgeoisie in the face of challenges stemming from modern industrial capitalism. The Klan sought to deny political rights to those whom it perceived as threats to that power. Indeed, one purpose of Klan racialism was to convince people that only a small, select group was "fit" for self-government. Imperial Wizard Simmons insisted, for example, that the supposed inability of Africans and most Asians to control their own affairs mandated carrying "the 'white man's burden' " well into the future. More generally, he described universal suffrage as "a very dangerous political doctrine." The ballot was not a right, but "a privilege," and neither blacks nor immigrants deserved it. And while they were hardly the primary target of such assertions, poorer whites were not immune. Simmons himself insinuated that both property and literacy qualifications should apply to all voters.[45]

His successor candidly announced that genuine democracy would impede the Klan's goals. "The Nordic can easily survive and rule," Evans explained, "if he holds for himself the advantages" secured by his forefathers. His supremacy would be lost, however, "if he surrenders those advantages" to immigrants and their children. The "Klansman's Creed" thus declared, "I believe my rights in this country are superior to those of foreigners."[46] African Americans and immigrants were thus the most immediate and aggrieved victims of what by the Klan's own admission was a wider attack on democracy conducted in defense of property and privilege. In this way, class perspectives, motives, and goals were at the nerve center of the Klan's racialism. Racism enabled Klansmen to reconcile conflicting impulses: on one hand, their regard for white popular sovereignty and their commitment to private property; on the other hand, their discomfort with growing economic concentration and their fear of genuine democracy in a society with a massive, lately quite militant, working class. Yet the Klan's construction of "race" was also shaped profoundly by gender-specific perspectives and motives.

Sexual themes saturated the Klan's racial agitation. So obsessively that it appeared intentional, Klansmen used bodily imagery to discuss race. Such terms as "proper blending of blood," "insoluble and indigestible" races, "mongrel population," "body politic," "pure and undefiled" blood, "racial pollution," and the like infested Klan lexicon.[47] But more important to the order's appeal were its oft-repeated allegations that men of other so-called races coveted native-born white women in various and distinctive ways.

In effect, Klansmen turned conflicts between classes and ethnic groups into rivalry over which men might possess which women.

As regards African Americans, the central argument the Klan put forward to oppose modification of Jim Crow was that any relaxation of its barriers would lead to racial "amalgamation." More specifically, reforms would lead to sex between black men and white women. Klan propagandists cited instances of multiracial social events and intermarriage in New York City as examples of the dangers of black electoral power. The order described the NAACP's mission as encouraging blacks around the country "to an open demand for social equality," inciting black men to "lust upon women of the white race," and telling them it was their "duty to marry a white woman." Voting rights, according to Imperial Wizard Evans, encouraged black men to sexually assault white women.[48]

Klansmen shared a sexual siege mentality with large numbers of their less strident white peers. By the late nineteenth century, large numbers of white Americans, particularly in the South, believed that black men had acquired an incorrigible desire to rape white women. This conviction grew so pervasive that rape came to be referred to in white society as the "new Negro crime." The most common and widely accepted—if utterly spurious—justification advanced for lynching was that it served to punish and prevent this alleged crime. Thus, Governor Clifford Walker sought to silence critics of lynching in the northern press with a challenge to report also "the unspeakable crime which provoked the violence." Walker then summoned up for his white listeners maudlin images of "the isolation of the Southern farm, leaving the women defenseless from the vagrant vagabond while their husbands toil for a living in the distant fields." More generally, supporters of racial discrimination employed the bogey of "social equality" as their primary and most effective argument. "Whenever, wherever, race relations are discussed," as Lillian Smith later observed, "sex moves arm in arm with the concept of segregation."[49]

Activists and historians alike have long struggled to make sense of the sexualization of racial conflict. The contrast between the rhetoric of chivalry and the reality of ruthless killing led many to interpret the sexual demonology of white supremacy as subterfuge. Writing in the 1920s, Walter White thus argued that sex charges were merely "a red herring" used to defame black victims and defuse white criticism. "Lynching has always been the means for protection, not of white women," he concluded, "but of

Stone Mountain, Georgia, the imposing granite butte outside Atlanta where the second Ku Klux Klan held its founding ceremony on Thanksgiving night in 1915. (Courtesy The Georgia Department of Archives and History)

One of the masterminds behind the spectacular growth of this male fraternal order in the early 1920s was a woman, Mary Elizabeth Tyler. Here, she poses while reading its publication *The Searchlight*. (Courtesy The Hargrett Rare Book and Manuscript Library, University of Georgia Libraries)

This 1925 ad for "Klan Day" at the Southeastern Fair shows how entrenched in white associational life the organization had become. (Courtesy Atlanta Historical Society)

The Imperial Palace in its heyday at 2621 Peachtree Road, one of the most fashionable streets in Atlanta. (Courtesy Atlanta Historical Society)

The Leo Frank case helped create the atmosphere in which the Ku Klux Klan could be revived. Here, gathered about Frank's suspended corpse, men in the crowd proudly wave to photographers. (Courtesy Georgia Department of Archives and History)

In the climactic scene of D. W. Griffith's 1915 epic film *The Birth of a Nation*, Reconstruction-era Klansmen prepared to lynch Gus (Walter Long), the film's black anti-hero. Censors removed a castration scene, but its point remained. The Klan used showings of the film to recruit new members. (Courtesy *Representations*, No. 9 (Winter 1976), p. 176)

Conscious of the power of symbol and the allure of ritual, the Klan devised elaborate rites such as this initiation ceremony to captivate members and supporters. (From *World's Work*, 1924)

IN PROPER HANDS

Although scholars have seen it necessary to emphasize one or another of the Klan's prejudices as its driving force, contemporary Klansmen felt no such need, as this anti-Semitic and anti-Catholic cartoon from a Klan promotional tract makes clear. (From *Heroes of the Fiery Cross*)

A view of Clayton Street in downtown Athens, where the Klan had its local headquarters in the late 1920s. (Courtesy The Hargrett Rare Book and Manuscript Library, University of Georgia Libraries)

Wartime regimentation and political repression largely wiped out alternative channels of popular dissent for ordinary white men, creating a vacuum the Klan came to fill. Here, troops march through the Five Points neighborhood of Atlanta in 1918. (Courtesy Georgia Department of Archives and History)

Yet the World War also opened a new era in Southern race relations; here, Augusta resident Frank Butler Black posed in his army uniform, a symbol of pride among many African-American servicepeople. (Courtesy Georgia Department of Archives and History)

The attenuation of male power and parental authority in the early twentieth century agitated many Klansmen. Here, the Georgia Young People's Suffrage Association spreads the word about votes for women in a 1920 parade. (Courtesy Georgia Department of Archives and History)

The Southern Manufacturing Company, one of the leading area mills, with workers' homes in background. This community would be a site of several night-riding incidents in the 1920s. (Courtesy The Hargrett Rare Book and Manuscript Library, University of Georgia Libraries)

This scene from a black neighborhood in Athens, probably from the 1910s or 1920s, reveals the poverty and deprivation of municipal services from which most blacks in the region suffered. (Courtesy The Hargrett Rare Book and Manuscript Library, University of Georgia Libraries)

Participants in the Fulton Bag and Cotton Mills strike of 1914–1915 march and publicize a mass meeting. One of the most pivotal and prolonged labor struggles of the area in the early twentieth century, this Atlanta strike dramatized the overt class conflicts that disturbed Klan members. (Courtesy National Archives and Southern Labor Archives, Georgia State University)

Yet this graduation photo of the first woman dentist in Athens—Ida Johnson Hiram (3rd row, 1st left)—also evokes the growing African-American middle class. Its members would play a key role in both the NAACP and the CIC. (Courtesy The Hargrett Rare Book and Manuscript Library, University of Georgia Libraries)

Women and child workers from the King Mill in Augusta in 1909. Wage-earning opened the way to new constructions of gender and sexuality among whites in the early twentieth century. Yet note also the black woman concealed from view in the left rear. Among the poorest-paid American industrial workers, white mill people could still take advantage of cheap black labor. (Photo by Lewis Hine. Courtesy The Hargrett Rare Book and Manuscript Library, University of Georgia Libraries)

By the 1910s, even the daughters of salaried employees and craftsmen began joining the labor force. These employees of the Southern Bell Telephone Company posed outside its new building in Athens in 1918. (Courtesy The Hargrett Rare Book and Manuscript Library, University of Georgia Libraries)

The defiant, precocious posturing of these Georgia mill workers, photographed in 1910, hints at how class was experienced by men and boys, too, in gendered ways. (Photo by Lewis Hine. Courtesy The Hargrett Rare Book and Manuscript Library, University of Georgia Libraries)

Capping his triumphant, populist campaign, Tom Watson in 1921 addresses a crowd in Thomson, Georgia, before leaving for Washington, D.C., to begin his term as U.S. senator. There, he would use his position to defend the second Ku Klux Klan from critics. (Courtesy Georgia Department of Archives and History)

The most common single occupation among Athens Klansmen was owner or proprietor of a small business; this drugstore belonged to one local member. (Courtesy The Hargrett Rare Book and Manuscript Library, University of Georgia Libraries)

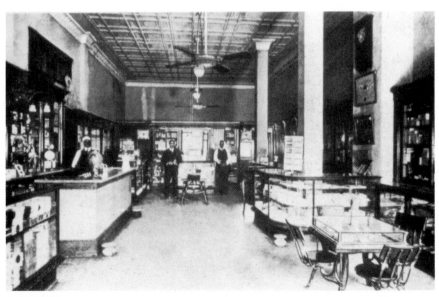

Small numbers and big obstacles notwithstanding, African Americans and immigrants were beginning to compete more directly with native-born white men. The E. D. Harris Drug Store was founded by an Athens black physician in 1910. (Courtesy The Hargrett Rare Book and Manuscript Library, University of Georgia Libraries)

"Costa's," an ice cream parlor presided over by an extended family of Greek immigrants, was very popular among Athens young people in the 1910s and 1920s. Indeed, it was one site of the emergence of the heterosocial youth culture that alarmed so many adults. (Courtesy The Hargrett Rare Book and Manuscript Library, University of Georgia Libraries)

VALDOSTA KLAN-OPEN AIR CEREMONIAL
KNIGHTS KU KLUX KLAN
VALDOSTA, GA. OCT. 10, 1922.

Klan organizers deftly tapped the potency of religious feeling to build their movement, as this open-air rally illustrates. (Courtesy Georgia Department of Archives and History)

Co-ed socializing at the University of Georgia in the 1920s. A crusade against young people's suggestive dancing and joy-riding in automobiles helped build the Athens Klan. (Courtesy The Hargrett Rare Book and Manuscript Library, University of Georgia Libraries)

First Christian Church, whose minister and his assistant served as exalted cyclops and kligrapp of the Athens Klan at mid-decade and helped to lead the city's social purity crusade. (Courtesy The Hargrett Rare Book and Manuscript Library, University of Georgia Libraries)

"Who In Th' 'el Said Vote For Dorsey ????"

A 1917 campaign cartoon from Tom Watson's newspaper. His racist attacks on Hugh Dorsey were amplified by the Klan in 1921 as part of a wider effort to subdue the nascent white racial liberalism embodied in such groups as the Commission on Interracial Cooperation. (Courtesy Atlanta Historical Society)

In this Marietta lynching from the 1920s, the perpetrators mock the corpse of their victim. Such gross white indifference to black suffering—these men obviously have no fear of identification or prosecution—helped make the Klan possible. (Courtesy Georgia Department of Archives and History)

Burned out of their home by Klan night-riders in 1949, this black woman and her son experienced the terror that many other families had in the 1920s. Then, no one even bothered to photograph the victims. (Courtesy Keystone Press Agency, New York, and Southern Labor Archives, Georgia State University)

A 1924 magazine article helps explain the confidence of Klan night-riders and the sparse documentation of their crimes. With so many politicians, law enforcement officers, and judicial personnel either cowed or complicit, prosecution of Klan vigilantes was unlikely. (From *Hearst's International Magazine*, 1924)

profits." Similarly, others have concluded that the cry of "social equality" was a ruse to safeguard economic privilege.[50]

Unquestionably, the evocation of sexual danger did shore up the white élite. But such interpretations fail to clear up the real enigma: why subordinate whites fell for the sleight of hand.[51] That is, how could such rhetoric rally the whites who seemingly stood to gain little? Because other kinds of power were also at stake. The charge of rape, as historian Jacquelyn Dowd Hall astutely observed, derived its power from the fact that it "was embedded . . . in the heart not only of American racism, but of [white] American attitudes toward women as well." At the same time as it enforced white supremacy, lynching also served to remind white women of their prescribed asexual, subordinate, and dependent roles in Southern society. To counterpoise economic concerns and sexual anxieties is thus to create a false dichotomy.[52]

Yet the evidence suggests that the prominence of sexual themes in the Klan's racism also had another important source. Male dominance and white supremacy shared common material roots in this setting. The key to the merging of economics and sexuality in Klan racism was the order's fealty to a vision of white petty-producer households in which women and youth were dominated by adult men. The Klan's racialism was thus grounded in the same domestic and social arrangements that its male supremacy and sexual conservatism issued from. Old traditions persisted in modified form in the way the male household head's control over the labor power and sexuality of his dependents helped ensure economic viability for contemporary farm and mill households. For all its irrational aspects, then, fear of female autonomy had a rational basis in the labor and service requirements of household survival. So, too, did control over women buttress racial hierarchy. Since racial affiliation was traced through the lineage of the mother, policing the borders of white society required the regulation of white women's sexuality. Not surprisingly, then, sexual relations between white women and black men became "the strongest taboo of the system."[53]

One of the commonest white supremacist campaigns waged by the Klan was in fact to end interracial intimacy. Around the country, state and local Klans promoted legislation to make marriage between blacks and whites a felony. There should be "no room in America," the Klan announced, for "any man or woman who believes in or teaches the mixing of our blood" with that of African

Americans. Yet if racial mixture itself had been the real fear, then the Klan should have included black women in its charges as often as black men, particularly since sexual contacts between white men and black women, usually forced, were far more common. Universal pronouncements were in fact the exception rather than the rule.[54]

The Klan never mentioned the rape of black women; it focussed almost exclusively on black men and their relations with white women. Whereas the rape of black women by white men confirmed the latter in their power over all blacks, male and female, and all women, black and white, sexual relations between white women and black men—particularly voluntary relations— defied white men's power over both groups. To rile up readers about the Al Smith campaign in 1928, for example, Klan publications employed front-page pictures of a black Tammany official in New York dictating a letter to his white secretary with a caption implying that his authority would translate into sexual access.[55]

In addition to proposing laws against interracial marriage, Klansmen worked in other ways to prevent or destroy such relationships. Southern chapters, for example, persecuted black men and white women who "violated social ethics," while Northern chapters warned prospective black grooms to cancel their weddings. The Klan press also printed notices of marriages between white women and black men to demonstrate "the necessity" of these legislative and vigilante efforts.[56] The image of passive white women put upon by aggressive black men was belied by the consensual nature of these marriages, however, suggesting that deeper concerns were at stake. In fact, intermarriage implied both an essential compatibility between blacks and whites and a voluntary social levelling that the Klan found intolerable.

Klansmen accused Jews, or rather Jewish men, of sexual offenses whose nature differed from those attributed to black men. One of the commonest was that Jews controlled the popular-culture industries alleged to be bringing moral ruin upon the nation's youth. The Klan's struggle to safeguard American families from the dangers of the "Jew-produced motion picture industry and the Jew-monopolized jazz music [and] sex publications" was thus adduced as a reason for Georgia men to join. Only "strict censorship," Imperial Wizard Evans insisted, could "keep the Jew-controlled stage and movies within even gunshot of decency."[57]

Behind such allegations were deep resentments at how economic and social change weakened white men's control over "their" women. These resentments found expression in an idiom at once populist and racist. The *Imperial Night Hawk* published the statement of a Protestant minister, supposedly a converted Jew, that ninety per cent of immigrant Jewish men believed "all Protestant girls are common property and can be bought for a song." Caleb Ridley likewise complained of "the total absence of respect for American girls by the great majority of foreign-born men and the younger generation of Jews," as did an Athens Klan lecturer. Another variation was to accuse Jewish employers of, as the *Kourier* put it, "produc[ing] by the slavery of under-paid girls working amid conditions and influences such as to make them easy prey to unvirtuous lives." Klan propagandists also alleged that Jewish men ran the "white slave" trade; it was, said a Georgia Klan lecturer, "in the hands of a 'Jewish trust' so to speak."[58]

The Klan also charged Jewish men with using their reputed economic power to seduce gentile women. "They bribe and buy those poor, uneducated girls," a Klan lecturer informed his Athens audience, "with fine silk underclothes, silk stockings, fine clocks, dresses and money." The notion that economic position gave Jewish employers sexual access to female employees had been central to the mass mobilization against Leo Frank that helped spur the formation of the second Klan. Now this speaker recalled Frank's alleged "lust . . . [for] the virtue of American girls" to make his case against Jews, while his Atlanta colleague cited Frank's case as evidence that Jews constituted "a national danger." Gentile Americans, proclaimed Blaine Mast, a Pennsylvania-based Klan writer, were "tired of the outrages inflicted upon innocent girls by Hebrew libertines." Mast offered as evidence for the "crimes and wrongs" of certain Jews the tale of a middle-aged, wealthy bachelor in his town. This man supposedly took gentile teenagers, "mostly working girls," in his car "to unfrequented spots and secluded places" and brought them to his rooms for "clandestine" trysts. After the man had "ruined" a young woman, Mast maintained, he discarded her like "an old worn-out coat."[59]

The dénouement of Mast's allegory showed how Klan propagandists also resented Jews for declining full assimilation into gentile society, a complaint in curious contrast with the order's opposite grievance against blacks. The *Searchlight* thus criticized what it construed as the "racial arrogance" of Jews, the product of a "superiority complex." "By deliberate election," Evans concurred, the Jew "is unassimilable." This alleged aloofness was one of the reasons adduced for the frequently made charge that Jews were "insoluble and indigestible." One aspect of Jewish "clannishness" that particularly irked Klansmen (imbuing the charge with overtones of wounded pride) was Jews' apparent lack of interest in marrying gentiles. "The descendants of Abraham," maintained Evans, "have denied their children the right of intermarriage with the Gentile."[60]

Comparison of the sexual demonology of anti-Semitism and of anti-black racism helps clarify the class and gender issues involved. In light of how Jews commonly stood surrogate in the Klan's world view for large capital, and African Americans for the propertyless population, it is notable that the order charged Jewish men with attempting to seduce gentile white women and black men with attempting to take them by force. In each case, the men's perceived position in the class structure was imagined to give them distinctive access to women of the Klan's group. Indeed, the method of the fantasized assailant varied according to his social position: the worker would take a woman by force; the owner, by money and cunning. In both cases, the method resembled that which his social group might use against the white proprietor in nonsexual power plays: for workers, a strike or rebellion; for employers, shrewd maneuvers. *All* the Klan's sexually related charges, in fact, focussed on the men of other groups. The lack of corresponding fears about female African Americans and Jews indicates that sexual jealousy was an important ingredient in white men's racism, as W. E. B. Du Bois had once suggested.[61] Since the Klan attacked black men for *marrying* white women and Jewish men for *not* doing so, moreover, it appears that the call for "racial purity" was not an absolute goal but rather a sign of deeper concerns.

Klansmen viewed women at some level as property; they also viewed them as symbols of power. From slavery forward, white men had taken sexual advantage of black women with no fear of legal reprisal. Such rapes served, among other ends, to humiliate black men by demonstrating to them—in the most searing way possible—their utter lack of social power and of the masculine "honor" whose prerequisite was the ability to control one's "own" women and protect them from the aggressions of other men.[62] Deeply conscious themselves of this tradition of using women as markers in symbolic power plays between men, Klansmen realized that other men could play the same game. The image of the black rapist was thus conjured out of Klansmen's fears of a militant claim to equality by a social subordinate: he would prove his dignity with their property even if he had to risk his life to get it.

Conversely, the image of the Jewish seducer connoted a humiliating display of command by a man with more money and power than they: he would use the womenfolk of subordinates to gratify his lust, yet scorn them as permanent partners.

In both fantasies, the association of "their" women with men of the "other" group denoted loss of position and prestige for Klansmen. In the one case, they imagined having to endure the dishonor historically visited on dispossessed, subordinate classes. In the other, the men of a formerly dispossessed, subordinate class symbolically raised themselves to Klansmen's level by gaining access to their women. In both cases, women were reified, turned into inanimate tools with which barriers were built up or brought down. The real point of reference was the middle-class man, caught between capital and labor and anxious about how the outcome of their rivalry would affect his power.

Gender was more than an emblem of race and class power, though; as these cases show, it gave these relations much of their charge. The Klan's commitment to WASP supremacy in fact converged with its dread of the growing autonomy displayed by the female relatives of its own constituents. One piece in the Klan press complained, not simply that "another white girl" had married a black man in New York, but that her parents "were helpless" to stop the union. The implication was that the Klan-proposed miscegenation law was necessary to restrain headstrong daughters over whom parents had lost control. Similarly, Alma White fretted over the immorality of young women who moved to the cities for work. Blaming their "fall" not on their own desires but on conniving Jewish men, White could avoid the troubling fact of their own sexual initiative.[63]

Finally, concerns about gender, race, and class came together in the Klan's eugenics program. Like many whites at the turn of the century, Klansmen deplored declining birthrates among native-born white women and urged them to bear more children to offset the numbers of foreign-born and African Americans. In such exhortations, commitment to an older model of male dominance converged with a desire to ward off democratic challenges from submerged groups. Thus, the Klan's scapegoating of immigrants for native-born Americans' growing inability to "support large families" assumed that having numerous children was indeed desirable, an assumption rooted in the male-dominated household economy. "Breeding better Americans in larger numbers" was likewise considered vital to maintaining position as a leading imperial power.[64]

The Klan shared a commitment to pseudo-scientific breeding with large numbers of contemporaries across the political spectrum. Still, there was a difference. The order's virulent racialism led it to apply eugenics in an especially dehumanizing, ghoulish manner. One Klan leader and minister thus maintained that "the methods employed in stock-raising" should be applied to human reproduction. He envisioned "elimination of the unfit" people and races from sexual activity and "development of the fit to the highest degree through the process of scientific study." Indeed, he went as far as to suggest the "segregating and isolating of certain males for the express purpose of developing a super race."[65] The Klan's pro-natalist program for white, native-born gentile women, like its sexualized racism generally, thus sought to fortify WASP supremacy, male dominance, and the rights of small property as integral components of one social system. But by far the most frightening combination of these commitments was that which came into play in lynchings of black men for alleged assaults on white women. Three such lynchings primed white residents in the Clarke County area for a decade of vigilantism in which elements of race, class, and gender overlapped and provided each other mutual support.

7

▲ ▲ ▲

Paramilitary Paternalism: The Politics of Terror

In September of 1917, the body of Rufus Moncrief, a thirty-year-old black man, was found riddled with bullets on a country road just south of Athens. Above his corpse was tacked a terse note: "You have assaulted one white girl but you will not assault another." The murder of Moncrief, carried out by just two carloads of men operating under cover of darkness, was mild compared with those that followed as the war aggravated social tensions. A year later, a posse of enraged white men pursued and caught Obe Cox, whom they accused of raping and murdering the wife of a prominent white farmer in neighboring Oglethorpe County. As several thousand spectators gathered, they chained Cox to an iron post, assembled a pyre beneath him, and burned him alive.[1]

Prior to these two cases, no lynching had occurred in Clarke County for more than forty years. So tranquil was it that Athens appeared a haven to many black Georgians fleeing harsher territory.[2] Then, suddenly, things changed. Between 1917 and 1921, four lynchings took place in the vicinity, each one involving an allegation of some threat to the purity or safety of a white woman. The Klan rode the crest of this wave of killings, encouraged as much by the tacit approval they received from most whites as by the acts themselves. In their wake, it set to work with its own campaign of intimidation and terror. Escalating the attacks on African Americans and including deviant whites among its targets, the Klan would test the limits of that community acceptance.

Whether local Klan members took part in the earlier lynchings is impossible to determine. Since even when they operated before the eyes of thousands, the lynchers were always described as "parties unknown," one cannot say for certain whether the rosters overlapped. But it is possible to understand, from their actions and words, why Klansmen practiced less lethal forms of vigilantism—most commonly, flogging men and sometimes women with thick rawhide straps under cover of darkness. The immediate goal of Klan violence was to terrify people out of engaging in particular kinds of behavior. As a collective effort to stave off or redirect changes in prevailing social relations, Klan violence was thus profoundly political. But methodical aggression like this needed cultural sanction.[3] That came from each of the strands in the Klan's world view: its reactionary populism, its racialism, its gender conventions, and its overall alarm about the state of society and government. Together, they worked to prompt and ennoble white male violence undertaken in defense of family and community. To put it another way, there were no significant restraining elements in Klan culture that might act to inhibit violence against outsiders to Klansmen's idea of community.

Finally, these episodes appear to have involved something else, more difficult to document. Assembling with their fellows to warn or whip a person who embodied the challenges of the day also probably stilled Klan members' own anxieties. At the same time that such violence served to subdue the propertyless, it also warned modernizing élites that ordinary white men were still a force to be reckoned with. In vigilantism, they acted out their rejection of remote government—whether directed by their social "betters" or their social "inferiors"—and dramatized instead a voluntary, local compact of white male household heads. Particularly when immune from punishment or reprisal, these rituals kept alive the vision of an older social order in which men like themselves still wielded paternalistic power. In Athens, events in the years before the Klan's rise made such impunity seem a good bet.

▲ ▲ ▲

As horrible as the killing of Obe Cox was, the next lynching was still more grisly. This one, in February of 1921, was "one of the most horrible in the history of the state," according to the *Atlanta Constitution*. The hunt began when a white farmer's wife was found murdered in neighboring Oconee County. Desperate to find a culprit, some focussed on John Lee Eberhart. The evidence connecting him to the crime was flimsy. Eberhart himself swore that he was innocent. But these details failed to deter the mob of Oconee and Clarke County men who abducted him from the Athens jail the night of his arrest, dragged him to a waiting car, and drove him to the scene of the crime. The fifty-odd police, sheriffs, and detectives in the vicinity of the courthouse made no effort to intervene. As an estimated three to seven thousand people raced to the scene by car and horseback to catch the spectacle, the mob chained Eberhart to a tree and piled wood beneath him. Taking their time, they tortured him to death by fire. Shortly after the murder, conclusive evidence surfaced that proved Eberhart was innocent.[4] But no matter.

By December of that year, whites were in such a furor and black life had become so devalued that a serious charge was no longer necessary to incite multiple murder. It took little to reactivate the frenzy. Out of work in the postwar depression, a young man named Aaron Birdsong approached "a prominent farmer" in Oconee County and asked him to rent him land or loan him a dollar. The farmer refused. When he was out of sight, so the reports went, Birdsong walked into the planter's house. His entry "frighten[ed] [the man's] wife and daughter." For this, Birdsong was to lose his life. A posse-cum-mob quickly assembled to hunt him down. But Birdsong did not allow the mob to take his life cheaply. Knowing what they had in mind, he shot the deputy sheriff who came for him and wounded a member of the mob. Having gained this brief reprieve, he tried to escape. Two respected local black men assisted his flight. Not only did they refuse to furnish information about Birdsong's direction to his pursuers, but they also supplied him with ammunition to keep them at bay. This solidarity so incensed the would-be lynchers that they murdered both men. One was shot from behind when he fled; the other was tortured to death. When the mob finally caught up with Birdsong, simple killing could not satiate its members. After shooting him, they exorcised his remains with fire.[5]

To the extent that they sought to incite terror among other African Americans, the lynchings succeeded. They certainly frightened black Athenians, who had hitherto felt secure that, as one doctor and NAACP member put it, "such a thing could not happen in 'Classic' Athens." Even some white citizens were appalled; the community began to show fault lines similar to those that appeared in response to Governor Dorsey's statement on the plight of blacks in Georgia. Following the murder of John Lee Eberhart, the

Athens Ministerial Association issued a condemnation of lynching, and many leading local men organized a Law Enforcement League to aid officials against lynch mobs.[6] The public fiction of white consensus was shattered; the future was open to contest.

It was at this juncture that the Klan set off on its own. By early 1921, according to a horrified black resident, the several hundred area whites in the Klan had assumed leadership of the offensive against African Americans. He accused them of practices like tying blacks to cars and dragging them along the roads, stealing their mail, driving them out of their homes and confiscating their goods, and, generally, "treetin the . . . negro worse than a dog."[7] At about the same time, dynamite destroyed a black church, two black schools, and a black lodge hall in Oconee County, where the lynching of Eberhart had taken place. As community meeting places that evinced participants' desire for education, for spiritual equality, and for economic advancement and general racial betterment, the targets were hardly coincidental. "It seems the more progress the Negro makes, the more his enemies increase," observed the editor of the *Atlanta Independent* in a discussion of such night-riding; "the more he develops and acquires education, property and wealth, the more he is opposed."[8]

Also set on fire was the cotton gin of R. E. Fullilove, a prominent Oconee County white man. Fullilove's offense, apparently, was too much sympathy with African Americans. Several months later, he received such serious anonymous death threats that he felt it necessary to post armed guards around his home. The source of their anger was his comment that blacks "were killed unjustly" and that lynching was wrong. Both the violence against blacks and the reprisals against white dissidents sent out messages to wider audiences: challenges to white supremacy, however modest, could be lethal. That the messages were heard was evident in the paucity of indictments and the absence of convictions, even though the police were said to have plenty of evidence on the culprits. No one complained, not even the organizations that spoken out for "law and order."[9] Having had their bluff called, élite opponents of mob violence retreated with barely a whimper.

The state's lenience emboldened the vigilantes. Poor whites now joined blacks in their scrutiny. In early March of 1922, five masked men visited Frank Kenney, a mill worker in East Athens, in the home he shared with his mother and two sisters. The intruders came to warn him about his conduct. Frank was having an affair with a married woman, Lillie Toole, on whom he spent most of the salary that otherwise would have gone to support his household, and he more than once abused his twenty-seven-year-old sister, Nora. The failure of earlier, informal attempts to change his ways precipitated the Klan's warning.[10] Frank ignored the warning.

Three weeks later, on the night of March 22, seven masked men broke into the Kenney home. Taking Frank from bed, they forced him in a waiting car and drove him out to the countryside. There they whipped him repeatedly and then left him to find his own way back to the city. Frank's kidnappers threatened to do worse if he did not change his ways. In effect, Kenney had exploited the prerogatives of male dominance. He used his power, not to maintain order in his household, but to safeguard his own profligate, abusive ways. He failed to provide for his womenfolk, he courted another man's wife, and he beat his adult sister when she implored him to change. The Klan in fact entered the case, according to Mrs. Kenney, on her daughter's request.[11]

The floggers gained confidence from several features of the case. The behavior that led to their intervention would make most of Kenney's contemporaries hesitate to publicly condemn his assailants for fear of being thought to condone it. Toole's own husband had threatened to kill the lovers, an action most would no doubt have excused as a crime of passion. Aware of the weakness of his position, Kenney did not press for prosecution, despite his certainty that the Klan was responsible. His reticence no doubt increased as he watched how the authorities handled his case. Local officials at first attributed the flogging to the Klan. Yet within days they began trying to exculpate it by redirecting the blame. They now said that Kenney was flogged "for paying too much attention to another man's wife"—as if this motive somehow proved the Klan's innocence. The case soon disappeared from the press and never resulted in any indictments. The Klan's repeated flogging, a few months later, of a man with chronic "domestic troubles" had a similarly anti-climactic outcome.[12] Both men had violated community notions of right. Although some residents may have disliked the Klan's methods, no one had the courage to challenge them directly in public. For the victims to have pressed for redress would have been to court worse trouble. Anyone who doubted that had only to look at how the Klan handled blacks who believed they had rights.

Ruthlessly—as Asbury McClusky and Odessa and Willie Peters were soon to learn. McClusky and the Peters couple lived in a small community of independent black farmers off the Dixie Highway between Athens and Winder, Georgia. Beginning with nothing, McClusky had acquired over twenty-five thousand dollars' worth of land and savings. Known to many prominent whites as "a good and useful citizen," McClusky had for years accommodated the etiquette of white supremacy by concealing his wealth and staying away from conflict. But in 1922, he did something that brought him to the Klan's attention: he came to the aid of a neighboring black family whose organ a white lender was trying to confiscate. Area Klansmen were already on edge; a few weeks before they had flogged a black Methodist minister whose "only offense," according to an outside investigator, was that he was "well educated and was teaching others of his race how to read and write." The night McClusky signed the bond to protect his neighbor's organ, a gang of masked men—the bailiff among them—came to his home. The men broke down the door and fired over thirty shots in the house, some of which ripped McClusky's arm to bits. As soon as they left, he escaped to seek safety in Atlanta.[13]

Infuriated by his survival and escape, Klansmen sought out other blacks to make their point. The following night, they came to the farm of Willie and Odessa Peters. In a struggle with his attackers, Mr. Peters wrestled one of their guns away, shot the owner in the stomach, and ran from the house. The men turned to Mrs. Peters, then pregnant, and beat her pitilessly. When finally they were through, they deposited her, unconscious, in a roadside ditch. Once she recovered, Mrs. Peters went to Atlanta to find her husband, who had been put in a fortified prison to safeguard him from lynching. Taking advantage of her fear, disorientation, and illiteracy, lawyers from one of Georgia's leading political families then swindled her out of the deed to her farm. Back home, Klan marauders drove out some of the remaining neighbors. The Peters never got their land back. His arm crippled and his spirit broken, McClusky sold his property at a loss and left the community he had worked so hard to build. The Klansmen involved suffered nothing. While some "good people" in the community looked askance at the Klan's actions, they also feared its power too much to speak out publicly or demand prosecution. As usual, the papers kept silent.[14]

Spurred on by their successes, the night riders grew more daring. In November, a note signed "K.K.K." was affixed to a black church in a hamlet outside Athens. The note warned "niggers" to leave the area or face the consequences. The sheriff maintained that the threat was merely a "joke" by young boys, a fabrication the Athens press eagerly repeated. Blacks, and some white employers worried about the loss of their labor supply, knew otherwise. Many no doubt remembered similar Klan-led efforts a few years back in counties north and east of Clarke. In these, such warning notes were the prelude to a nightmarish campaign of harassment and terror that included crowds of white men shooting into black homes on an almost nightly basis. The point of it all was to so harass and terrify African-American residents that they would abandon their land and belongings and flee the area. Hundreds did. Soon after the note was posted outside Athens, groups of masked night-riders abducted and flogged area black residents in a series of raids. Terrified, some 250 African Americans held a mass meeting with fifty "white farmers" from Barrow Country to appeal to "law-abiding white citizens" for protection.[15]

The appeal yielded little. Klan representatives were soon winning "thunderous applause" from meetings called to condemn the night-riding practiced by their organization. By the spring of 1923, the siege had spread across many rural counties in North Georgia. Cold calculation joined hysteria as Klan marauders confiscated the property left behind by owners in their determined haste to seek safety. Flight from the Clarke County area, which depleted the black farm population by more than half, in fact peaked in the years of these vigilante attacks. By October of 1923, Georgia blacks were heading north at the rate of 1,500 a week; how many went to escape such terror we will never know. In any event, the "unrest" and "exodus" resulting from the attacks proved sufficient to widen the private breach among whites into a quasi-public split. On the initiative of Judia Jackson Harris of the Athens Teacher Training and Industrial Institute and the local Interracial Committee, a few prominent residents denounced what the most forthright among them, a white female teacher, called the "reign of terror" so many of the area's African Americans now lived under.[16]

As they disrupted and destroyed black lives, Klansmen persisted in their efforts to police white ones. While it is hard to tell what the Klan members who participated in each particular vigilante episode thought, it is possible to tease out something of their consciousness and modus operandi from cases about which evidence survives. Klan floggers looked to violence as a tool to enforce hierarchy on recalcitrant community members. In the case of McClusky and the Peters couple, this intent was obvious. The floggers wanted to punish these African Americans for their success and to break the community solidarity successful resistance depended on. In the Kenney case, the way the Klan came to the aid of a female victim tended to distract from the ultimate purpose of the effort: to fortify the chain of white male dominance by mending weak links. Thus, while some white men might find themselves disciplined for neglecting their duties and abusing their power, others might come to the Klan's attention for failing to exert sufficient control over their wives.[17]

This was evident in Klansmen's notice to *Arnold Moss*, a poor farmer in rural Athens. "We give you fair warning that we have been called upon to make investigation in you and your wives affairs," they wrote, admonishing him "to keep her off [the] streets. . . . You Be [ruler] of your home. But treat your wife right. We mean she must respect you and home and children." As if to emphasize their regard for mitigating factors, the Klan authors also stated their awareness that *Moss* was "a hard working man" and that he had "cut out drinking." In effect, he had met two of their criteria for appropriate male conduct; if he would only compel his wife to act properly, there would be "no more said or done."[18]

In other cases, the method for shoring up men's power was less direct, as in several episodes in which Klansmen sought to control women by using violence against suitors disapproved of by the women's male relatives. In 1927, for example, the Athens Klan looked into the case of Miss *Carrie Ivey*, who had recently returned to town. The daughter of an Athens mechanic, *Ivey* was a "New Woman" who had moved to Bartow, Florida, and taken a job as a stenographer. There, according to third parties, she had "been ruined" by the married brother of her employer. The investigation was requested by the Exalted Cyclops of Bartow, who intimated that the chapter would take vigilante action against *Ivey's* suitor if the rumors were true.[19]

A more dramatic illustration took place in Royston, about twenty-five miles north of Athens. There, some fifty Klansmen conducted an armed, robed "raid" in March of 1926 against a man involved with another's wife. This time, one Klan member, a farmer, was killed, and another, a druggist, was wounded when their intended victim defended himself. Fellow Klan members portrayed their attempt to compel the woman's fidelity by attacking her lover as a valiant effort to maintain "the Chastity of our ladies." Describing their fallen brother as a "Christian gentleman," they solicited the Athens Klan for funds to build a monument to this martyr to chivalrous manhood.[20]

The Royston case illustrated what Klan spokesmen implied but rarely stated outright: that maintaining female chastity was a service white men provided for other white men. To have heeded women's own wishes would have contradicted the purpose of the enterprise. By punishing the men with whom women defied respectable standards, Klansmen tried to thwart women's own aspirations. Their raids sought to exact a kind of subordinate female behavior that ordinary economic and community pressure seemed unable any more to ensure. The focus on men in these attacks revealed graphically how the ideology of chivalry reduced women to possessions, passive pawns men warred over. Indeed, in these cases, the Klan made clear its belief that women's sexual comportment *ought* to be subject to the control of men with proprietary rights over them.

As they concentrated on immediate issues, however, Klan floggers probably had more abstract questions at least dimly in view. In effect, they engaged in a kind of collective fantasy of their band of white brothers as a surrogate state to replace the one which they had such doubts about. Perhaps it is not surprising, then, that critics should complain of "self-constituted authorities" or "invisible government," for that was precisely the Klan's project. Its forays, in fact, involved the ritual enactment of the movement's notions of reasonable government. This was particularly true where the intended victims were white, since attacks on fellow whites required more in the way of rationalization. Symbolically asserting their unilateral claim to public power, floggers self-consciously acted the parts of prosecutors, judges, and hangmen. The rites of violence thus usually included a prior investigation, a practice that prompted to Athens townspeople to complain that "spying and reporting" on neighbors was turning Athens into "a warring camp."[21] While providing the proceedings with an aura of legality, warnings that preceded violence and mock trials for the victims also confirmed to practitioners their own righteousness and allayed doubts about their ongoing authority over community life.

White residents understood that the script mattered to the Klan, which is why those who appealed to it for aid affirmed their willingness to have their cases "investigated" before it took action. Male petitioners acted as their own counsel in the kangaroo-court proceedings. They sometimes buttressed their briefs with the names of fellow white male witnesses whose word could help convince the judges. When *W. A. Lease* wrote to the Klan for help in retrieving his adulterous wife, he thus appended a list of endorsements from his hometown, including those of a justice of the peace, the postmaster, and two ministers. Similarly, *C. W. Hale,* a worker at the Cord Mill outside Athens, wrote to the organization the day after Klansmen had paid him a night-riding visit in 1925. His point in writing was to redirect the blame for his family's problems. "The trouble," he asserted, "is with my wife." To convince the Klan of his wife's recalcitrance, *Hale* secured the signatures of three male neighbors.[22] These men appreciated the common logic that united scattered acts of intimidation and violence, and tried to make it serve their ends.

▲ ▲ ▲

In the attacks it mounted on outsiders to, and transgressors of, its vision of community, in fact, the Athens Klan was hardly anomalous. The vigilante practices of local Klansmen found support in the ideology articulated by national Klan leaders and by representatives of the movement in other states and localities. In forums for members and sympathizers, Klan leaders sometimes even used vigilantism as a selling point to distinguish their organization from others. Where Klan chapters existed, the *Searchlight* boasted, "there's a will . . . to stamp out the sort of crime that is no less criminal because it is without the reach of the legally constituted authorities."[23] A promotional novel used "the Vigilance Committee" and "the white caps" as synonyms for the Klan, and described how it "avenged" various "unlawful acts" through vigilante means. Imperial Wizard Simmons, for his part, warned that "if needed we have a great invisible and mysterious force that will strike terror into the hearts of lawbreakers." Lest doubts remained about the meaning of such assertions, Simmons' successor, Hiram Evans, proclaimed that his organization, like its Reconstruction-era namesake, would "restor[e] control."[24]

But Klansmen did not use violence simply because it was effective. They used it because they believed they had a *right* to use it. They derived this sense of entitlement from many different elements of their ideology, whose combined effect was to grant permission for such violence. The most obvious support for vigilantism came from the Klan's reactionary populist world view. Klansmen trusted established public institutions less than their more élite contemporaries, whom they also suspected of ambitions to take over the state. Rather than look to it for redress, the Klansmen sought to build a white-sheeted militia to enforce their values and combat threats to their standing.[25]

Klan leaders could justify their acts under the rubric of fighting "tyranny," a term they used in its original, highly specific meaning in the republican tradition. Thus used, it denoted infringement upon the rights of the property-holding, white, male citizenry, or failure to respect their collective will. Claiming Thomas Jefferson and Patrick Henry as his inspiration, Simmons asserted that, throughout history, the "Ku Klux spirit" had defended "freedom" in the face of "oppression," which included unjust laws and court decisions as well as acts in civil society. In Simmons' interpretation, this spirit had animated both the Boston Tea Party and the Reconstruction Klan. The alleged tyranny came, in one case, from the minions of the King; in the other case, from the enfranchisement of propertyless former slaves. Another Klan propagandist, the Grand Dragon of Pennsylvania, argued that selective and active intolerance was necessary to defend "liberty."[26]

In their effort to enlist the core rhetoric of American political culture in the service of vigilantism, Klan leaders adapted a tradition common to earlier American vigilante movements from whose record they took heart. These movements, according to their leading scholar, had developed by the nineteenth century a common "ideology of vigilantism." The concept of "popular sovereignty" provided the framework to justify the vigilantes' activities. Since their government was of, by, and for the people, citizens were entitled to act in its stead when they represented the popular will. Indeed, vigilantes invoked the Jeffersonian notion of the citizenry's right to defy the state if it ignored the will of the people. Vigilantes also portrayed their victims as aggressors and their own acts as community self-defense. Finally, the simple but eminently republican motive of thrift came into play: as taxpayers, vigilante leaders found their methods cheaper than those of established authorities. The latter would be of no small concern when perceived crime waves coincided with economic recession and tight government budgets. Like the Klan, moreover, these earlier episodes were most often practiced by those with a stake in society: propertied white men, generally adult heads of households. Their leaders generally came from among the élite, their rank and file from the middle orders of the community, and their targets from the poor or lower classes.[27]

In the Klan's case, middle-class standing, or at least identification, led members to feel distinctively entitled to intervene in disputes of the so-called selfish classes above and beneath them with paramilitary means if need be. Whether America would survive the "period before us now of stress and strain," Gould explained, depended on the will and organization of the middle class. "The Klan literally is once more the embattled American farmer and artisan," proclaimed Evans, "coordinated into a disciplined and growing army." The Bonapartist flavor of this language, one contemporary French visitor to the United States recognized, was no doubt intentional: " 'honest men' are those on the side of order who can be relied upon to strike hard when necessary"—in the case of Klansmen, "to take the law into their own hands should the government prove inadequate." Indeed, Klan leaders proffered their movement as "a stabilizing power" against "the unrest threatening to undermine our American civilization." Klan propagandists insisted that redress could come from no quarter but theirs: political parties, courts, churches, and private organizations were all either too corrupt, too weak, or too compromised to do the job.[28]

Some of the clearest intimations of the Klan's intent to suppress labor struggle by force if necessary came from its highest national officers—those who built and directed the movement. Thus, after invoking the specter of a workers' revolution in the United States, Simmons said that the Klan insisted of "Americans of all classes who now prepare their minds for civil war that they *must and shall* make peace." "A vast majority of our farming people and middle classes are ready to *demand*," he asserted, of "the wage-working population . . . that America and Americanism can solve their great problem without rebellion." "Let none mistake our purpose," he warned, "we shall prevent war." In other words, if it came to it, Klansmen themselves would intervene to suppress such "rebellion." Simmons' successor elaborated. "The Klan has no other purpose" than "the breaking down of all barriers between classes" and "the creation of a united America." But, Evans said, "vigorous and organized resistance" prevented the peaceful achievement of this. "The next and immediate objective of the Klan," therefore, was "to make sure that the patient will take the medicine."[29] Only force could compel labor to give up class loyalties; only the Klan would administer it with the requisite determination.

Further support for that militancy came from evangelical Protestantism. Defending those who had burned a black man at the stake, the president of the Georgia State Senate in 1921 admonished critics of the lynchers that "the wages of sin is death." Just as Klansmen used tendentious readings of the Bible to corroborate their notions of justice, so they co-opted Christianity's foremost symbol to sanctify their vigilante practices. In Athens as throughout the nation, the paramount sign of the Klan was the burning cross. In klavern ceremonies, it was ritually described as "the emblem of that sincere, unselfish devotedness of all klansmen to the sacred purpose and principles we have espoused." In short, it served to convince members of the righteousness of their cause. Movement leaders exploited its authority with greatest skill at Klan rallies. Usually held in open fields after dark, the rallies would be illuminated by the blaze of a wooden cross, sheathed in oil-soaked cloth and set afire at a key moment, and sometimes as tall as seventy feet. No other source of light was permitted, not even matches. "It is not difficult," as one minister briefly in the Klan observed, "to see what effect such a surrounding will have on the mind of one who is easily excited." That effect would be compounded by the assertions of Klan speakers that this icon—a source of terror to its victims—was a reminder to members to pattern their lives after Jesus Christ, "to serve and sacrifice for the right."[30]

When Klan leaders invoked Christ as the model members should aspire to, they hardly had in mind liberal Christians' Prince of Peace. Rather, the man Klansmen were to emulate was the potent and vengeful Redeemer who "purged the temple with a whip of chords." Evans thus disputed "the usual picture" of Christ as "effeminate"; the Bible showed, rather, his essential "virility." "Jesus," said Evans, "was fit": "a robust, toil-marked young man who had conserved both His physical and mental strength." American Protestantism, for its part, was "certainly neither soft nor lax." Rather, it was "a thing of rugged steel . . . forged in the terrific stress . . . of wresting a continent from savages and from the wilderness." Only "the red-blooded and virile" could rise to the challenge of Christian America.[31]

Klansmen fused this conception of a virile, manly faith with the long tradition of Protestant nationalism in the United States that designated Americans as God's "chosen" people to produce a fearsome, military rendition of Protestantism. It was, after all, an axiom needing no proof to Klan members that Christianity was a "white man's religion" that presumed "white supremacy." The organization's leaders frequently described themselves as a Protestant "army," and used the language of "battle" to describe their mission against their "enemies." "Onward Christian Soldiers" became a Klan anthem. Their "fighting instinct," Evans boasted, was the pre-eminent "racial characteristic" of American men.[32]

The ardent masculinism at work here gained force from another critical legitimating mechanism in vigilantism: men's right
to control and protect their households. Republicanism's conviction that the future of the republic lay in orderly families and virtu-
ous citizens became, for the Klan, a justification for the white,
male citizenry to compel obedience to their moral standards on
their own when the state failed them. Time and again, Klan spokespeople asserted that extralegal forays were needed to defend their
homes, just as their predecessors in the Reconstruction era had
needed to protect the home and white womanhood. "If the law
will not protect our homes, we must," proclaimed an Oklahoma
minister in defense of the Klan in 1922. "These are strong words,"
he admitted, "but the times call for stern methods." "The reason
the Ku Klux Klan have organized is because our homes have been
threatened," a Klan-affiliated evangelist in Colorado concurred.
Georgia's Grand Dragon referred to morals-related night-riding as
the Klan's "purity of homes" drive; no clear boundaries differentiated legal from illegal methods.[33]

The persuasiveness of home protection as a rationale for vigilantism came from the way it blended republican premises with
notions of male honor. And male honor, with its associated rhetoric of chivalry, was central to the allure of the second Klan. The
word "HONOR" formed the foundation for the seal of the Imperial
Palace; above it stood two Klan "Knights" on rearing stallions. As
a "Knight" of the KKK, each member was bound to the code of
chivalry, hence the "protection" of white womanhood. Indeed, the
member was to be "a MAN who values HONOR more than life."
In this exaltation of honor, the Klan tapped deep roots. The impera-
tives of honor had supported the rule of Southern patriarchs, both
planters and yeomen, over their dependents in the nineteenth-
century South. Slaves, male and female, and white women and
children were denied the honor all free white male proprietors
could lay claim to. "Honor" thus, as historian Edward Ayers concluded, "presupposes undisguised hierarchy."[34]

It also presupposed a propensity for violence. Offended honor
acted as a hair trigger for many Klansmen. A man's home was his
castle, after all; it "must be protected," by violence if necessary.
When a man's control over "his" women and children was challenged or his honor affronted in some other manner, particularly
where it involved female sexuality, the price of maintaining that
honor was personal violence. Honor thus contained a built-in imperative to violence, since a man only possessed honor insofar as

his equals acknowledged it. To seek redress from the state was to concede the loss of honor and accept public shame. A man who could not defend his own honor was no man at all. One writer in Texas demonstrated this logic when he demanded of an anti-Klan judge what he "would do if some wretch should insult [his] daughter." The writer inquired whether the judge "would prosecute or . . . have the real manhood to make it a personal matter—to take the law into your own hands?"[35] Where honor was involved, "real men" spurned the courts and sought personal vengeance.

In Klan quarters, the line between honor and sadism easily wore thin, as fervent masculinism sharpened the proclivity to violence. While Klansmen ridiculed the "sissy," they held up "real he-men" as role models.[36] Some Klan floggers evidently took real pleasure in their power over their terrified victims and in the violence they inflicted. Again and again, Klansmen and their sympathizers described night-riding as going "to have some fun." In this "fun," sexual currents frequently gave violence an erotic charge. More than a few Klan flogging teams stripped their victims or pulled down their pants and underwear for the whippings, which sometimes continued until blood flowed freely—in one case, even after the terrified victim had defecated all over himself. In so shaming victims, these rituals of degradation thereby also symbolically divested them of power and exorcised the threat they had posed to the traditions Klansmen would uphold.[37]

The view that vigilantism expressed the Klan's core beliefs and values also finds support in the way that scattered episodes of Klan vigilantism conformed to common patterns. For reasons that will become clearer later, it is impossible to estimate reliably either the frequency of the incidents or the relative proportions of the alleged offenses precipitating them. But, according to knowledgeable contemporaries, the victims appeared to be evenly divided between blacks and whites. "While the flogging of white persons has received the major part of newspaper attention," commented the CIC in 1927, "the victims of flogging have probably been equally divided between whites and Negroes."[38] If the data do not permit exact calculations, they do clarify how vigilantism served to defend older relations of power and to compel obedience to lower-middle-class white codes of conduct. Evidence of this can be found in the way the Klan attacked different social groups for different reasons.

Native-born whites, in other parts of the United States as in Athens, were most often flogged for morals offenses. As the leading subjects in the Klan's vision of society, adult white men also numbered heavily among the victims, attacked for offenses such as bootlegging, adultery with other men's wives or daughters, failure to support their families, chronic abuse of their womenfolk, or unwillingness to marry young women whom they'd had intercourse with. Family values converged with fiscal conservatism in Klansmen's worry about the waning of men's sense of duty to their families. In Athens and across the nation, Klansmen engaged in extralegal efforts to compel delinquent husbands to support their wives and children, tracing the whereabouts of deserters, issuing them warnings, and flogging those who failed to comply.[39]

White women, whom the Klan's gender ideology tended to portray as passive appendages to men, usually suffered the order's violence only when they acted independently in ways subversive of this ideology. Their offenses included selling whiskey, adultery, prostitution, and severely neglecting their children. White children, whom Klansmen accorded less moral responsibility than women, were almost never flogged by the Klan; instead, it held adults responsible for their indiscretions.[40]

Yet, whereas the Klan used vigilantism to buttress rigidly demarcated gender roles among whites, it almost never attacked blacks for intraracial morals offenses. The only morals allegations it appears the Klan flogged blacks for were interracial infractions, such as selling moonshine to whites or having sex or marrying across the color line.[41] Since one function of the Klan's conservative moral code was to solidify white supremacy by constituting "respectable" whites as a cohesive community distinct from blacks, the order's lack of interest in regulating morality within the black community made sense.

Indeed, the Klan worked diligently to deny blacks access to white models of gender. Klansmen made no pretense of offering even nominal protection to black women; they, like immigrant women, were by definition beyond the pale of the Klan's vaunted chivalry. Like black men, they, too, could be intimidated, kicked, beaten, or whipped; not even old age or pregnancy would shield them. The order also tried to deprive black men of the very signs of masculinity it encouraged white men to display. Thus, Klansmen were liable to attack black men for standing up for their rights as citizens, for amassing property, for protecting their kinfolk, or for defending their homes.[42]

Indeed, most Klan attacks on African Americans stemmed from their assertions of their own rights or defense of those of other blacks. Efforts to register and vote by Southern blacks particularly irked the Klan. It employed parades and vigilante violence to "intimidate" prospective black voters and "make them more amenable to influence," in the euphemistic words of one former Klan leader. S. S. Mincey, a seventy-year-old black Republican leader in south central Georgia, was thus kidnapped and flogged to death in 1930 for his political activities; others received threats because of their protests against segregation or their participation in the NAACP.[43]

Resistance to laboring on whites' terms—referred to by the Klan as "failure to work"—could also result in persecution. Still other African Americans suffered Klan vengeance for offering economic competition to whites or having achieved success on their own. The order set fire to black businesses and drove their owners from Waycross, Georgia, in 1922; it whipped black farm owners in other parts of the state to make them sell their property at a loss.[44] In Springfield, Missouri, local Klansmen threatened a black physician on the grounds that he was performing abortions on white women, a baseless charge to camouflage their real complaint: the man's wife was a leading NAACP activist, and together the couple had "accumulated good property."[45]

Finally, no matter how upright their morals, whites who defied the Klan's injunctions to class harmony might also suffer its wrath. As early as 1918, local Klan chapters burned crosses, kidnapped and beat or flogged union leaders, and ran activists out of town to prevent labor organizing and strikes.[46] By 1923, an informed journalist could conclude that "the Klan is nearly everywhere and always an enemy of organized labor." Its record led even the politically timorous American Federation of Labor to adopt resolutions condemning the activities of masked organizations and the Ku Klux Klan specifically.[47]

▲ ▲ ▲

But if the argument so far advanced is true—if vigilante violence was Klan thought in action—then one must wonder why more evidence of such activities is not easily accessible. After all, it might be argued, even the hundreds of episodes reported in the national press seem few compared to the numbers enrolled in the Klan. Some recent studies have argued essentially this point: the Klan of this era, the authors say, was not particularly violent. Indeed, they explicitly distinguish it on these grounds from the Klans built during Reconstruction and the civil rights movement.[48]

The problems with such arguments are many. First of all, they display the same fallacious emphasis on quantification that led a few historians in the early 1970s to play down the extent and significance of whipping as a means of control by slaveholders in the antebellum South. As Herbert Gutman pointed out in a biting critique of such efforts, not every slave had to be whipped regularly for the owners to make their will felt. Similarly, an episode of Klan flogging affected, not just the individual victim, but every member of the community who learned of it. Evoking this violence as they did, even warnings could instill sufficient terror to make people change their behavior.[49] More generally, though, the problem with attempts to characterize the second Klan as nonviolent is their epistomological innocence: their assumption that, if evidence of violence does not appear in the usual sources—newspapers in particular—then it must not have occurred. Such interpretations have also tended to accept Klan leaders' public proclamations against "lawlessness."

Klan leaders did, after all, repeatedly deny that their movement engaged in extralegal activity, and the historian cannot simply ignore these disavowals.[50] But to accept them at face value would be naïve and dangerous. The same men, after all, regularly denied—sometimes in the same breath—that the Reconstruction-era Klan was violent or that their own organization was racist or anti-Semitic. The occasions of such denials were also important. They usually came after the public revelation of some vigilante activity attributed to the Klan. As part of an effort to evade prosecution, the disavowals were obviously self-interested.[51]

A close reading of the wording of the denials suggests further cause for skepticism. Klan speakers and publications often purposefully qualified their affirmations of respect for the law. A guide to organizing Klan chapters spoke of how the Klan's investigating committee should aid law enforcement personnel "in the *proper* performance of their legal duties." The Klan's Constitution bound members "to protect and defend the [U.S.] Constitution . . . and all laws passed *in conformity thereto,* and to protect the States and the people thereof *from all invasion of their rights from any sources whatsover.*" In other words, some laws were constitutional, therefore worthy of defense; others, it would seem, could be defied in the name of the people. Which was which, presumably, was up to the Klan.[52]

Here the exclusivity of the Klan's vision of American citizenship played a critical enabling role. For the movement did not believe that anyone other than native-born, Protestant white people had the right to a say in American law-making; blacks, Jews, and Catholics had no place in public affairs. This is what Klan leaders meant when they said their movement had only "one purpose": "to put the Government of the United States into the hands of none but American citizens." Laws passed in the interests of other groups could thus be represented as necessarily illegitimate, while regulation of the activities of such "aliens" by true citizens was by definition lawful. African Americans were particularly vulnerable in this line of reasoning because the assumption that they were outsiders to American democracy and unworthy of due process of law was so ingrained in the dominant culture.[53] Philosophically, then, Klansmen would have had little difficulty justifying vigilantism as law-abiding conduct.

In practical terms, their organization was also well suited to supporting and protecting such activities. Why don masks and robes if there was nothing to hide? "It is not logical," as one former member observed, "to believe that if the intent and purpose is good and for the general welfare, the members will be ashamed, or afraid, to be known." The order's commitment to strict confidentiality could also serve clandestine ends. Members were warned to keep quiet, especially about "Secret Work." "Just remember," concluded one such counsel, "even a sucker would not get caught if he kept his mouth shut."[54] Nothing was left to chance. Members were tutored in "complete obligation," "secrecy and obedience," and "loyalty." As one manual for organizers bluntly put the code: "he who violates the standard ought to be made to pay the price." The warnings were not idle. When Klan secrets leaked in Athens, the organization saw the danger as sufficiently serious that it hired a private investigator to find out how and by whom. Those who violated their oath of loyalty to the Klan could suffer expulsion, social ostracism, and sometimes violence or even the threat of assassination.[55]

Some evidence exists, moreover, that top Klan leaders actively fostered vigilante activities. Former Klan members reported that officials from the Imperial Palace—Simmons and Evans included—tutored regional warlords in the how-tos of "rough stuff." For their part, members of the Simmons block, once deposed, charged that their former organization took upon itself "the purpose of regulating men and women at the whipping post, blackmail, murder, and arson."[56] The testimony of disgruntled members should of course be handled with care. Still, the charges were remarkable both for their unanimity—despite coming from widely scattered parts of the country—and for their singularity. It is difficult to imagine the same things being said of any other contemporary organization, no matter how much it had disappointed its members.

Similarities in the methods employed by Klan floggers in different areas lend further credibility to accusations that the practice was institutionalized. The commonest pattern was for the Klan to discover the offending behavior either through its own "intelligence"-gathering apparatus or through a complaint made by a third party. The case would then be turned over the chapter Klokann, or "investigating committee" head. If this body found the accused guilty, the first "corrective" act was usually a warning designed to intimidate: a note, a night-time visit, or a cross-burning or parade at their home. If the offenders failed to oblige, a group of Klansmen, often robed and wearing black masks—known internally as "the wrecking crew"—would abduct them from their homes under the cover of night. After taking them to a secluded site, usually a spot outside city limits, Klansmen would flog their victims with as many as fifty lashes with a thick leather strap.[57]

Scattered evidence also suggests, however, that not all Klan members had access to knowledge of these activities. Some Atlanta-area members said that they had belonged for several years before they became aware of the local "secret committee" that carried out flogging operations. The Exalted Cyclops was fully aware of its activities, however; he appointed its chief—sometimes a different one for each action—and passed on complaints for action. The members of the committee were known only to its head—who drafted them—and the Exalted Cyclops.[58]

On the rare occasions when Klan members were indicted or their methods exposed, Klan leaders followed what seemed a prefabricated script. As soon as arrests were made, national and state Klan officials set to work to thwart prosecution with public denials of culpability and efforts to suppress testimony and destroy evidence. While publicly condemning extralegal violence, the Klan quickly hired lawyers for its accused members. Its representatives also tried to intimidate witnesses, and sometimes journalists and even judges as well. The Imperial Office also sent its own "investigators" to stymie inquiry in the guise of aiding it. If a case did reach trial, Klan officials instructed subpoenaed members to lie on the stand—as they themselves did to the press—and tried to ensure juries stacked with their fellows.[59] Here honor worked as insulation; historically, the rule of honor had helped to solidify military bodies against civilian interference. If none of these subterfuges succeeded, the Klan might protect itself with the specious claim that the local "secret" committees acted autonomously, without the formal authorization of the parent order.[60]

These methods almost never failed. So efficacious were they that the governor of Oklahoma found he had to declare martial law and institute military tribunals in order to successfully prosecute Klan vigilantism. But he soon found himself impeached by the Klan-dominated state legislature. Not until 1987, in fact, was the Ku Klux Klan as a body ever successfully held accountable for the violence its members perpetrated in service to its goals.[61]

Seen in this light, the relative lack of documentation of Klan vigilantism is much less puzzling. Indeed, the gaps in the record appear rather as yet another illustration of the Klan's remarkable power. That power included the capacity to shape—and distort— the historical record by keeping victims silent and journalists and courts at bay. Indeed, the Klan seems to have practiced violence most commonly where its members could get away with it—and, thus, where evidence of it would be least likely to survive.

Klan leaders were not stupid; they took care to prevent haphazard violence.[62] To have engaged in vigilantism wherever they had members would have been to invite destruction of their movement. After all, there were parts of the country where its chapters could not even assemble publicly without risking counterattack. In many areas of the Northeast, for example, Catholics in particular mobilized fierce counter-demonstrations against the Klan.[63] In such areas, Klan violence would have invited physical reprisal. Such people also cast ballots and sat on juries. Klans in areas in which Catholics, African Americans, or Jews were a significant part of the electorate, or where white non-members became aroused, might find themselves facing hostile courts or being subjected to restrictive laws, such as the anti-mask legislation proposed in several states and passed in a few.[64] In short, where Klansmen and their sympathizers were in a minority, to have antagonized opponents would have invited bad publicity, prosecution, political battles, and possible retaliation.

A successful campaign of terror required conditions found in a different kind of setting: a restricted electorate, compliant politicians, a cowed press, and the active complicity of law enforcement officials. This environment was most common in the South. Hence it is not surprising that Southern Klans practiced the lion's share of vigilantism.[65] They did it because they could. In the South, particularly in small towns and rural areas, the potential gains from vigilantism outweighed the costs. The groups who most barred the Klan's way in other areas of the country lacked power here: immigrants, because of their small numbers; workers, because of the Southern ruling class's fierce opposition to labor organization; and blacks, because of the coercive apparatus of Jim Crow. The only force left with any power to oppose the Klan, then, was the small fraction of the Southern élite that disdained it, men and women not notable for stiff resistance to injustice.

Operating in such an environment, many Southern public officials, perhaps most, gave tacit endorsement to Klan vigilantism, especially in the unsettled years immediately following the war. As a result, those responsible were seldom indicted and almost never convicted. In Georgia and Alabama, for example, governors Clifford Walker and Bibb Graves conspired from their executive offices to avert legal punishment for their fellow Klan members. Walker commuted the sentence of one of the only Georgia Klansmen indicted for vigilante activity. For his part, Walker's predecessor, Thomas Hardwick, assigned to the investigation of one flogging a sheriff and a police chief the victim had already identified as among his assailants.[66]

Some evidence in fact suggests that the Klan enrolled or enlisted the support of law enforcement officials—police, prosecutors, justices of the peace, and judges—before initiating vigilante activity, presumably to stave off indictments.[67] Such men, after all, had a history in the South at least of habitual violence against blacks who asserted their rights. Their off-duty practices were but one end of a continuum of repression; "extralegal force supplements and supports the legal action if it is considered uncertain or inadequate." Imperial Wizard Simmons even stated once, in defending the Klan from charges of vigilantism against blacks, that "there is never a stand taken unless an officer of the law supervise[s] K.K." In cases that did come to light, time after time it surfaced also that police or other law enforcement personnel had either participated in the violence, suggested that the Klan undertake it, belonged to the order, or condoned its activities more indirectly.[68] Even Southern police who did not actually belong themselves were unlikely to disapprove of Klan violence. "They liked them," one Atlanta non-member on the police force recalled of his fellow officers' attitudes to the Klan; "they didn't have nothing against the Ku Klux." Some victims, having recognized law enforcement officials among their assailants, understandably believed prosecution futile.[69]

It was. Since particularly in less populous rural counties,

native-born white property-holders (those likeliest to sympathize with the Klan even if they did not belong) generally staffed the juries, the chances of conviction were scant in any case. Moreover, the few indicted Klansmen could look forward, in some states, to having fellow members decide their appeals in state supreme courts. "Experience seems to indicate," observed the Commission on Interracial Cooperation in 1927, "that in most American communities the members of lynching or flogging bands need have little or no fear of prosecution."[70]

The rarity of prosecution resulted also from the vulnerability of the victims singled out by Klan floggers. Black residents especially appeared convinced that efforts to seek legal redress locally might only bring worse trouble upon them. Some would appeal to national organizations in Washington, D.C., without even bothering to try to obtain relief from local authorities. Their dismal appraisal of the prospects for redress through ordinary channels was well-founded in a state like Georgia, where addressing a black female witness with the respectful title of "Miss" could land a black man in jail for contempt of court.[71]

While black victims of the Klan had no hope of justice, most white victims had little more. Indeed, "through fear or shame," few of the Klan's white victims reported to legal authorities.[72] "It is not considered an honor to be whipped by masked men," as one sardonic Georgian put it in 1927. A North Carolinian opponent of the Klan later explained that much of its support derived from a public consensus that "generally speaking the [white] people that they punished had a whole lot lacking in their character and they deserved some punishment." According to him, non-Klan white residents would point to people "leading these immoral lives, and they've been doing it for ten years and the children out there are suffering and nothing's being done about it. So the Klan did something about it; they put the whip to them."[73] Neighbors like these were unlikely to indict or convict.

The most instructive illustration, however, of the factors conspiring to limit the documentary record of Klan violence comes from the case of a multiple lynching in Aiken, South Carolina. As a lynching, it was unusual as an episode of Klan vigilantism in the 1920s, which most commonly involved intimidation or violence short of murder. Yet, as a case of uncommon crime that would seem difficult to cover up, its resolution speaks eloquently about how less noteworthy incidents would be handled. The case was that of the Lowmans, a black family who lived in the countryside outside Aiken. In the spring of 1926, a sheriff and his deputies came to the Lowman home to arrest one of the sons on suspicion—later proved baseless—of selling whiskey. Finding him missing, the sheriff hit the young man's sister, Bertha, in the mouth; when Mrs. Lowman came to her daughter's rescue, he shot her through the heart. Hearing the shots, the Lowman men ran in from the fields. A scuffle and weapon fire ensued, in which the sheriff was killed. The bullet most likely came from the gun of one of his own deputies, but the three Lowman siblings were arrested all the same. A "farcical" trial in a courtroom packed with armed Klan members resulted in death sentences for the two brothers and life imprisonment for their sister. Outraged, a black lawyer from Columbia, South Carolina, appealed the case to the State Supreme Court and won them a retrial. One of the brothers was acquitted as a result. Upon hearing of the acquittal, a mob began to gather in Aiken. Knowing what was coming, white men and women from as far as ninety miles away began to converge on the site. The mob marched on the jail and carried the prisoners outside town to the site where the crowd, now numbering almost a thousand, had gathered. There, Bertha Lowman "begged so piteously for her life that members of the mob had a hard time killing her." But they did. The mob leaders freed Bertha and her brothers and told them to run—only to fire dozens of shots into their backs when they did, to the glee of the crowd.

No one was indicted for the willful murder of Clarence, Damon, and Bertha Lowman. The coroner's jury charged the crime to "parties unknown." The governor paid lip service to the need to apprehend the guilty parties, but did nothing. Investigating for the NAACP, Walter White found out why. The dead sheriff and the deputies who first visited the Lowman home were all Klansmen. So was the new sheriff, who handed over his prisoners over to the mob without any struggle, and the Aiken attorney—and newly elected state representative—at whose office the murder was planned. The mob itself was organized, led, and staffed by the Klan—although it tried to enlist non-members so as to be able to maintain later, if need be, that the lynching "was not a Klan affair." White supplied the governor with the names, addresses, and occupations of the killers in the mob, all of whom belonged to the Klan. Fourteen of the twenty-four men were some kind of law enforcement personnel. Among the spectators at the lynching were the president and vice-president of one local manufacturing company, superintendents from three plants, and an overseer from another—along with three of the governor's own cousins. Not surprisingly, in these circumstances, the lynchers were never prosecuted. That fact alone should convince us of the dangers of uncritical reliance on public sources in writing the history of the Klan. If Walter White had not investigated, historians today would never have heard of the Lowman case, much less known of the leading part the Klan played in it. The official record would have had us conclude that the culprits were "parties unknown."[74] Rarely does a story so vividly illustrate the maxim that history is written by the victors.

In short, the decision about whether to use violence was for the Klan a tactical one, one answered in the affirmative most often south of the Mason-Dixon line. Yet, even in areas where it seemed too risky to try at the moment, the possibility of violence at some later date loomed. For Klan culture generated a propensity to vigilantism like an acorn does an oak; all the seed needed to grow was nourishment from good soil. Vigilante violence was the concentrated expression of that culture, of the brutal determination to maintain inherited hierarchies of race, class, and gender that Klansmen sought to conceal with a mask of chivalry. "The only reason that men are not tarred and feathered, whipped and driven from their homes and deprived of their constitutional rights in Ohio," observed one astute contemporary, "is because the klan is not strong enough. When it is," he predicted, "those things will come."[75]

III

▲▲▲

MEANINGS

8

▲ ▲ ▲

Conclusion:
The Second Klan
in Wider Perspective

This story should convince us of one thing: that there are not
two Germanys, a good one and a bad one, but only one. . . .
Wicked Germany is merely good Germany gone astray, good
Germany in misfortune, in guilt, and ruin. For that reason it is
quite impossible for one born there simply to renounce the
wicked, guilty Germany and to declare, "I am the good, the
noble, the just Germany in the white robe; I leave it to you to
exterminate the wicked one."

—Thomas Mann[1]

The Klan never did become strong enough to attempt in Ohio what
it had in Georgia. Within a decade of its founding, the Knights of
the Ku Klux Klan was on the wane. Nationally, the Klan's power
crested in the years 1924 and 1925. With state and local variations,
it began to flag thereafter. By 1926, observers around the country
were reporting smaller numbers and dwindling influence.[2] Following a brief revival when Catholic New Yorker Al Smith ran for
President, the decline became terminal in 1928. A later, short-lived
resuscitation in the mid-'thirties proved insufficient to keep creditors at bay. With initiation and dues receipts dwindling, Klan officers sold the Imperial Palace in 1936. Eight years later, the order
quietly dissolved, unable to pay the taxes it had evaded over the
years.[3]

As in other parts of the country, in Georgia the Klan began to ebb after mid-decade. In 1926, it proved unable to unseat Superior Court Judge James K. Hines, a Klan critic whom it had declared war upon. Nor did it prevent the state from going to Al Smith in the presidential election of 1928. One observer estimated that from 156,000 in 1925, the Klan's membership in Georgia plummeted to 1,400 in 1930. If the figures exaggerated the extent of the decline, they nonetheless captured its pace. During the 1928 presidential campaign, the Athens chapter experienced a brief revival; one hundred lapsed members sought reinstatement, and almost that many new ones signed up. But thereafter, the numbers resumed their downward slide. By 1930, Klansmen had given up their headquarters in downtown Athens, and met instead at the Masonic Hall in nearby Bogart.[4]

Historians have explained the demise of the Klan in a variety of ways, usually a combination. Many have pointed to internal problems such as hypocritical leaders and factionalism. Others have cited effective opposition from the press, civic leaders, or residents in particular locales. Some have argued that the very success of Klan politicians in winning office bred demoralization, as once in power they failed to furnish the dramatic changes they had promised. Similarly, some point to the subsiding of local problems such as crime, or to how experience exposed the falsity of Klan claims.[5] One writer has even suggested that the Klan's very nature doomed it.[6]

If that particular assertion seems wishful thinking, many of the other observations are apt. But as full explanations, they beg some important questions. Opposition from the press and local élites may well have spurred decline in some areas, for example, but why did a corresponding decline take place in areas without such indigenous opposition? Similarly, the notion that the Klan so contradicted core American values that defeat was foreordained runs aground in the face of evidence showing, not only a great deal in the way of shared values, but also a lack of contrition in Klan members even years later. In interviews, many affirmed their willingness to "do it again" if they felt conditions warranted a resurgence of their movement.[7]

But the main problem with prevailing accounts of the Klan's decline is the parochial vision that serves as their starting point. State or local in conception, almost none of the scholarly studies on the Klan, particularly those in recent years, examines the American movement in its international setting.[8] They make no effort to come to terms with the Klan as an expression of what European historian Arno Mayer refers to as "the General Crisis and Thirty Years' War of the twentieth century." Bounded on one end by the First World War and on the other by the Second World War, this epoch was marked by pervasive social change and political crisis, above all by the contest between right and left, "the ideological struggle," as Mayer sums it up, "between fascism and bolshevism." That match ultimately yielded the regimes of Mussolini in Italy, of Franco in Spain, and of Hitler in Germany.[9] Any account of the second Klan that fails to consider it in the light of these contemporaneous movements is bound to yield a limited understanding of its place in history.

▲ ▲ ▲

If historians have largely overlooked the common ground occupied by European fascist movements and the Klan of the 1920s, many contemporaries did not. "A 'legitimate offspring of the Facista movement in Italy' " was how small farmers in Oklahoma's Farmer-Labor Union described the Klan of their day; Atlanta's black newspaper referred to Klansmen as "American Fascisti." Some pundits concurred. "Fascist in inspiration" in the view of a contemporary French writer sojourning in the United States, the Klan was "mob rule in favour of order." Arthur Corning White also argued that the Klan was a homegrown variant of European fascism. He observed that, unlike traditional élite-based conservatism, Mussolini's movement and the Klan both had their roots in the petite bourgeoisie. The Klan's largely middle-class members, White argued, were roused in good part by a deep-seated economic malaise. They saw in the Klan a force to end the debilitating conflict between labor and capital, by violent means if need be. "That the Klan in the long run will resort to brickbats instead of ballots," White concluded, "seems as certain as the setting sun."[10]

Some Klan spokespeople recognized the kinship between their movement and those of Mussolini and Hitler. The *Searchlight* in 1922 described Mussolini's rise to power as "a sign of political health in Italy and a guarantee against the crazy and experimental forms of government with which Russia is afflicted." The *Imperial Night Hawk*, for its part, asserted that Mussolini's fight to crush "communism and anarchy" was "an entirely worthy cause." A California Klan minister coupled the Klan as "a nationalistic movement" with "the Black Shirts of Italy and the Grey Shirts of Germany." The Reverend Charles Jefferson of New York stated the relationship most aptly. In postwar Italy, he said, "things got into a mess" and the established politicians flailed ineffectually until "a strong man" stepped in and took charge. "The Ku Klux Klan," he explained, "is the Mussolini of America," the organizational expression of the "vast volume of discontent in this country with things as they are."[11]

When, after the onset of the Great Depression, organizations emerged in the United States that openly identified with fascists across the sea, moreover, the Klan came to their aid. Many Klan figures cooperated with and helped build these organizations, including the Georgia Klan leaders Walter A. Sims, Caleb Ridley, J. O. Wood, and James R. Venable. The reason for the crossover is not hard to find. The Mussolini-inspired American Order of Fascisti, or Black Shirts, which enlisted several of these Klan stalwarts, proclaimed its commitment to white supremacy, its opposition to both communism and conventional conservatism, and its dedication to organizing "the great middle class." It campaigned to solve white unemployment by taking jobs from blacks, and defended racist murders. Arthur Bell, Grand Dragon of the New Jersey Klan for two decades and later vice-president of the Nazi-identified German-American Bund, characterized the relationship between the two thus in 1940: "The principles of the Bund and the principles of the Klan are the same."[12]

Much more than American historians have realized, in fact, Klan ideology shared common features with its Nazi contemporary.[13] Like the Klan, Hitler embraced a reactionary populism that blended outspoken resentment of established élites with vitriolic anti-communism. Regularly exaggerating the power and prospects of the left, he was obsessed with destroying the organized workers' movement. Yet his was a counterrevolutionary ideology for a modern age of mass politics. National Socialism spoke in the name of the middle classes, of the "masses" as against the "classes" it excoriated for pulling society apart. Like the Klan's criticisms of America's economic system, his attacks on Germany's were superficial, and his positive proposals vague. Sidetracking basic questions of class power, they focussed on exploitative relations of exchange; in particular, the alleged machinations of finance capital and the corrupting reign of "Mammon."[14]

His detestation of Marxism as "the concentrated intellectual essence of today's universal world-concept," led Hitler to reject much of the rest of the heritage of the French revolution and the Enlightenment. In place of reason, he exalted passion and advocated a propaganda geared to the emotions. He recognized and exploited the power of symbolism and such rituals as torchlight parades. In place of careful analysis, he pushed conspiratorial explanations. In his view, "part of a great leader's genius was to make even widely separated adversaries appear as if they belonged to one category."[15] In place of a parliamentary democracy he declared mired in corruption and soft on communism, he called for a strong man to take charge with a plebiscitary popular mandate. In place of the rule of law, he substituted the calculated paramilitary terror of the Brown Shirts. As one German historian wrote: "The Nazi answer to the problem of class division was to abolish its expression by force." And, finally, in place of the notion of fundamental human equality, "the idea of the majority," Hitler substituted "the aristocratic principle within the people": strident nationalism and murderous racism.[16]

Indeed, as they were for the Klan, nationalism and racialism were National Socialism's means of countering the class divisions it so abhorred. Race was posited as the wellspring of human culture and history. Demonizing Jews, in particular, as the source of all the purported evils of modernity—from materialism, to Bolshevism, to changes in sexual mores and mass culture—Hitler positioned them at the core of his revolutionary counterrevolutionism, a lightning rod to give it mass appeal.[17]

Like the Klan's, the Nazis' anti-Semitism was steeped in sexual themes. It both preyed on and promoted anxieties about changes in women's roles and sexuality. As part and parcel of their antipathy to modern social relations and attitudes, the Nazis execrated the transformations occurring in gender roles and relations between the sexes and generations and upheld conventional family values. Hitler denounced the selfishness and sensuality of the New Woman and urged upon German womanhood a nostalgic version of the nineteenth-century respectable middle-class housewife, an intensely familial and communally oriented woman who always put the needs of her husband and children ahead of her own. Pushing women to make motherhood the center of their lives, he also condemned birth control and promised to outlaw abortion. And, all the homoerotic undercurrents flowing among his band of brothers notwithstanding, he condemned homosexuality as evidence of modern degeneracy.[18]

Not only in its world view, but also in its dynamics as a social movement, the Klan had much in common with German National Socialism and Italian Fascism. All three movements emerged from the crucible of world war, and grew in times of economic difficulty, class polarization, and political impasse. Each mobilized men and women from a broad spectrum of the population, but had particular attraction for the petite bourgeoisie. Notwithstanding their popular mass basis, however, each of these movements also enlisted the active backing or toleration of important members of the established élite, and gained strength from the legitimacy thus bestowed. They also exerted particular appeal for members of the police and armed forces, who in turn provided aid and cover for the movements' extralegal terror. Finally, all three movements had similar organizational styles in their conscious emphasis on the irrational, on liturgical rituals, and on public displays of power.[19]

Of course, to highlight the family resemblance is not to assert that these movements were similar in all respects. Italian fascism, for example, did not rely on anti-Semitism and other racialism the way that the Klan and National Socialism did.[20] And the Klan differed in some important ways from those two prototypical fascist movements. Not least, the Klan's class politics were more ambiguous. Here the Klan bore the marks of its birthplace. Operating in a nation with much lower levels of working-class organization and consciousness, the Klan sometimes even posed as the friend of "organized labor" in the face of common enemies.[21] More commonly, the Klan took advantage of the deep racial, ethnic, and skill divisions in the American working class to advance its project, especially in the South. It preyed on the narrow craft-union consciousness of native-born, Protestant, white, skilled workers in efforts to turn them against black, foreign-born, and radical workers. And it managed to win the support of some, most notably in conservative Jim Crow locals that, like the Klan, opposed industrial unionism.[22] The Klan's greatest successes among trade unionists, in fact, tended to follow disastrous defeats, sometimes involving strikebreakers from other ethnic groups, which left some native-born white workers casting about for scapegoats and alternatives to class-based politics.[23]

The particular constellation of class and race in the American South also made for some unusual conflicts within the Klan and between the Klan and local élites. Of all sections of the American capitalist class, the South's planter class appeared most akin to the peripheral élites who provided critical early support to European fascism.[24] And, yet, things were more complicated than they might at first appear. While Klan vigilantism against rural African Americans usually served the interests of white planters who wanted a cheap and pliant labor force, on occasion it went so far as to threaten the very basis of production by driving out the workers who performed it. When masked terror led to black exodus, in fact, planters sometimes came to believe things had gone too far. Often enough, however, they then found it difficult to restrain their erstwhile allies. White small farmers and tenants probably enjoyed this opportunity to outdo planters in proving their loyalty to the "white race" while in the process enhancing their own immediate bargaining position by limiting labor supply. Planter representative J. J. Brown, past president of the Georgia Farmers' Union and then state Secretary of Agriculture, thus once found himself confronted with the odd task of urging his fellow Klan members to "do all they could to hold the Negro[e]s in the South."[25]

There were other contrasts as well. Klansmen seemed less inclined than their Continental peers to welcome the idea of dictatorship, even if that was the conclusion their leaders' screeds against impotent government pointed to. Klan spokespeople sometimes criticized Mussolini, not only for being an "ally of the Pope" or a rival nationalist who might appeal to Italian Americans, but also for being a despotic ruler.[26] Internally, the unruliness of the Klan's rank and file plagued its leaders from early on. Having been recruited with an ideology that stressed popular sovereignty for native-born Protestant whites, many members came to resent the discipline that Klan leaders sought to impose. The ability of such fractious members to win support from their fellows in Klan ranks contributed mightily to the hothouse environment that produced offshoot organizations on a regular basis. Other aspects of the movement similarly complicate the analogy with European fascism. However hedged in formulation and belied in practice, Klan leaders' declared regard for the Constitution meant that they at least feared appearing to deviate from it. Finally, the Klan's reverence for Protestantism also distinguished it. Seeing established churches as competing centers of belief or power, Nazis were far less inclined to accept them.[27] Future research will undoubtedly uncover more contrasts.

The point is thus not to argue for absolute homology. It is rather to insist that the Klan was not a movement *sui generis:* it had enough in common with contemporary European mass movements of far right to make for meaningful comparison. The family resemblance between the Klan and classic fascism, however, puts the problem of interpreting the Klan's demise into a whole new light.

▲ ▲ ▲

Once the Klan is viewed in transnational perspective, a more chilling hypothesis emerges about why Klan strength waned so quickly right across the country after mid-decade. The causes usually adduced for this decline may be incidental to the simple fact that circumstances in the United States never reached the point that they did in the nations where fascism ultimately triumphed. After all, as late as the elections of 1928, the Nazis took only 2.6 percent of the total vote and were seen as "a minor, and declining splinter party." By contrast, in 1924, a Klan write-in candidate for mayor was able to attract more than one-third of the vote in Detroit, the fourth-largest American city. Had the Depression not hit Germany as hard as it subsequently did, National Socialism might today be dismissed as the Klan sometimes is: a historical curiosity whose doom was foreordained.[28]

In the United States, on the other hand, the social conditions that once fueled Klan growth had largely abated by mid-decade. In the nation at large, the postwar recession gave way to boom and renewed growth by 1923. The economic crisis loomed larger and longer in the South and in farming regions of the Midwest than in the industrial North, not as dependent on the "sick" industries of agriculture, textiles, and mining. Yet, even in the South, the sense of economic apocalypse had faded by mid-decade. By 1925, the regional economy had rebounded, and the press was reporting with palpable relief the revival of crop values and textile demand.[29]

The pitched class conflict of 1919–1921 also rapidly abated, as employers, with the aid of the government Red Scare and new technology, regained the upper hand. By 1924, the offensive against labor and the left had largely succeeded. For the first time in American history, unions failed to grow during a time of relative prosperity. On the contrary, the proportion of the labor force in unions dropped from over twelve percent in 1920 to eight percent in 1930, as the number of union members fell from five million to less than three and a half million. In proportion to the wage-earning population, it was the weakest labor movement of any of the leading Western industrial powers. Socialist Party membership in the nation as a whole dropped drastically from 110,000 in 1919 to 12,000 four years later; the Communist Party, for its part, never amassed more than 20,000 members at any point in the decade. By contrast, the German Communist Party in the 'twenties—the smaller of the country's main left-wing parties—enlisted a membership that ran as high as 380,000, and drew as many as three million votes.[30]

Like the struggle for "industrial democracy" in the United States, the wartime and postwar offensive for racial equity had clearly run aground by a few years into the new decade. Despite determined, often heroic, efforts, blacks had been unable to attract enough white support to dismantle any of the apparatus of white supremacy, as the federal government's unwillingness to pass even mild anti-lynching legislation attested. The NAACP lost almost 200 branches by 1923, and over 70,000 members—or more than two-thirds of its 1919 roster—over the decade. The hemorrhage was especially severe in the South, where both challenge and resistance had been greatest.[31]

Locally, the discouragement African Americans felt was evident in that the Athens NAACP, which inspired such hopes at its founding, ceased functioning by mid-decade. By then, even the Atlanta chapter, the boldest in the state, had lapsed into inactivity. Nationally, black disillusionment found expression in the massive following attracted by nationalist leader Marcus Garvey. His advocacy of black pride and program of self-determination bore witness to his followers' aspirations, much as his insistence on the impossibility of achieving racial equity in the United States did their despair. The discouragement of blacks, like that of labor, no doubt undercut the urgency many of the Klan's followers and sympathizers had hitherto felt.[32]

The decline of that challenge from below also helps explain the changed attitudes of members of the governing élite towards the Klan. Many public officials, business leaders, and opinion makers had proved more than willing to turn a blind eye to the Klan's activities in early years, just as their counterparts in Italy and Germany did. Occasional pieties about letting the law take its course notwithstanding, they looked away as Klan shock troops did the dirty work of enforcing the hierarchies of class, race, and sex. Only after the challenge from below had been successfully staved off did large numbers come to identify the Klan as a threat. When they did, it was sometimes less for its unsavory features than for its mass following. Across the country, élite critics used phrases such as "the menace of such Jacobins," and "the incubator of anarchy and the aide de camp of Lenine and Trotsky" to denounce the Klan, or equated it with the radical wing of organized labor.[33]

Without the extra charge that came from association with la-

bor militancy and black struggle, the changes in gender and generational relations came to seem less threatening. As important as
feminism, the so-called sexual revolution, and the spread of the
commercial leisure industry had been in winning the Klan a mass
following, neither their continued spread nor the ongoing and
widespread defiance of Prohibition proved sufficient to keep Klan
members mobilized after the other challenges had receded. Here,
in fact, if in few other areas, the Klan defeat was unambiguous.
The persistence of bootlegging and gambling, the popularity of
dancing and movie-going, and young women's enthusiasm for the
social freedom and sensual pleasures symbolized by the flapper all
made clear the tenuousness of the family values the Klan stood for.

In short, on most fronts, Klansmen could feel, if not triumphant, at least relieved by mid-decade. As the sharp polarizations
of the postwar years abated, their movement must have come to
seem like overkill to all but the most devoted. Without extreme
conditions, extreme measures enjoyed less legitimacy. But that
change in circumstances leaves open, for us, an unsettling question: what if the interwar social crisis *had* reached the scale in
America that it did in Italy or Germany? How might native-born,
middle-class whites have reacted?

We can never know the answer to that question. But we do
know enough to make easy optimism untenable. Indeed, what
emerges most forcefully from this study of the Klan is the wealth
of cultural material at hand that a movement like the Klan could
build on. Under conditions of economic uncertainty, sharply contested social relations, and political impasse, assumptions about
class, race, gender, and state power so ordinary as to appear "common sense" to most WASP Americans could be refashioned and
harnessed to the building of a virulent reactionary politics able to
mobilize millions. What appears distinctive about the Klan is less
the specific ideas it stood for than the way it synthesized them and
proffered itself as, in the words of contemporary Walter White,
their "direct-action expression."[34] But those ideas themselves, at
least in more understated form, had a long-standing and widely accepted place in the dominant culture.

Of course, to say this is not to imply that all elements in
American culture worked to the Klan's advantage, or even that all
of those that did necessarily did so, that they could not have led to
different conclusions. Here I am much persuaded by the accounts
of European fascism that reject fatalistic readings of the proclivities
of the petite bourgeoisie and stress, instead, contingency: the de-

gree of organization of anti-fascist forces and the political choices made by their leaders mattered very much.[35] We simply have no way of knowing how the Klan would have fared had not the Red Scare and the American Plan for industry eliminated left-wing popular alternatives to the status quo. In any event, even in the narrowed political spectrum of the 1920s, Klan leaders confronted some ideas and values that defied their ambitions. Majority rule, religious tolerance, and regard for the rule of law, for example, all had significant, if not majoritarian, followings.

Nevertheless, what seems most striking in this story is the adaptability of many conventional American sensibilities to a reactionary populist project. The core elements of Klan ideology were not as aberrant as one might imagine. Generations of observers of American culture, for example, have remarked on what one writer calls "the imperial middle": the pervasive assumption that America had always been and should stay a middle-class society, and the corollary denial that other classes had valid interests of their own. In times of pitched struggle, such as that which followed World War I, those axioms could easily slide into an insistence that class conflict was illegitimate, even treasonous, and should be suppressed. The unusual sway of individualism, moreover, made the United States fertile terrain for racialist explanations of why some people succeeded and others failed, explanations that ranged from Manifest Destiny to Social Darwinism and eugenics. Indeed, historians have lately become more aware of how, from the time of the republic's founding, American ideas of middle-class standing and citizenship rights were coded in racially exclusive ways. So, too, was American middle-class consciousness molded and galvanized, from its very origins, by notions of appropriate gender roles and moral respectability.[36] The evangelical strain so prominent in American culture could also play its part in establishing an indigenous form for the apocalyptic, anti-rational, and Manichean emphases of fascist thought.

Finally, each of these elements in turn contributed to the unusual sanction American culture conferred on extralegal violence in defense of prevailing relations of power. This country was, after all, as James Weldon Johnson once reminded an impervious Congress, "the only spot, civilized or uncivilized, on the face of the globe where a human being can be burned alive with immunity." It would not be such a long leap from the many American varieties of vigilantism—not only lynching, but also white-capping, anti-labor citizens' committees, and, more generally, the veneration of rough-and-ready frontier "justice" in popular culture—to the "politics of the piazza" characteristic of fascism in its mobilization phase.[37]

If this assessment is valid, if core elements in American culture were more amenable to a fascist sensibility than is usually recognized, the question inevitably follows of why the Great Depression turned out as it did in the United States. In fact, mass middle-class movements of the far right did develop in the 1930s, although not under the aegis of the Klan, perhaps then too spent from its earlier performance. Some of these movements—like William Dudley Pelley's Silver Shirts—were openly sympathetic to fascism. Others—like the followings of Huey Long and Father Coughlin in the early 1930s—are more difficult to classify. But all enlisted a predominantly petit-bourgeois following and mobilized around the core axioms discussed here.[38] Perhaps, in fact, given these cultural traditions and political mobilizations, the question we should be asking is not the common one of why American reform was not more radical in the 1930s, but why it happened in the relatively benign ways it did at all.

That question is too big for this book, and no single answer to it will do. But among the reasons ordinarily cited by historians for why the United States bypassed the road some of its great-power peers took, one may call for special emphasis at a time in which trade unions have become targets of scorn for nearly all points on the political spectrum: the fact that, in contrast to the 1920s, a strong and inclusive working-class movement was able to pose an alternative to both the far right and the discredited status quo. As civil rights movement veteran Anne Braden once observed, the times when the Klan has failed to grow are as instructive as those when it has. In neither the 1930s nor the 1960s did it make much headway, she argues, because in each period "strong mass movements advocated real answers to social and economic problems" at the same time as "there was a strong offensive against the ideology of racism."[39]

If this analysis has merit, then the irony is acute. That phenomenon deemed least American by the dominant culture from the founding of the republic forward—class struggle waged by the propertyless, many of them African Americans, Catholics, and recent immigrants—may have contributed more than we will ever know to keeping reactionary populist movements at bay in the United States during the Great Depression. Perhaps, after all, it was those with the least "stake in society" who had the most stake in defending democracy.

▲ ▲ ▲

Appendix on Names

The author-unfriendly state of current United States law has led me to change some of the names used in this book. I have no desire to protect these individuals or their families' reputations. On the contrary, one subtheme of this book is how white Americans' refusal to face the truth about the first Klan helped make the second Klan possible. But neither am I interested in scandal. Since my purposes are scholarly and political rather than sensational, I have decided that it would not detract from them to alter names to conceal identities. The principles followed were these. Where individuals' membership in or contact with the Klan was reported by the press or in the courts, and is thus already a matter of public record, I have used real names. Where the membership or contact was revealed only in the unpublished Klan manuscript, I have created pseudonyms whose fictitious character is signalled by italics. Where the names must also appear as correspondents in the notes, however, I have kept the initials constant and used them in lieu of full names in citations. Those who need to know the actual identities of people in this study for scholarly purposes may contact me through my publisher. Others should consult the Klan manuscript for themselves.

▲ ▲ ▲

Appendix on Database

The information about Athens Klan members in this work comes from a database built from a variety of sources. The Athens Klan manuscript and the local newspaper yielded an initial list of members. This group was then pared down to exclude anyone whose name was listed only once (to reduce the possibility of accidental misidentification), those whose initials were shared by more than one resident, and those whom the sources listed as residents of a county other than Athens, since tracking members through public records for the five counties adjoining Clarke County as well would have been an insuperable task. Such pruning resulted in a final list of 543 members.

These individuals were then traced through a variety of sources: the Clarke County federal manuscript censuses for 1910 and 1920; Athens city directories for the years 1904, 1916–17, and 1926–27; county tax digests for the years 1900, 1910, 1918, 1921, and 1927; Superior Court records; the local press; and records of local churches and voluntary organizations held by the Georgia Department of Archives and History and the Hargrett Rare Book and Manuscript Library at the University of Georgia. Among the sources of the data on class position, the most unusual in historiography of the Klan and the most rewarding for the analysis offered here were tax digests, which recorded annually the value of the assets held by all Georgia households. These digests made it possible to portray Klan members' economic standing in a more concrete way than other studies, using only city directories and the census, have been able to. Also, by tracing members over the course of two decades, one may develop a more dynamic and relational understanding of class than is common in the literature on the Klan.

As do any historical data, however, these suffer flaws and omissions. Of the 543 members on the final list, 125 could not be found in other Clarke County public records—presumably because they resided outside county limits or only came to the area in the 'twenties—leaving 418 on whom some information was available. Data could not be found on each variable for every member, so percentages cited in the text are adjusted to the variable at hand. The database also probably understates the participation of certain groups. Farmers of all tenures—planters, working landowners, and tenants—probably joined the Klan in greater numbers than my data suggest. City directories only included farmers near city limits, and Clarke County was in any case more urban than most contemporary Southern counties. Similarly, transients would be underenumerated in the public records, as would mill workers in villages outlying Clarke County. Finally, since internal Klan records for the early 'twenties are not available, the data probably understate the participation of high-status, wealthy men, who, at least in other areas of the country, tended to drop out before mid-decade. These weaknesses tend to rule each other out, however: since certain high, low and middle categories are all probably undercounted, the overall profile should be fairly accurate.

▲ ▲ ▲

Abbreviations
Used in Notes

ABH	*Athens Banner Herald*
ACHF	Athens-Clarke County Heritage Foundation
ACLU	American Civil Liberties Union Archives
ADB	*Athens Daily Banner*
ADH	*Athens Daily Herald*
AGC	Asa Griggs Candler Papers
AHR	*American Historical Review*
AHS	Atlanta Historical Society
AK	Ku Klux Klan, Athens Chapter No. 5, Papers
BUGA	*Bulletin of the University of Georgia*
CES	*Columbus Enquirer-Sun*
CCL	Georgia, Commissioner of Commerce and Labor
CCSC	Clarke County, Superior Court
CDF	Cameron Douglas Flanigen Papers
CIC	Commission on Interracial Cooperation Papers
CNM	Carlton-Newton-Mell Collection
CSB	Charles S. Bryan Papers
DCB	David Crenshaw Barrow Papers
DU	Special Collections Library, Duke University
ECB	Eugene Cunningham Branson Papers

EDC	Incoming Correspondence, Records of the Executive Department, Georgia Department of Archives and History
EU	Special Collections, Robert W. Woodruff Library, Emory University
FLT	Frances Long Taylor Papers, UGA
FSAA	Federal Surveillance of Afro-Americans
FWP	Federal Writers' Project Papers
GDAH	Georgia Department of Archives and History
GHQ	*Georgia Historical Quarterly*
HGG	Hugh Gladney Grant Papers
HJA	Henry J. Allen Papers
HS	Hoke Smith Papers, Russell Library
JAH	*Journal of American History*
JJB	John James Blaine Papers
JLH	Julian Larose Harris Papers
JMB, AHS	Joseph Mackey Brown Papers, Atlanta Historical Society
JMB, UGA	Joseph Mackey Brown Papers, UGA
JRH	James Robert Hamilton Papers
JSH	*Journal of Southern History*
JWB	Josiah William Bailey Papers
LAC	Living Atlanta Collection
LH	*Labor History*
NAACP	National Association for the Advancement of Colored People Papers
NEH	Nathaniel E. Harris Section, Executive Department Correspondence
NSMC	Nina Scudder Memorial Collection
NYPL	New York Public Library
NYT	*New York Times*
PCM	University of Georgia, Prudential Committee Minutes
RBR	Richard B. Russell, Jr., Papers
REM	Ralph Emerson McGill Papers
RL	Russell Library, University of Georgia–Athens
RLF	Rebecca Latimer Felton Collection
RLR	Robert L. Rodgers Papers

SHC	Southern Historical Collection
SHSW	State Historical Society of Wisconsin
TEW	Thomas Edward Watson Papers, Southern Historical Collection
TINCF	Tuskegee Institute News Clippings File
UA	University Archives, University of Georgia, Athens
UGA	University of Georgia Libraries, Hargrett Rare Book and Manuscript Library
USDA	United States Department of Agriculture
WAC	Bishop Warren Akin Candler Papers
WTA	William Tate Archives

▲ ▲ ▲

Notes

Introduction

1. W. E. B. DuBois, "The Shape of Fear," *North American Review* 223 (1926), 294–95.

2. Reports of the second Klan's numerical strength vary from a historian's low estimate of 1.5 million to the organization's own claim of five million men in 1926. See, respectively, Kenneth T. Jackson, *The Ku Klux Klan in the City, 1915–1930* (New York, 1967), xii, 236; *New York Herald Tribune*, Sept. 15, 1926, in series C, box 317, National Association for the Advancement of Colored People, Papers, Manuscripts Division, Library of Congress, Washington, D.C. (hereafter, NAACP).

3. For overviews of the Klan's successive incarnations, see David M. Chalmers, *Hooded Americanism: The History of the Ku Klux Klan*, third ed. (New York, 1981); Arnold S. Rice, *The Ku Klux Klan in American Politics* (Washington, D.C., 1962); Wyn Craig Wade, *The Fiery Cross: The Ku Klux Klan in America* (New York, 1987). On Duke, see *The Emergence of David Duke and the Politics of Race*, ed. Douglas Rose (Chapel Hill, 1992).

4. Quote from Stanley Frost, *The Challenge of the Klan*, 6–7. Recent published studies that apply the methods of the new social history include Robert Alan Goldberg, *Hooded Empire: The Ku Klux Klan in Colorado* (Urbana, 1981); Larry R. Gerlach, *Blazing Crosses in Zion: The Ku Klux Klan in Utah* (Logan, Utah, 1982); Shawn Lay, *War, Revolution and the Ku Klux Klan: A Study of Intolerance in a Border City* (El Paso, 1985); William D. Jenkins, *Steel Valley Klan: The Ku Klux Klan in Ohio's Mahoning Valley* (Kent, Ohio, 1990); Leonard J. Moore, *Citizen Klansmen: The Ku Klux Klan in Indiana, 1921–1928* (Chapel Hill, 1991). For a sam-

pling of the new work, see Shawn Lay, ed., *The Invisible Empire in the West: Toward a New Historical Appraisal of the Ku Klux Klan of the 1920s* (Urbana, 1992). Among the leading older works on the Klan are, in addition to those already cited, John Moffat Mecklin, *The Ku Klux Klan: A Study of the American Mind* (1923; reprint, New York, 1963); Emerson H. Loucks, *The Ku Klux Klan in Pennsylvania: A Study in Nativism* (Harrisburg, Penn., 1936); Charles Alexander, *The Ku Klux Klan in the Southwest* (Lexington, 1965); idem, *Crusade for Conformity: The Ku Klux Klan in Texas, 1920–1930* (Houston, 1962); Jackson, *Ku Klux Klan in the City*. Some scholars whose primary focus lay elsewhere have also offered influential interpretations of the second Klan. See, in particular, Richard Hofstadter, *The Age of Reform, from Bryan to F.D.R.* (New York, 1955); John Higham, *Strangers in the Land: Patterns of American Nativism, 1860–1925* (1963; reprint, New York, 1974); Seymour Martin Lipset and Earl Raab, *The Politics of Unreason: Right Wing Extremism in America, 1790–1970* (New York, 1970); Stanley Coben, *Rebellion Against Victorianism: The Impetus for Change in 1920s America* (New York, 1991).

5. Important recent arguments to this effect include Paula Giddings, *When and Where I Enter: The Impact of Black Womanhood on Race and Sex in America* (New York, 1984); Bell Hooks, *From Margin to Center,* (Boston, 1984); Hazel Carby, *Reconstructing Womanhood: The Emergence of the Afro-American Woman Novelist* (New York, 1987); Elsa Barkley Brown, "Womanist Consciousness: Maggie Lena Walker and the Independent Order of Saint Luke," *Signs* 14 (Spring 1989); Evelyn Brooks Higginbottom, "African-American Women's History and the Metalanguage of Race," *Signs* 17 (Winter 1992), 251–74; and the synthesis by Patricia Hill Collins, *Black Feminist Thought: Knowledge, Consciousness, and the Politics of Empowerment* (New York, 1990). For a groundbreaking study of female involvement in the Klan, see Kathleen M. Blee, *Women of the Klan: Racism and Gender in the 1920s* (Berkeley, 1991). My work, in contrast, applies gender analysis to men.

6. Quotes from Leonard J. Moore, "Historical Interpretation of the 1920's Klan: The Traditional View and the Populist Revision," *Journal of Social History* 24 (Winter 1990), 348, 352. Moore's is the most extreme statement of this tendency; more nuanced versions include Goldberg, *Hooded Empire*; Gerlach, *Blazing Crosses in Zion*; Lay, *War, Revolution and the Ku Klux Klan*; Christopher N. Cocoltchos, "The Invisible Government and the Viable Community: The Ku Klux Klan in Orange County, California, During the 1920's" (Ph.D. diss., University of California, Los Angeles, 1979). Interestingly, the Reconstruction-era Ku Klux Klan, whose violence and racism no scholar today would dispute, also sought to regulate community morality. See Charles Flynn, *White Land, Black Labor: Caste and Class in Late Nineteenth-Century Georgia* (Baton Rouge, 1983), 31–56; also U.S. Congress, Senate, *Testimony Taken by the Joint Select Committee to Inquire into the Condition of Affairs in the Late Insurrec-*

tionary States, Georgia, vols. 1 and 2, 42nd Congress, 2nd session (Washington, D.C., 1872).

7. For an incisive review of the literature about race-coding in contemporary American politics, see Adolph Reed, Jr., and Julian Bond, "Equality: Why We Can't Wait," *Nation* 253 (9 Dec. 1991), 733–37. For analysis of how such signs work, see Murray Edelman, *The Symbolic Uses of Politics* (Urbana, 1967).

8. On the concept of reactionary populism, see Nancy MacLean, "The Leo Frank Case Reconsidered: Gender and Sexual Politics in the Making of Reactionary Populism," *Journal of American History* 78 (Dec. 1991), esp. 920–21; also Ronald P. Formisano, *Boston Against Busing: Race, Class, and Ethnicity in the 1960s and 1970s* (Chapel Hill, 1990).

9. The treatments of republican ideology that have most influenced my thinking include Edmund S. Morgan, *American Slavery, American Freedom: The Ordeal of Colonial Virginia* (New York, 1975); Gordon Wood, *The Creation of the American Republic, 1776–1787* (Chapel Hill, 1969); Isaac Kramnick, *Republicanism and Bourgeois Radicalism: Political Ideology in Late Eighteenth-Century England and America* (Ithaca, 1990); Eric Foner, *Tom Paine and Revolutionary America* (New York, 1976); Steven Hahn, *The Roots of Southern Populism: Yeomen Farmers and the Transformation of the Georgia Upcountry, 1850–1890* (New York, 1983); Sean Wilentz, *Chants Democratic: New York City and the Rise of the American Working Class, 1788–1850* (New York, 1984); Edward L. Ayers, *Vengeance and Justice: Crime and Punishment in the Nineteenth-Century American South* (New York, 1984), esp. 40–45. Of these, Kramnick is most attentive to the reactionary elements of republicanism vis-à-vis the liberalism of its day and to the way, notwithstanding their contrariety, both streams continued to influence the political discourse of Americans (see esp. 1–40, 294). The common tendency of recent scholars to assume people in the past must have accepted either republicanism *or* liberalism is also criticized by Daniel T. Rodgers, *Contested Truths: Keywords in American Politics Since Independence* (New York, 1987), esp. 9–10.

On republican thought's presuppositions about race, see Morgan, *American Slavery, American Freedom;* J. William Harris, *Plain Folk and Gentry in a Slave Society: White Liberty and Black Slavery in Augusta's Hinterlands* (Middletown, 1985). On gender, see Linda Kerber, *Women of the Republic: Intellect and Ideology in Revolutionary America* (New York, 1986); Joan Gunderson, "Independence, Citizenship, and the American Revolution," *Signs* 13 (1987), 59–77; Ruth H. Bloch, "The Gendered Meanings of Virtue in Revolutionary America," ibid., 37–58; Stephanie McCurry, "The Two Faces of Republicanism: Gender and Proslavery Politics in Antebellum South Carolina," *JAH* 78 (March 1992), 1,245–64.

10. W. E. Burghardt Du Bois, "Georgia: Invisible Empire State," in *These United States: A Symposium,* vol. I, ed. Ernest Gruening (New York, 1924), 322–45.

11. For scholarly treatments of Southern Klan chapters' vigilante activities in these years, see Alexander, *Crusade for Conformity*; idem, *Ku Klux Klan in the Southwest*; Roger Kent Hux, "The Ku Klux Klan in Macon, 1922–1925" (M.A. thesis, University of Georgia, 1968); idem, "The Ku Klux Klan in Macon, 1919–1925," *GHQ* 62 (1978), 155–68; idem, "The Ku Klux Klan and Collective Violence in Horry County, South Carolina, 1922–1925," *South Carolina Historical Magazine* 85 (July 1984), 211–19.

12. Examples of this kind of reasoning in scholarship on the Klan include Alexander, *Crusade for Conformity*, 6; Hux, "Klan in Macon," 66; Gerlach, *Blazing Crosses in Zion*, 138–39.

13. Among the outstanding models of such inquiry are Morgan, *American Slavery, American Freedom*; and Barbara J. Fields, "Ideology and Race in American History," in J. Morgan Kousser and James M. McPherson, eds., *Region, Race, and Reconstruction* (New York, 1982). In discussions of Klan ideology, I use the word "racialist" in preference to "racist" in order to emphasize that we are not looking simply at attitudes, but rather at expressions of a systematic world view with racial categories as one of its core elements.

Chapter 1

1. Minutes, 15 Sept. 1925, box 2, Ku Klux Klan, Athens Chapter No. 5, Papers, Hargrett Rare Book and Manuscript Library, University of Georgia Libraries, Athens, Georgia (hereafter, AK and UGA, respectively). All names italicized in the text are fictitious. See Appendix on Names for explanation.

2. Ibid. For policy and depictions of dangerous members, see Knights of the Ku Klux Klan, Department of Realms, *Klan Building: An Outline of Proven Klan Methods for Successfully Applying the Art of Klancraft in Building and Operating Local Klans* (Atlanta, n.d.), 10. For a former Atlanta member's recollections of how such reports were treated, see Sanders Ivey interview by Cliff Kuhn, transcript, p. 30, box 37, Living Atlanta Collection, Atlanta Historical Society (hereafter, LAC and AHS, respectively).

3. See CDM to Governor L. G. Hardman, 15 March 1927, Record Group 1, Incoming Correspondence, 1927–1931, Records of the Executive Department, Georgia Department of Archives and History (hereafter, EDC and GDAH, respectively). Local Klan members were identified through the Athens Klan Papers and, in some cases, the press. Further information about individual members was culled from a wide variety of sources, including the local press, the federal manuscript census, state tax digests, city directories, superior court minutes, and publications by or about local churches and civic organizations. The database is described further in the Appendix. Subsequent references to members that draw on the database will only cite sources in the case of direct quotes, newsworthy events, or information gleaned from manuscript sources. Readers who want more

extensive documentation for the points made in this book are urged to consult my doctoral dissertation, "Behind the Mask of Chivalry: Gender, Race, and Class in the Making of the Ku Klux Klan of the 1920s in Georgia" (Ph.D. diss., University of Wisconsin–Madison, 1989).

4. C. Lewis Fowler, *The Ku Klux Klan: Its Origin, Meaning and Scope of Operation* (Atlanta, 1922) 12; undated *ABH* clipping, Vertical Files, Biography, s.v. "M. G. Michael," UGA.

5. See Kenneth Coleman and Charles Stephen Gurr, eds., *Dictionary of Georgia Biography* (Athens, 1983), s.v. "Simmons, William J."; Clement C. Moseley, "Invisible Empire: The History of the Ku Klux Klan in Twentieth-Century Georgia" (Ph.D. diss., University of Georgia, 1968), 7; Jackson, *Klan in the City*, 4–8.

6. See Coleman and Gurr, eds., *Dictionary of Georgia Biography*, s.v. "Tyler, Mary Elizabeth Grew"; Moseley, "Invisible Empire," 12, 8; Wilma Dykeman and James Stokely, *Seeds of Southern Change: The Life of Will Alexander* (Chicago, 1962), 101; Jackson, *Klan in the City*, 11–12.

7. Clarke, as quoted in Jackson, *Klan in the City*, 10; see also Chalmers, *Hooded Americanism*, 32–33; Dykeman and Stokely, *Seeds of Southern Change*, 97–98.

8. *Athens Daily Herald*, 24 Dec. 1915, p. 1 (hereafter, *ADH*); George Tindall, *The Emergence of the New South, 1913–1945*, (Baton Rouge, 1967), 53; Frances Long Taylor, "The Negroes of Clarke County, Georgia, During the Great War," *BUGA*, 29, n. 8 (Sept. 1919), 45; Jack E. Parr interview, 2 Dec. 1976, transcript, p. 3, Athens–Clarke County Heritage Foundation, Oral History Tapes, UGA (hereafter, *ACHF*); S. V. Sanford to Joseph Mackey Brown, 12 July 1917, Joseph Mackey Brown Papers (restricted section), UGA (hereafter cited as JMB, UGA); C. D. Flanigen to R. B. Nevitt, 9 Nov. 1917, box 2, Cameron Douglas Flanigen Papers, UGA (hereafter, CDF); typed minutes from meeting of the ministers of Athens, 10 Oct. 1917, box 5, Carlton-Newton-Mell Collection, UGA (hereafter, CNM); *Athens Daily Banner*, 9 Jan. 1921, p. 2 (hereafter, *ADB*).

9. Hiram Wesley Evans, *The Klan of Tomorrow and the Klan Spiritual* (n.p., 1924), 20; Jackson, *Klan in the City*, 29; Knights of the Ku Klux Klan, *Official Message of the Emperor of the Invisible Empire of the Knights of the Ku Klux Klan to the Initial Session of the Imperial Klonvocation* (Atlanta, 1922), 5; *ADH*, 24 Dec. 1915, p. 1. For the rituals, see Alvin J. Schmidt, *Fraternal Organizations* (Westport, 1980), 196–201. For members' enthusiasm for the rituals, see correspondence in box 1, AK.

10. Norman Fredric Weaver, "The Knights of the Ku Klux Klan in Wisconsin, Indiana, Ohio and Michigan" (Ph.D. diss., University of Wisconsin, 1954), 29; Gerlach, *Blazing Crosses*, 136; Knights of the Ku Klux Klan, Realm of Georgia, *Official Document*, Jan.–Mar. 1928, 7; Robert L. Duffus, "Ancestry and End of the Ku Klux Klan," *World's Work*, 46 (1923), 532; Carl Hutcheson to Thomas E. Watson, 12 July 1921, box 12, Thomas E. Watson Papers, Southern Historical Collection, Library of the Univer-

sity of North Carolina at Chapel Hill (hereafter, TEW and SHC, respectively); *Searchlight,* 24 March 1923; *Imperial Night Hawk,* 24 Oct. 1923; Winfield Jones, *Story of the Ku Klux Klan* (Washington, D.C., 1921), 93.

11. Marion Monteval, *The Klan Inside Out* (1924; reprint, Westport, 1970), 53; Edward Vance Toy, "The Ku Klux Klan in Oregon: Its Program and Purpose" (M.A. thesis, University of Oregon, 1959), 39, 58; Jackson, *Klan in the City,* 10. For Clarke County, see Minutes, May 4, 1925, box 2, AK; also Samuel Green to H. K. Brackett, 24 June 1930, box 1, AK; *Athens Banner Herald,* 6 Aug. 1930, pp. 1, 5 (hereafter, *ABH*); *ADH,* 24 Dec. 1915, p. 1; *ADH,* 3 April 1915, p. 1.

12. *ADH,* 28 Dec. 1915, p. 1; *ABH,* 12 Nov. 1928, p. 1.

13. *ABH,* 3 July 1930, pp. 1, 6; Lipset and Raab, *Politics of Unreason,* 125. See also Gerlach, *Blazing Crosses,* 38, 134; Jackson, *Klan in the City,* 10, 18–19, 29–30, table 4; Goldberg, *Hooded Empire,* 35–37, 65.

14. *Kourier,* Dec. 1924, p. 26; Jackson, *Klan in the City,* 64, 150; Cocoltchos, "Invisible Government," 138, 163; Goldberg, *Hooded Empire,* 47, 168; Gerlach, *Blazing Crosses,* 45; Lipset and Raab, *Politics of Unreason,* 123–24; Toy, "Klan in Oregon," 134–36; Jenkins, *Steel Valley Klan,* esp. ix–xi. Percentages for Clarke County are adjusted. These numbers no doubt underestimate Klansmen's church membership, since rosters were not available for all churches, and they were not always complete. The distribution of Protestant church members in the general Clarke County white population roughly corresponded to that in the Klan. United States, Bureau of the Census, *Census of Religious Bodies, 1926,* Part I, Summary and Detailed Tables (Washington, 1930), 589, 592.

15. *ABH,* 8 Jan. 1928, p. 8; *ADB,* 15 Nov. 1921, p. 5.

16. Robert Moats Miller, "A Note on the Relationship between the Protestant Churches and the Revival of the Ku Klux Klan," *JSH* 22 (Aug. 1956), 356; *Imperial Night Hawk,* 21 May 1924, p. 8.

17. *ABH,* 15 Jan. 1925, pp. 1, 8; *ABH,* 2 Dec. 1927, p. 1; *ABH,* 29 Oct. 1925, p. 4.

18. *Searchlight,* 17 Nov. 1923, p. 1; *Kligrapp's Quarterly Report* forms in box 3, AK; *Searchlight,* 17 Nov. 1923, p. 1.

19. Jackson, *Klan in the City,* 37; correspondence in preparation for a 1924 Athens rally in box 1, AK; "Special Bulletin," 31 Dec. 1926, loc cit.

20. Jackson, *Klan in the City,* xii, 15, 237; *Imperial Night Hawk,* 28 March 1923, p. 8; Kenneth Coleman, ed., *A History of Georgia* (Athens, 1977), 293; New York *Herald-Tribune,* 15 Sept. 1926, in series C, box 317, NAACP.

21. Quote from Frost, *Challenge of the Klan,* 1; Loucks, *Klan in Pennsylvania,* 147; Knights of the Ku Klux Klan, *Papers Read at the Meeting of the Grand Dragons of the Knights of the Ku Klux Klan* (Atlanta, 1923), 97–98.

22. See, for example, Goldberg, *Hooded Empire*, 38–40, 48; and Gerlach, *Blazing Crosses*, 133.

23. For analysis of these organizations, see Blee, *Women of the Klan*. In Athens, see Klan, Minutes, 18 May 1926, AK; B. E. V. to Knights of the KKK, 7 June 1927, box 1, AK; *ABH*, 2 May 1926. Unfortunately, no other records of the Athens WKKK survive.

24. Ticket to "Great Klan Barbecue," folder 6, Ku Klux Klan Miscellany Collection, UGA; Klan, Georgia, *Official Document*, 23 Dec. 1926, p. 4; Minutes, 29 Sept. 1925, box 2, AK; ibid., Jan. 19, 1926, AK; Mrs. S. A. et al. to "Officers and Members of Athens Klan No. 5," 1 Feb. 1926, box 1, AK.

25. Klan, Georgia, *Official Document*, Nov. 1926, p. 2; Klan, *Klan Building*, 4, 5, 7 (italics in originals).

26. Quoted in Jackson, *Klan in the City*, 257. See also Leonard Dinnerstein, *The Leo Frank Case* (New York, 1968).

27. Coleman, *History of Georgia*, 292–93; William G. Shepherd, "Fighting the K.K.K. on Its Home Grounds," *Leslie's* 133 (15 Oct. 1921), 511.

28. For a deft analysis, see Michael Rogin, " 'The Sword Became a Flashing Vision'; D. W. Griffith's *The Birth of a Nation*," *Representations* 9 (Winter 1985), 150–95.

29. Quoted in Jackson, *Klan of the City*, 3–4; Robert L. Zangrando, *The NAACP Crusade Against Lynching, 1909–1950* (Philadelphia, 1980), 33–34; John Dittmer, *Black Georgia in the Progressive Era* (Urbana, 1977), 185–86.

30. Henry Lincoln Johnson to Nathaniel E. Harris., 9 Dec. 1915, box 234, EDC; C. C. Pope to James Weldon Johnson, 17 Jan. 1921, series C, box 312, NAACP. See also *New York Call*, 23 March 1921, in v. 190, American Civil Liberties Union Archives, Seeley G. Mudd Manuscript Library, Princeton University (hereafter, ACLU); *Atlanta Independent*, Jan. 20, 1921, p. 4; *Atlanta Independent*, May 1, 1924, p. 1.

31. *Catholic World*, Jan. 1923, p. 443; *New York Times*, 7 Sept. 1923, p. 17 (hereafter, *NYT*); Weaver, "Klan in Wisconsin," 244–48; depositions in John James Blaine Papers, Manuscripts Section, State Historical Society of Wisconsin (hereafter, JJB and SHSW, respectively); also Weaver, "Klan in Wisconsin," 255–64; Jackson, *Klan in the City*, 168.

32. *NYT*, 25 July 1926, II, 2, 3; *NYT*, 17 Nov. 1922, p. 4; *NYT*, 17 June 1921, p. 15; *NYT*, 5 Oct. 1921, p. 18; *NYT*, 3 April 1923, p. 2; *NYT*, 1 Nov. 1924, in series C, box 317, NAACP. See also Philip S. Foner, *Organized Labor and the Black Worker, 1619–1981* (New York, 1982), 166n.

33. Dorsey quoted in Horace C. Patterson and Gilbert C. Fite, *Opponents of War, 1917–1918* (Seattle, 1968), 223; Casefile OG 196145, RG 65, FBI, frame 536, reel 10, FSAA (italics mine).

34. See Robert K. Murray, *Red Scare: A Study of National Hysteria, 1919–1920* (New York, 1964); William Preston, Jr., *Aliens and Dissenters:*

Federal Suppression of Radicals, 1903–1933 (New York, 1963); Nell Irvin Painter, *Standing at Armageddon: The United States, 1877–1919* (New York, 1987), esp. 388.

35. Copy of the petition in series I, frames 725–27, reel 7, CIC; undated *ABH* clipping, frame 291, reel 14, *Federal Surveillance of Afro-Americans (1917–1925): The First World War, the Red Scare, and the Garvey Movement*, ed. Theodore Kornweibel (Frederick, Maryland, 1985), (hereafter, FSAA). While the historian is hard pressed to uncover evidence of active local opposition, the Klan itself later claimed that in Athens it was "BEING FOUGHT TO THE LAST DITCH," an opposition it blamed on "Catholics and Jews in the city." *Searchlight,* 5 April 1924, p. 8; ibid., 21 June 1924, pp. 1, 8.

36. See Klan, Georgia, *Official Document,* March 1927, p. 2; J. Q. Jett to R. L. Felton, box 8, Rebecca Latimer Felton Papers, UGA (hereafter, RLF); CIC, *Annual Report,* 1922–23, series 1, frames 442 and 555, reel 4, CIC.

37. Will W. Alexander to Marion M. Jackson, 4 June 1926, series I, frames 481–82, reel 7, CIC; *ABH,* 7 Dec. 1927.

38. Telegram from Herbert Bayard Swope to David C. Barrow, 3 May 1924, box 26, David Crenshaw Barrow Papers, University Archives, Main Library, University of Georgia, and Barrow's response, ibid. (Hereafter, DCB and UA, respectively); Thomas G. Dyer, *The University of Georgia: A Bicentennial History, 1785–1985* (Athens, 1985), 180–82; University of Georgia, Prudential Committee Minutes and Miscellaneous Papers, 1922, UA (hereafter, PCM); notes on resolution, 30 Jan. 1921, box 1, PCM. For a sense of the criteria considered relevant in faculty appointments, see R. P. Brooks to C. M. Snelling, 12 April 1926, box 21, Robert Preston Brooks Papers, UGA.

39. John Wade to Mrs. J. H. Harris, 4 Aug. 1924, box 6, JLH; Minutes, 17 Oct. 1923, University of Georgia Demosthenian Society, UA.

40. For examples of Klan legislative projects, see Klan, *Official Monthly Bulletin,* 1 Dec. 1926, p. 1; Klan, Georgia, *Official Document,* Dec. 1926, p. 1; ibid., Oct.–Nov. 1927.

41. N. B. Forrest to JPM, 9 April 1926, box 1, AK; Athens Klan, Minutes, 20 April 1926, box 2, AK; JPM to N. B. Forrest, 29 Nov. 1925, box 1, AK. See also J. Mills Thornton III, "Alabama Politics, J. Thomas Heflin, and the Expulsion Movement of 1929," *Alabama Review* 22 (April 1968), 90; Weaver, "Klan in Wisconsin," 118; Blee, *Women of the Klan,* 148–49.

42. See *Athens Banner,* 8 Sept. 1922, p. 1 (hereafter, AB); N. B. Forrest to JPM, 3 Dec. 1925, box 1, AK; *Imperial Night Hawk,* 2 July 1924, p. 1; Klan, Georgia, *Official Document,* May–June, 1927, p. 2; Anon. to "Dear Sir," 14 June 1922, Pre-Gub. series, box 1, Richard B. Russell, Jr., Papers, Russell Library, University of Georgia–Athens (hereafter, RBR and RL, respectively); Monteval, *Klan Inside Out,* 103–6, 172; Lem A. Dever, *Masks*

Off! Confessions of an Imperial Klansmen, 2nd ed. (n.p., 1925), 9, 35; Dykeman and Stokely, *Seeds of Southern Change,* 109.

43. *NYT,* 18 Nov. 1923, II, 1; Moseley, "Invisible Empire," 38, 40–41, 45, 61–64; J. Q. Jett to Rebecca Latimer Felton, [1924], box 8, RLF; William F. Mugleston, "Julian Harris, the Georgia Press, and the Ku Klux Klan," *GHQ* 59 (1975), 291–92; Edward Young Clarke to "Dear Sir," [c. 1921], Pre-Gub. series, box 1, RBR; Press Release, 12 Jan. 1923, series C, box 313, NAACP.

44. *NYT,* 23 Sept. 1923, II, 1; *Atlanta Independent,* 21 Dec. 1922, 1.; *NYT,* 10 Dec. 1922, in series C, box 315, NAACP; *NYT,* 1 Nov. 1923, 1; Klan, *Official Monthly Bulletin,* 1 Dec. 1926; Goldberg, *Hooded Empire,* xi.

45. N. B. Forrest to JPM, 9 April 1926, box 1, AK; Moseley, "Invisible Empire," 71–72; *New York Herald Tribune,* 28 April 1924, in series C, box 316, NAACP; *NYT,* 21 March 1924, p. 3; *New York World,* Oct. [n.d.] 1924, in series C, box 317, NAACP; *Imperial Night Hawk,* 2 July 1929, p. 1.

46. Moseley, "Invisible Empire," 42–43; *NYT,* 14 Sept. 1922, in series C, box 314, NAACP; *NYT,* 13 Nov. 1922, 16. For one Georgia newspaper editor's subsequent counsel to Hoke Smith to avoid this "hornet's nest" during the gubernatorial race of 1926, advice Smith took, see Ben A. Neal to Hoke Smith, 26 Jan. 1926, box 9, Hoke Smith Papers, RL (hereafter, HS); *NYT,* 18 Nov. 1923, II, 1. See also *NYT,* 7 July 1924, in series C, box 316, NAACP. In Alabama, according to one non-member politician, the entire "ruling class" in state politics belonged to the Klan. *Diary,* 1928–1930, pp. 2, 72, box 12, Hugh Gladney Grant Papers, DU (hereafter HGG); see also Jackson, *Klan in the City,* 38.

47. Quoted in Clifford M. Kuhn, Harlon E. Joye, and E. Bernard West, *Living Atlanta: An Oral History of the City, 1914–1948* (Athens, 1990), 313–14.

48. *ABH,* p. 4 July 1926, p. 4; Moseley, "Invisible Empire," 82–83; Mugleston, "Julian Harris," 284–90; Robert Preston Brooks, "What Is Worth Fighting for in American Life?" typescript of speech, box 20, Brooks Papers; Mrs. James Cornelison interview, transcript, p. 6, ACHF.

49. *ABH,* 30 July 1926, p. 8; J. R. Hamilton, Address to the Grand Jury, 3 Oct. 1921, James Robert Hamilton Papers, SHC (hereafter, JRH).

50. Loucks, *Klan in Pennsylvania,* 62–69; Jackson, *Klan in the City,* 7; Klan, Georgia, *Official Document,* March 1927, p. 3; ibid., April 1927, 1.

51. Klan, Georgia, *Official Document,* Nov. 1926, pp. 2, 4; Evans, *Klan of Tomorrow,* 25; Klan, *Klan Building,* 7, 8; also Loucks, *Klan in Pennsylvania,* 62–69.

52. Knights of the Ku Klux Klan, *Thirty Three Questions Answered* (Atlanta, 1924), 16; Klan, Georgia, *Official Document,* Oct. 1926, p. 1.

53. Klan, Georgia, *Official Document,* March 1927, p. 4; ibid., Jan.–Mar. 1928, p. 4; R. A. Patton, "A Klan Reign of Terror," *Current His-*

tory 28 (April 1928), 54; "Confession of Luke Trimble," p. 4, box 60, Ralph Emerson McGill Papers, EU (hereafter, REM); Robert L. Duffus, "The Ku Klux Klan in the Middle West," *World's Work* 46 (1923), 367–68; Donald W. Stewart to Gov. H. J. Allen, 7 July 1922, series B, box 48, Henry J. Allen Papers, Library of Congress, Washington, D.C. (hereafter, HJA); *NYT,* 14 Sept. 1924, section VIII, 12.

54. Howard Washington Odum, *An American Epoch: Southern Portraiture in the National Picture* (New York, 1930), 321; John A. Cobb to Mrs. M. A. L. Erwin, 16 July 1924, box 89, Howell Cobb Papers, UA.

55. "Application for Citizenship" forms, box 2, AK; Ku Klux Klan, *Kloran,* 5th ed. (Atlanta, 1916), 28, 40; Klan, Georgia, *Official Document,* Nov. 1926, p. 4; Klan, *Klan Building,* 10; *Atlanta Independent,* Nov. 15, 1923; J. Q. Jett to R. L. Felton, [1924], box 8, RLF. See also *NYT,* 2 Sept. 1924, p. 21; *NYT,* 11 April 1925, p. 9.

56. Alfred Vagts, *A History of Militarism, Civilian and Military* (New York, 1959), quote on 13–14. For the growth of militarism among fraternalists by the late nineteenth century, see Mary Ann Clawson, *Constructing Brotherhood: Class, Gender, and Fraternalism* (Princeton, 1989), 233–38.

57. See the "Application for Citizenship" forms in box 2, AK; Klan, *Klan Building,* 4, 13, 14. The Klan also idolized Theodore Roosevelt, the foremost civilian militarist in the nation's history. See *Kourier,* Dec. 1927, p. 15; Evans, *Klan of Tomorrow,* 14. For martial men, see Robert L. Duffus, "Ancestry and End of the Ku Klux Klan," *World's Work* 46 (1923), 534; Chester T. Crowell, "The Collapse of Constitutional Government," *Independent* 109 (9 Dec. 1922), 334; *NYT,* 18 March 1928, p. 2; *NYT,* 30 March 1928, p. 7; Samuel Streetsman to Gov. Henry J. Allen, 1 Nov. 1922, series B, box 49, HJA; Lipset and Raab, *Politics of Unreason,* 124.

Chapter 2

1. Quoted in N. Gordon Levin, Jr., *Woodrow Wilson and World Politics: America's Response to War and Revolution* (New York, 1968), 132.

2. For recognition by contemporaries of the transformation in Southern life, see Robert Preston Brooks, *Georgia Studies: Selected Writings* (Athens, 1952), esp. 224; William Joseph Robertson, *The Changing South* (New York, 1927). More generally, see, on developments in the South, Gavin Wright, *Old South, New South: Revolutions in the Southern Economy Since the Civil War* (New York, 1986), 52, 62; Daniel Joseph Singal, *The War Within: From Victorian to Modernist Thought in the South, 1919–1945* (Chapel Hill, 1982); and in the nation at large, Robert H. Wiebe, *The Search for Order, 1870–1920* (New York, 1967); Walter Dean Burnham, *The Current Crisis in American Politics* (New York, 1982), esp. 138; Louis Galambos and Joseph Pratt, *Rise of the Corporate Common-*

wealth: U.S. Business and Public Policy in the 20th Century (New York, 1988), esp. 39–99; Stephen Skowronek, *Building a New American State: The Expansion of National Administrative Capacities, 1877–1920* (Cambridge, Mass., 1982), esp. 163–284.

3. Frank Jackson Huffman, "Old South, New South: Continuity and Change in a Georgia County, 1850–1880" (Ph.D. diss., Yale University, 1974), 138–39; Barrow et al., *History of Athens,* 49, 99.

4. Quote from Thomas Jackson Woofter, Jr., "The Negroes of Athens, Georgia," *BUGA,* 14, n. 4 (Dec. 1913), 13. See also Institute for the Study of Georgia Problems, "Survey of Athens and Clarke County, Georgia; Part I," *BUGA,* v. 44, n. 3 (March 1944), 1–3; Earnest B. Wilson, "The Water Supply of the Negro," *BUGA,* v. 31, n. 3a (March 1931), 34–35.

5. "Oldest Barber in Town," interview by Grace McCune, 28 March 1939, transcript, p. 5, box 13, Federal Writers' Project Papers, SHC (hereafter, FWP); Jack Thomas interview, transcript, p. 1, box 2, ACHF. For a sense of Athens in 1920, see American Red Cross, *Report of Survey: Athens and Clarke County* (Athens, [1920]).

6. *ADB,* 19 Feb. 1921, in series C, box 355, NAACP; *ADB,* 8 Sept. 1921, p. 4.

7. Jurgen Kuczynski, "The Merchant and His Wage Earning Customers in the South and in the North," *American Federationist,* 35 (Nov. 1928), 1,340; Georgia, Commissioner of Commerce and Labor, *Annual Report* (1920), 3, 10 (hereafter, CCL); ibid. (1917), 4; ibid. (1918), 4; ibid. (1922), 5; Taylor, "Negroes of Clarke County," 36–37; Robert Preston Brooks, "A Local Study of the Race Problem," *Political Science Quarterly* 26 (June 1911), 218–19.

8. On these efforts, see Jacquelyn Dowd Hall, Robert Korstad, and James Leloudis, "Cotton Mill People: Work, Community, and Protest in the Textile South, 1880–1940," *AHR* 91 (1986), 266; Herbert Jay Layne, *The Cotton Mill Worker in the Twentieth Century* (New York, 1944), 203–6; Bernard Yabroff and Ann J. Herlihy, "History of Work Stoppages in Textile Industries," *Monthly Labor Review* 76 (April 1953), 369–70; Mercer Griffin Evans, "The History of the Organized Labor Movement in Georgia" (Ph.D. diss., University of Chicago, 1929), esp. 32, 90–95, 348–49; George L. Googe, "Organizing the Workers of Georgia," *American Federationist* 35 (Nov. 1928), 1,326.

9. *ADH,* 22 Jan. 1914, p. 1; *ADH,* 24 March 1914, p. 1; *ADH,* 2 Oct. 1916, p. 1; *ADH,* 8 Jan. 1914, p. 4; *ADB,* 28 April 1920; "Life of a Retired Mill Worker," interview by Sadie B. Hornsby, 10 Jan. 1939, transcript, pp. 13, 8, box 11, FWP; *ADH,* 25 Jan. 1916, p. 4; *ADB,* 21 June 1921, p. 1.

10. Quoted in Brooks, "Local Study of the Race Problem," 212–13, 218–19; Willard Range, *A Century of Georgia Agriculture, 1850–1950* (Athens, 1954), 259–61; H. S. O'Kelley, "Sanitary Conditions Among the Negroes of Athens, Georgia," *BUGA,* 18, n. 7 (July 1918), 23; Jacqueline

Jones, *Labor of Love, Labor of Sorrow: Black Women, Work, and the Family, from Slavery to the Present* (New York, 1986), 130–31, 133; Robertson, *Changing South*, 130.

11. T. W. Singleton to Joseph Mackey Brown, 14 June 1913, box 6, Joseph Mackey Brown Papers, Atlanta Historical Society (hereafter, JMB and AHS, respectively).

12. Wilson Jefferson to Mr. James Weldon Johnson, 3 May 1917, series G, box 45, NAACP. See also Thomas Jackson Woofter, *Negro Migration: Changes in Rural Organization and Population of the Cotton Belt* (New York, 1920), 15, 139.

13. Quoted in Taylor, "Negroes of Clarke County," 39, 40–41, also 35–36, 43, 51–54. See also Range, *A Century of Georgia Agriculture*, 266–67; Tindall, *Emergence of the New South*, 61.

14. Case file 10218–117, RG 165, frames 545–49, reel 19, FSAA. See also *Atlanta Independent*, 28 Aug. 1918, frame 704, reel 221, Tuskegee Institute News Clipping File, Hollis Burke Frissell Library, Tuskegee Institute, Alabama (hereafter, TINCF).

15. For exceptions to this neglect, see Raymond Wolters, *The New Negro on Campus: Black College Rebellions of the 1920s* (Princeton, 1975); Neil R. McMillen, *Dark Journey: Black Mississippians in the Age of Jim Crow* (Urbana, Ill., 1989), 302–17.

16. Wilson Jefferson to Mr. James Weldon Johnson, 3 May 1917, series G, box 45, NAACP. Tindall, *Emergence of the New South*, 159. See also the branch files of the NAACP (series G).

17. Application for Charter, Athens Branch, NAACP, 7 Feb. 1917, series G, box 43, NAACP; [Walter F.] White to John R. Shillady, 9 July 1918, series G, box 49, NAACP; G. C. Callaway to Roy Nash, 3 June 1917, series G, box 43, NAACP.

18. Branch report, Atlanta NAACP, 3 April 1919, series G, box 49, NAACP; ibid, 20 June 1921, loc. cit. Synopsis of Work of NAACP, Atlanta Branch, 1919, loc. cit.; T. K. Gibson to James Weldon Johnson, 7 March 1919, box 43, loc. cit. See also Atlanta Branch circular, 12 Feb. 1919, loc. cit.; Georgia, *House Journal* (1920), 452.

19. Dykeman and Stokely, *Seeds of Southern Change*, 55; John Michael Matthews, "Studies in Race Relations in Georgia, 1890–1930" (Ph.D. diss., Duke University, 1970), 157; John William Fanning, "Negro Migration: A Study of the Exodus of the Negroes between 1920 and 1925 from Middle Georgia Counties," *BUGA* 30, no. 8b (June 1930), 32–33.

20. Case file 10218–319, RG 165, NARC, frame 254, reel 21, FSAA; Case file 10218-289, RG 165, frames 175–76, loc. cit.; Case file 10218-311, RG 165, frame 241, loc. cit.

21. Taylor, "Negroes of Athens," 54; Case file BS 203677, RG 65, frames 872-916, reel 7, FSAA.

22. Typescript history, "Origin and Purpose," [n.d.], series I, frames 908–9, reel 4, CIC; also *Progress in Race Relations: Survey of the Work of*

the CIC, 1924–1925, frame 516, loc. cit.; Alexander quoted in Dykeman and Stokely, *Seeds of Southern Change,* 80, also 11. For overviews of the CIC, see ibid., 52–96; Morton Sosna, *In Search of the Silent South: Southern Liberals and the Race Issue* (New York, 1977), 20–41.

23. CIC, "Annual Report, 1922," series I, frame 418, reel 4, CIC; Kuhn et al., *Living Atlanta,* 165–66; Prudential Committee, Minutes, 14 Feb. 1927, PCM; Dyer, *University of Georgia,* 178–82.

24. Quoted in Dykeman and Stokely, *Seeds of Southern Change,* 95; also 81–96. See also Kuhn et al., *Living Atlanta,* 249; Jacquelyn Dowd Hall, *Revolt Against Chivalry: Jessie Daniel Ames and the Women's Campaign Against Lynching* (New York, 1979).

25. Annie Laura Ragsdale, "The History of Co-Education at the University of Georgia, 1918–1945" (M.A. thesis, University of Georgia, 1948) [no page numbers in text]; Dyer, *University of Georgia,* 170–73; *ADH,* 17 Jan. 1914, p. 2; *ADH,* 22 April 1914, p. 8; *ADH,* 29 April 1914, p. 2. More generally, see A. Elizabeth Taylor, "Revival and Development of the Woman Suffrage Movement in Georgia," *GHQ* 42 (1958), esp. 345–46; Ann Firor Scott, *The Southern Lady: From Pedestal to Politics* (Chicago, 1970).

26. *ADH,* 8 July 1914, p. 4; A. Elizabeth Taylor, "The Last Phase of the Woman Suffrage Movement in Georgia," *GHQ* 43 (March 1959), 16, 18; *ADH,* 12 June 1916, p. 4.

27. Ragsdale, "Co-Education"; Dyer, *University of Georgia,* 172–73. For like sentiments now common elsewhere in the South, see Robertson, *Changing South,* 128–29; nationally, see Nancy Cott, *The Grounding of Modern Feminism* (New Haven, 1987), 11–50.

28. "Mildred Lawson," [pseud.], interview by S. B. Hornsby, transcript, p. 7, box 12, FWP; Martha Comer, interview, transcript, box 1, ACHF, 3; also "I Maids for the Co-Eds," interview by S. B. Hornsby, 14–16 March 1939, transcript, p. 7, box 11, FWP.

29. *ADH,* 5 June 1921, II, p. 4; *CES,* 2 July 1920, p. 4; *ADB,* 23 July 1921, p. 4; *ADB,* 13 Aug. 1921, p. 5. For analogous fears elsewhere, see Carroll Smith-Rosenberg, "The New Woman as Androgyne," in *Disorderly Conduct: Visions of Gender in Victorian America* (New York, 1985), 250–52.

30. *ADB,* 25 Aug. 1921, p. 5; *ABH,* 7 Feb. 1930, 1; also *ADB,* 25 Feb. 1920, 4.

31. "I Maids for the Co-Eds," transcript, p. 11, FWP. See also "Mildred Lawson," [pseud.], transcript, loc. cit.; "Julie Nickerson," [pseud.], interview by L. T. Bradley, transcript, p. 9, box 11, loc. cit.

32. Elaine Tyler May, *Great Expectations: Marriage and Divorce in Post-Victorian America* (Chicago, 1980), 2; Department of Commerce, Bureau of the Census, *Marriage and Divorce, 1916* (Washington, D.C., 1919), 11, 36, 40; idem, *Marriage and Divorce, 1922* (Washington, D.C., 1925), 75.

33. *ADH,* 28 March 1914, p. 4; *ABH,* 1 April 1927, p. 4; *ABH,* 10 July 1923, p. 4; *ADB,* 20 July 1919, p. 3; *ADB,* 13 April 1920, p. 1.

34. For the related shift from Victorianism to modernism in Southern culture, see Singal, *War Within,* esp. 9, 83, 111, 318. For gender as a sign in a wider discourse of ambivalence about modernity, see Cott, *Grounding of Modern Feminism,* 271.

35. Barrow, *History of Athens,* 98; Coleman, *History of Georgia,* 235, 269; Howard A. Turner and L. D. Howell, "Condition of Farmers in a White-Farmer Area of the Cotton Piedmont, 1924–1926," U.S. Department of Agriculture, Circular No. 78 (Washington, D.C., 1929), 20–21; U.S. Department of Agriculture, brochure, "Clarke County Farm Statistics, 1900–1955," p. 2, s.v. "Clarke County," File II, Record Group 4, GDAH; Range, *Century of Georgia Agriculture,* 267.

36. Allison Davis, Burleigh B. Gardner, and Mary R. Gardner, *Deep South: A Social Anthropological Study of Caste and Class* (Chicago, 1941), 73–75, 84–89, 91, 327–28, 409–10, 413; Walter B. Hill, "Rural Survey of Clarke County, Georgia, with Special Reference to the Negroes," *BUGA,* 15, n. 3 (March 1915), 23; Huffman, "Old South," 221; Jones, *Labor of Love,* 90–91; Turner and Howell, "Condition of Farmers," 16, 26, 37; Wright, *Old South, New South,* 55; Joseph and James Costa interview, transcript, pp. 1–2, box 1, ACHF.

37. Quote from "Life on 'Happy Top,'" interview by S. B. Hornsby, transcript, p. 5, box 11, FWP. See also Layne, *Cotton Mill Worker,* 65, 129; Lois MacDonald, *Southern Mill Hills: A Study of Social and Economic Forces in Certain Mill Villages* (New York, 1928), 27; Douglas DeNatale, "Traditional Culture and Community in a Piedmont Textile Mill Village" (M.A. thesis, University of North Carolina, Chapel Hill, 1980), 19, 27.

38. D. W. Dyal to Thomas E. Watson, 6 June 1922, box 14, TEW, SHC. See also "Life of a Retired Mill Worker," transcript, pp. 14, 15, 19; John Morland, *Millways of Kent* (Chapel Hill, 1958), 42, 67–68; "Life History of Mrs. Ann Waldrop," [pseud.], interview by S. B. Hornsby, 3 Jan. 1939, transcript, p. 15, box 11, FWP.

39. Testimony of Walter Francis White, U.S. Congress, Senate, Committee of the Judiciary, *Punishment for the Crime of Lynching,* 73rd Congress, 2nd session, 20–21 Feb. 1934, p. 10; Walter White, *Rope and Faggot: A Biography of Judge Lynch* (New York, 1929), 232, 234, 255; CIC, *Annual Report* (1922), series I, frame 421, reel 4, CIC.

40. Arthur F. Raper, *Preface to Peasantry: A Tale of Two Black Belt Counties* (Chapel Hill, 1936), 172; the Rev. R. D. Ponder to Editor, *The Crisis,* 31 May 1920, series G, box 43, NAACP; see also S. S. Humbert to Mr. J. W. Johnson, 10 Aug. 1922, loc. cit.

41. *ABH,* 11 Feb. 1923, p. 1; *ABH,* 11 June 1922, p. 4; *ABH,* 15 April 1927, p. 1; *NYT,* 22 Nov. 1920, p. 8.

42. Quoted in Dale Newman, "Work and Community Life in a Southern Town," *LH* 19 (Spring 1978), 212. See also Marjorie A. Potwin, *Cotton*

Mill People of the Piedmont: A Study in Social Change (New York, 1927),
102–4, 108; MacDonald, *Southern Mill Hills*, 31, 44–45; Hall et al., "Cotton Mill People," 250.

43. For examples, see David Montgomery, "Violence and the Struggle
for Unions in the South, 1880–1930," *Perspectives on the American South*
1 (1981), 35–47; Tom Tippett, *When Southern Labor Stirs* (New York,
1931), 64–65, 87, 106; Irving Bernstein, *The Lean Years* (Boston, 1960), 17,
23, 41; Evans, "Organized Labor," 229.

44. See *ADB*, 11 May 1922, p. 5; *ADH*, 4 Aug. 1914, p. 8; *ABH*, 16
May 1927, p. 1; *ABH*, 11 July 1926, p. 10; Morland, *Millways of Kent*,
90–91; J. Wayne Flynt, "Folks Like Us: The Southern Poor White Family,"
in *The Web of Southern Social Relations*, ed. Walter J. Fraser, Jr., et. al.
(Athens, 1985), 229–30; Kuhn et al., *Living Atlanta*, 131, 136–37. More
generally, see Linda Gordon, *Heroes of Their Own Lives: The Politics and
History of Family Violence; Boston, 1880–1960* (New York, 1988), 250–88.

45. For class structure and power in the South in these years, see
Wright, *Old South, New South*; David L. Carlton, *Mill and Town in South
Carolina, 1880–1920* (Baton Rouge, 1982), 118–19; Paul D. Escott, *Many
Excellent People: Power and Privilege in North Carolina, 1850–1900*
(Chapel Hill, 1985); C. Vann Woodward, *Origins of the New South,
1877–1913* (Baton Rouge, 1951). For Georgia see Numan V. Bartley, *The
Creation of Modern Georgia* (Athens, 1983), 103–46.

46. See Robert Preston Brooks, *The Agrarian Revolution in Georgia,
1865–1912* (Madison, 1914), 52–53, 63–68; Harold D. Woodman, "Post-
Civil War Southern Agriculture and the Law," *Agricultural History* 53
(Jan. 1979), 319–37; idem, "Postbellum Social Change and Its Effects on
Marketing the South's Cotton Crop," *Agricultural History* 56 (Jan. 1982),
215–30.

47. United States Department of the Interior, Census Office, *Tenth
Census of the United States* (Washington, D.C., 1883), vol. 3, *Agriculture*,
252; ibid., *Fourteenth Census of the United States* (1920), pt. 2, vol. 6,
p. 334. Hereafter cited as *Census*, with decennial year. On commercial
production and the yeomanry's demise, see Hahn, *Roots of Southern Populism*; Coleman, *History of Georgia*, 266.

48. Wright, *Old South, New South*, 111–12; Range, *Century of Georgia Agriculture*, 265; Coleman, *History of Georgia*, 259; *Census* (1920), pt.
2, vol. 6, *Agriculture*, 297–98, 306–7.

49. Hall et al., "Cotton Mill People," 245; *Census* (1910), *Supplement
for Georgia*, 592, 608; *Census* (1920), vol. 3, *Population*, 206, 222; ibid.
(1890), 758; *Census* (1880), *Manufactures*, 104, 114; ibid. (1919), vol. 9, p.
265; Barrow, *History of Athens*, 98; *ADB*, 3 March 1922, 3. The dominance of town life in Clarke County differentiated it from the South as a
whole where three in every four people remained in the countryside, although only one in two in farming. See I. A. Newby, *The South : A History*
(n.p., 1978), 294, 382.

50. Morland, *Millways of Kent*, 23–24; MacDonald, *Southern Mill Hills*, 49, 52; CCL, *Annual Report* (1921), 4; *Census* (1920), vol. 9, *Manufactures*, 270; "Life of a Retired Mill Worker," transcript, p. 16.

51. Ralph C. Patrick, Jr., "A Cultural Approach to Social Stratification" (Ph.D. diss., Harvard University, 1953), 297, 300–304; Morland, *Millways of Kent*, passim; Bess Beatty, "Textile Labor in the North Carolina Piedmont: Mill Owner Images and Mill Worker Response," *LH* 25 (Fall 1984), 496; Kuhn et al., *Living Atlanta*, 35.

52. Jack Thomas, interview, transcript, p. 1, ACHF; Morland, *Millways of Kent*, 31, 35, 39, 43–44, 202, 215; Jennings J. Rhyne, *Some Southern Cotton Mill Workers and Their Villages* (Chapel Hill, 1930), 206–7, 34; MacDonald, *Southern Mill Hills*, 75–76, 113–115; Harriet L. Herring, *Passing of the Mill Village* (Chapel Hill, 1949), 57; Carlton, *Mill and Town*, 165, 170. More generally, see Jacquelyn Dowd Hall et al., *Like a Family: The Making of a Southern Cotton Mill World* (Chapel Hill, 1987), esp. chapters 3–6.

53. Carlton, *Mill and Town*, 157; Morland, *Millways of Kent*, ix, 80, 183; Bernstein, *Lean Years*, 8; Dale Newman, "Work and Community Life in a Southern Town," *LH* 19 (Spring 1978), 207; MacDonald, *Southern Mill Hills*, 83, 99, 103, 132–134, 150; Patrick, "Cultural Analysis," 296–97; Ina Cooper to Aaron Cohen, 24 March 1917, box 2, CDF.

54. Bernstein, *Lean Years*, 9; MacDonald, *Southern Mill Hills*, 28, 56, 96, 128–29; Harriet L. Herring, *Welfare Work in Mill Villages* (Chapel Hill, 1929), 28–29; Layne, *Cotton Mill Worker*, 58–59; Ina Cooper to Aaron Cohen, 24 March 1917, box 2, CDF; Cooper to C. D. Flanigen, 3 Feb. 1917, loc. cit.; saying quoted in MacDonald, *Southern Mill Hills*, 78.

55. Woofter, *Negro Migration*, 55. More generally, see *Census* (1920), *Agriculture*, pt. 2, vol. 6, pp. 297, 306–7; Wright, *Old South, New South*, 101, 104–7, 119.

56. Davis et al., *Deep South*, 55; Dittmer, *Black Georgia in the Progressive Era*, 27; Woofter, "Negroes of Athens," 7, 8, 12; "Principle of Grammar School Thirty-Three," interview by S. B. Hornsby, 27 July 1939, transcript, pp. 4–7, 10, box 11, FWP; *ADH*, 17 Nov. 1914, 1.

57. M. Ashby Jones to "Dear Brother," 24 May 1921, box 3, Robert L. Rodgers Papers, GDAH (hereafter, RLR). On the roots and content of the CIC, see Blaine A. Brownell, *The Urban Ethos in the South, 1920–1930* (Baton Rouge, 1975), 151–53; Sosna, *In Search of the Silent South*, 26–27; Dykeman and Stokely, *Seeds of Southern Change*, 53–54, 64–65; Cynthia Neverdon-Morton, *Afro-American Women of the South and the Advancement of the Race, 1895–1925* (Knoxville, 1989).

58. Huffman, "Old South," 86; DeNatale, "Traditional Culture," 19, 27; *Census* (1920), *Manufactures*, 267; ibid., (1930), *Population*, pt. 1, p. 510; Bessie Mell Industrial Home, Minutes, 1901–1909, box 6, Nina Scudder Memorial Collection, UGA (hereafter, NSMC).

59. On inheritance and authority, see Davis, *Deep South*, 95–99; John

Bondurant, Sr., interview, transcript, p. 1, ACHF; Turner and Howell, "Condition of Farmers," 26, 30, 37–39. On the flight of rural youth, see *ADH*, 14 Jan. 1914, p. 4; Brownell, *Urban Ethos in the South*, 74–75, 79–80; Orie Latham Hatcher, *Rural Girls in the City for Work* (Richmond, 1930), 41, 83–84. On child labor, see Coleman, *History of Georgia*, 306; Bernstein, *Lean Years*, 57.

60. For "filiarchy," see Gilman M. Ostrander, *American Civilization in the First Machine Age: 1890–1940* (New York, 1970), esp. 9–13, 237–73. On challenges to Victorianism as a factor in the Klan's rise, see Coben, *Rebellion Against Victorianism*, 136–56; also Singal, *The War Within*, 318. For declining church participation among young people, see Edmund D. Brunner, *Church Life in the Rural South* (1923; reprint, New York, 1969), 75.

61. See *ADH*, 9 Nov. 1914, p. 4; Grand Jury Report, 17 Jan. 1922, p. 391, Clarke County Superior Court, Minutes (hereafter, CCSC); "Life on 'Happy Top,' " interview by S. B. Hornsby, 17 March 1939, transcript, pp. 5–6, 17, FWP. Elsewhere, see Robertson, *The Changing South*, 114, 126–27; John D'Emilio and Estelle B. Freedman, *Intimate Matters: A History of Sexuality in America* (New York, 1988), 171–235.

62. Dyer, *University of Georgia*, 178–82; William F. Mugleston, "The Press and Student Activism at the University of Georgia in the 1920s," *GHQ* 64 (Fall 1980), 214–52; William Tate, Diary, vol. I, 16 Jan. 1925, William Tate Archives, UA; also, ibid., 14 Jan. 1925; ibid., 20 March 1925.

63. Bartley, *Creation of Modern Georgia*, 75–87; Coleman, *History of Georgia*, esp. 218–22; Evans, "Organized Labor," 24–26; Jonathan Garlock, comp., *A Guide to Local Assemblies of the Knights of Labor* (Westport, 1982), 53.

64. On Georgia Populism, see C. Vann Woodward, *Tom Watson: Agrarian Rebel* (1938; reprint, New York, 1977); Hahn, *Roots of Southern Populism*; William F. Holmes, "The Southern Farmers' Alliance: the Georgia Experience," *GHQ* 72 (Winter 1988), 627–52.

65. Bartley, *Creation of Modern Georgia*, 96–102; Coleman, *History of Georgia*, 300–301; Alex Matthews Arnett, *The Populist Movement in Georgia* (New York, 1922), 185; CCSC, Registration of Charters, Book 2, 25 Jan. 1890 (pp. 1–2), and 11 Dec. 1894 (pp. 98–99), GDAH.

66. See Escott, *Many Excellent People*, esp. 241, 253–61; Woodward, *Origins of the New South*, 212; Coleman, *History of Georgia*, 248, 277; J. Morgan Kousser, *The Shaping of Southern Politics: Suffrage Restriction and the Establishment of the One-Party South, 1880–1910* (New Haven, 1974).

67. Russell Korobkin, "The Politics of Disfranchisement in Georgia," *GHQ* 74 (Spring 1990), 20–58; Matthews, "Studies in Race Relations," 116–17; Thomas E. Watson, *Socialists and Socialism* (Thomson, Ga., 1910). On Populism as a movement representing the interests of middle-range farmers and cool towards large planters and propertyless rural people

alike, see Holmes, "Southern Farmers' Alliance," 646–51; David Montgomery, "On Goodwyn's Populists," *Marxist Perspectives* 1 (Spring 1978), 166–73; James Green, "Populism, Socialism, and the Promise of Democracy," *Radical History Review* 24 (1980), 7–40.

68. Institute for the Study of Georgia Problems, "Survey of Athens and Clarke County," 4–8. On the commercial-civic élite and municipal reform, see Brownell, *Urban Ethos in the South*, 47, 157–58, 172; Kuhn et al., *Living Atlanta*, 89, 311; Coleman, *History of Georgia*, 303–4, 308.

69. Ina Cooper to Aaron Cohen, 24 March 1917, box 2, CDF; Brownell, *Urban Ethos in the South*, 54; Dykeman, *Seeds of Southern Change*, 101; Escott, *Many Excellent People*, 260–67; Wright, *Old South, New South*, 123, 176–77, 259.

70. "I've Been Drifting," interview by S. B. Hornsby, 21 July 1939, transcript, p. 14, box 12, FWP; Ralph Wardlow, "Negro Suffrage in Georgia, 1867–1930," *BUGA*, vol. 33, n. 2a (Sept. 1932), 78–79.

71. See Samuel P. Hays, "The Politics of Reform in Municipal Government in the Progressive Era," in *American Political History as Social Analysis* (Knoxville, 1980); James Weinstein, *The Corporate Ideal in the Liberal State, 1900–1918* (Boston, 1968), 92–116; Tindall, *Emergence of the New South*, 143, 145; Skowronek, *Building a New American State*, esp. 165, 211, 286–87; Burnham, *Current Crisis in American Politics*, 110, also 86–87, 97, 109.

72. Tindall, *Emergence of the New South*, 111–13; Harry Hodgson to E. C. Branson, 28 October 1922, box 6, Eugene Cunningham Branson Papers, SHC (hereafter, ECB); Fanning, "Negro Migration," 17–21; USDA, "Clarke County Farm Statistics," box 9, File II, GDAH.

73. H. G. Smith to Hoke Smith, 28 Dec. 1920, box 5, HS; Harry Hodgson to E. C. Branson, 28 Oct. 1922, box 6, ECB; T. E. Fillyow to TEW, 8 August 1921, box 12, TEW; T. Z. Daniel to TEW, 23 June 1921, loc. cit.; see also F. M. Hughes to TEW, loc. cit.

74. *ADB*, 25 Aug. 1921, p. 6; MacDonald, *Southern Mill Hills*, 83–84, 104; Bernstein, *Lean Years*, 3, 9; "Life on "Happy Top,'" transcript, p. 17.

75. Georgia, *Senate Journal* (1922), 49; R. L. Woodruff to Hoke Smith, 11 December 1920, box 5, HS; J. J. Brown quoted in Fite, *American Farmers*, 34–35; W. F. Bowe, Jr., to TEW, 24 Dec. 1921, box 13, TEW; *ADB*, 11 July 1922, p. 1. On the growth of vagrancy, see E. I. Price to Superintendent M. L. Brittain, 8 Feb. 1921, box 2, CDF; Brownell, *Urban Ethos in the South*, 107; Kuhn et al., *Living Atlanta*, 89.

76. Yabroff and Herlihy, "Work Stoppages," 368; David Montgomery, "Struggle for Unions," 35–47; Layne, *Cotton Mill Worker*, 206; nationally, see Bernstein, *Lean Years*, 144–49, 200; Robert W. Dunn, *The Americanization of Labor: The Employers' Offensive Against the Trade Unions* (New York, 1927), esp. 21–33.

77. Evans, "Organized Labor," 32; Coleman, *History of Georgia*, 272; Bernstein, *Lean Years*, 63–64, 84, 89, 149, 242; Murray, *Red Scare*,

263–81; Frank Stricker, "Affluence for Whom? Another Look at Prosperity and the Working Classes in the 1920s," *LH* 24 (1983), 5–33.

78. *NYT*, 11 Oct. 1920, p. 20; J. S. Hale to Hoke Smith, 7 Dec. 1920, box 5, HS; J. L. Cartledge to TEW, 28 Dec. 1921, TEW; Helen Dortch Longstreet to Rebecca Latimer Felton, 20 Oct. 1923, box 7, RLF. See also Turner and Howell, "Condition of Farmers," 34–35; Range, *Century of Georgia Agriculture*, 251; Fite, *American Farmers*, 34.

79. Thomas J. Shackelford to Hoke Smith, 29 Dec. 1920, box 5, HS; the Rev. R. M. Walker to TEW, 14 Jan. 1922, box 13, TEW; Ernest W. Fanning to TEW, box 12, loc. cit. See also L. R. Driskell to TEW, 6 Sept. 1922, box 15, loc. cit.; MacDonald, *Southern Mill Hills*, 71, 144.

80. August Andrae [illeg.] to TEW, 2 June 1921, box 12, TEW, SHC; Thomas W. Hardwick to Rebecca Latimer Felton, 16 Oct. 1918, box 5, RLF; Thomas W. Hardwick to RLF, 12 Jan. 1918, loc. cit. See also TEW to Robert L. Rodgers, 19 Aug. [n.d.], box 5, RLR; Dr. Grace Kirkland to Rebecca Latimer Felton, 11 Oct. 1920, box 6, RLF.

81. TEW to Rebecca Latimer Felton, 4 June 1921, box 8, RLF, UGA; Helen Dortch Longstreet to Hoke Smith, 3 May 1923, box 17, HS.

82. For the election results, see Edward L. Cashin, "Thomas E. Watson and the Catholic Layman's Association of Georgia" (Ph.D. diss., Fordham University, 1962), 235. For Watson's causes, see Woodward, *Tom Watson*, 473–86.

83. R. M. Johnson to TEW, 8 Dec. 1921, box 13, TEW; two Valdosta telegrams to TEW dated 27 Jan. 1922, loc. cit.; J. S. Dean to TEW, 7 Feb. 1922, loc. cit. Citations in the text are limited to Georgia, but Watson's papers from these years contain communications of support from all over the nation.

84. J. S. Hale to Hoke Smith, 7 Dec. 1920, box 5, HS. "Shall we be yoked into slavery to the moneypower?" queried another Georgian. August Andrae [illeg.] to TEW, 2 June 1921, box 12, TEW.

85. The best treatment of Watson's evolution can be found in Woodward, *Tom Watson*. See also Barton C. Shaw, *The Wool-Hat Boys : Georgia's Populist Party* (Baton Rouge, 1984), esp. 46, 101, 149–153. For a similar trajectory, see "Sketch of Robert L. Rodgers," box 5, RLR. This is not to imply that there was a linear movement from Populism to the Klan; on the contrary, Georgia Populists no doubt divided after the 1890s, as did their counterparts elsewhere. At least one former state Populist leader, Judge James K. Hines, became an outspoken opponent of the Klan. See *ABH*, 5 Nov. 1928, p. 5.

86. For his own denial coupled with endorsement, see Watson to F. J. Barbee, 26 May 1921, box 11, TEW. See also Z. Tiffy to TEW, 9 Sept. 1921, box 12, loc. cit., in which the writer informed Watson of a rumor that he was "a chartered member of the Klan and empowered to solicit membership of the U.S. Senators." The rumor is plausible, as the Klan reportedly had a secret section for prominent politicians.

87. See Woodward, *Tom Watson*, 463–86; Benjamin M. Blackman to TEW, 19 Dec. 1921, box 13, TEW; Carl Hutchinson to TEW, 12 July 1921, box 12, TEW; Rice, *Klan in American Politics*, 59–60; Sam H. Campbell to TEW, 16 Sept. 1922, box 15, TEW; J. J. Brown to TEW, 21 Jan. 1921, box 11, TEW; W. A. Sims to Watson, 24 Dec. 1921, box 13, TEW; Cashin, "Thomas E. Watson," passim; Sam H. Campbell to TEW, 16 Sept. 1922, box 15, TEW.

88. Caleb A. Ridley in *Searchlight*, 7 Oct. 1922, 4; ibid., 21 July 1923, 7; Jefferson Davis Klan No. 3 to TEW, 7 Oct. 1921, box 13, TEW; telegram from Clarke to Mrs. TEW, 27 Sept. 1922, box 15, TEW; A. D. Kean to Sen. TEW, 23 March 1922, box 14, TEW; A. S. Whitfield to TEW, 9 Sept. 1922, box 15, TEW, 88.

89. For Watson's critics in the Frank case, see Dinnerstein, *Leo Frank Case*, passim; for the transition to anti-Klan efforts, see Shepherd, "Fighting the K.K.K. on Its Home Grounds," 508–11, 526; Robert L. Duffus, "Counter-Mining the Ku Klux Klan," *World's Work*, 46 (July 1923), 275–84. On Athens, see Paul Hunnicutt to Hoke Smith, 12 Aug. 1920, box 32, HS; Moseley, "Invisible Empire," 71–72; *ADB*, July 20, 1920, 4; telegram to W. T. Anderson from J. H. Beusse et al., 5 Aug. 1920, box 32, HS.

90. Thomas Walter Reed, *David Crenshaw Barrow* (Athens, 1935), 71, 183–84; article reprinted in *ABH*, 21 Feb. 1926, 10. For concern over the danger to outside investment, see form letter from M. Ashby Jones, for CIC, May 24, 1921, box 3, RLR; W. M. Bryant to Hoke Smith, Feb. 4, 1926, box 9, HS; Howell C. Erwin to Smith, Dec. 23, 1925, loc. cit.

91. Charles Tilly, "Collective Violence in European Perspective," in Graham and Gurr, *History of Violence in America*, 41. See also E. P. Thompson, "The Moral Economy of the English Crowd in the Eighteenth Century," *Past & Present* 50 (Feb. 1971), 76–136; George Rude, *The Crowd in History: A Study of Popular Disturbances in France and England, 1730–1848*, rev. ed. (London, 1981). For an American example of this approach, see Robert P. Ingalls, "Lynching and Establishment Violence in Tampa, 1858–1935," *JSH* 53 (Nov. 1987), 613–44.

92. On the Regulators, see Rachel N. Klein, "Ordering the Backcountry: The South Carolina Regulation," *William and Mary Quarterly*, 3rd series, 38 (Oct. 1981), 661–80. On White Caps, see Ayers, *Vengeance and Justice*, 255–61; Richard Maxwell Brown, "Historical Patterns of Violence in America," in Graham and Gurr, *History of Violence in America*, 70–71; James O. Nall, *The Tobacco Night-Riders of Kentucky and Tennessee, 1905–1909* (Louisville, 1939); E. W. Crozier, *The White-Caps: A History of the Organization in Sevier County* (1899; reprint; n.p., 1963); Paul J. Vanderwood, *Nightriders of Reelfoot Lake* (Memphis, 1969); William F. Holmes, "Whitecapping in Georgia: Carroll and Houston Counties, 1893," *GHQ* 64 (1980), 388–404; idem, "Moonshining and Collective Violence: Georgia, 1889–1895," *JAH* 67 (1980), 589–611.

Chapter 3

1. "The Red, White, and Blue Barber Shop," interview by S. B. Hornsby, 30 March 1939, transcript, p. 14, box 12, FWP; also "Life on 'Happy Top,' " transcript, passim; "A Visit with Aunt Jerry," interview by Grace McCune, 13 Jan. 1939, transcript, pp. 8–10, FWP.

2. "The Red, White, and Blue Barber Shop," transcript, p. 17, passim; also "Life on 'Happy Top,' " transcript, p. 16.

3. Frost, *Challenge of the Klan*, 6–7. One 1930 study reported that the highest stratum of mill workers was the homeowners, who suffered little from the social segregation or stigma that their non-owning co-workers did. Rhyne, *Southern Cotton Mill*, 49. The figure in the text is an adjusted percentage, reflecting the 327 members for whom this data was available. Where members were dependents in the period under consideration, the variable reflects their parents' home ownership.

4. For similar patterns elsewhere, see Jackson, *Klan in the City*, 64, 150; Alexander, *Crusade for Conformity*, 28–30; Cocoltchos, "Invisible Government," 138, 163; Goldberg, *Hooded Empire*, 47, 168; Gerlach, *Blazing Crosses*, 45; Lipset and Raab, *Politics of Unreason*, 123–24; Toy, "Klan in Oregon," 134–36.

5. See Appendix for explanation of how the data were compiled and why they probably understate rural participation.

6. See *Census* (1920), *Agriculture*, pt. 2, vol. 6, pp. 306–7; *Census of Agriculture: 1925*, pt. II, *The Southern States*, 407. On the old and new middle class, see C. Wright Mills, *White Collar: The American Middle Classes* (1951; reprint, New York, 1976); Jurgen Kocka, *White-Collar Workers in America , 1890–1940: A Social-Political History in International Perspective* (Beverly Hills, 1980).

7. In the dynamic class analysis used in this study, "dominant occupation" indicates direction. Thus, the dominant occupation of a man who worked in the mills as a youth but later acquired a farm would be registered as farmer; conversely, that of a farm laborer who became a clerk would be clerk. For observations of declining élite participation by mid-decade, see Robert S. Lynd and Helen Merrell Lynd, *Middletown: A Study in Contemporary American Culture* (New York, 1929), 481–84; Goldberg, *Hooded Empire*, 45.

8. See, e.g., Floyd Adams interview, 15 Dec. 1976, transcript, box 1, ACHF; B. L. Flanigen interview, 11 Nov. 1976, transcript, loc. cit. I used $7,000 in assets as a baseline for major proprietors, this being the average among rural landowners able to hire tenants, according to Turner and Howell: see "Condition of Farmers in a White-Farmer Area of the Cotton Piedmont," 28–29.

9. *ADH*, 1 July 1914, p. 2; *ADH*, 19 March 1914, p. 1. Copy of petition in series I, frames 725–27, reel 7, Commission on Interracial Cooperation Papers, originals held by Robert W. Woodruff Library, Atlanta Univer-

sity, Atlanta, Georgia (microfilm edition: Ann Arbor, 1984). Hereafter, CIC.

10. Patrick, "Cultural Analysis," 69–70, 219–20, 299, 322, 335–36, 394; Davis et al., *Deep South*, 75–79, 108; Hall, "Disorderly Women," 374–76; Morland, *Millways of Kent*, 65, 167–68, 175–80; "Red, White, and Blue Barber Shop," transcript, pp. 2–3.

11. This discussion builds on a classic USDA study of a neighboring Piedmont farming county in order to make meaningful inferences from assets about relative class standing. That study found the average net assets of sharecroppers to be $77; of renters, $440; of landowning farmers with no tenants, $3,160; and of landowning farmers with one or more tenants, $7,164. See Turner and Howell, "Condition of Farmers," 28–29. On property requirements for suffrage, see Wardlow, "Negro Suffrage in Georgia," 81–83.

12. When corrected to include only those born before 1890, for example, the mean assets for 1921 rose to $3,189, while for those born after 1890, the corresponding figure was $678.

13. The proportion of Klan members in unskilled, menial labor dropped from seventeen percent in 1910 to two percent in 1926–27. The proportions in skilled trades and lower-level white-collar employment, in contrast, grew from three percent and fifteen percent, respectively, in 1900, to twenty percent and thirty-four percent in 1926–27. The absolute numbers of those whose occupation could be identified in the earlier years were relatively small, however—forty in 1900 and 104 in 1910.

14. The mean assets of Klan owners or managers of small businesses were valued at $1,348 in 1918, $2,022 in 1921, and $1,965 in 1927.

15. "Oldest Barber in Town," transcript, pp. 2–5; Woofter, "Negroes of Athens," 38; idem, *Negro Migration*, 54.

16. "Grocery Store," interview by S. B. Hornsby, 3 Aug. 1939, transcript, pp. 1–7, FWP; "Waiting Room in a Bus Station," interview by Grace McCune, 27 and 29 March 1939, transcript, p. 6, box 13, loc. cit. The "average" referred to was that specified in Turner and Howell, "Condition of Farmers."

17. These professors were not included in the data base since their membership could not be verified. See Appendix for explanation.

18. *ABH*, 29 July 1926, p. 5; *ABH*, 1 Aug. 1926, p. 7; *ABH*, 16 May 1926, p. 6.

19. Hiram Wesley Evans, "The Ballots Behind the Ku Klux Klan," *World's Work* 55 (1927–28), 246; *Searchlight*, 30 Sept. 1922, p. 4; ibid., 11 Aug. 1923, p. 2.

20. Quotes from Klan, *Papers Read at the Meeting*, 86; *Searchlight*, 16 Jan. 1922, p. 3. For Simmons and Mary Elizabeth Tyler, see entries in Coleman and Gurr, *Dictionary of Georgia Biography*. For other leading Klansmen, see the "Biographical Introduction" to Leroy Amos Curry, *The Ku Klux Klan Under the Searchlight* (Kansas, 1924).

21. On "rapid, uneven expansion and abrupt collapse [as] the motor force behind small owner protest" in another context, see Philip G. Nord, *Paris Shopkeepers and the Politics of Resentment* (Princeton, 1986), 484.

22. Georgia, State Tax Commission, *Annual Report*, 1921, quote from p. 4, loss figures on p. 10. Generalizations in text based on this and subsequent *Annual Reports*.

23. Ibid., 1925–1926, p. 4. On the growth of outside capital in the South in the 1920s, see Wright, *Old South, New South*, 63–64. On local business, see G. H. Bell interview, transcript, pp. 3–5, ACHF; B. L. Flanigen interview, transcript, pp. 5–7, loc. cit.; Mrs. James Cornelison interview, transcript, p. 5, loc. cit. See also Galambos and Pratt, *Rise of the Corporate Commonwealth*, esp. 89.

24. "Oldest Barber in Town," transcript, 11–13; "Grocery Store," transcript, 1–8; also "From Farm to Filling Station," interview by S. B. Hornsby, transcript, pp. 5–9, box 11, loc. cit. On the disproportionate toll of regulation on small business, see Weinstein, *Corporate Ideal in the Liberal State*, 252.

25. The total number this information was available for was 180.

26. See, for example, Goldberg, *Hooded Empire*, 35–37, 45–48, 64–65, 174–78; Cocoltchos, "Invisible Government," 614; Gerlach, *Blazing Crosses*, 77, 133; Moore, *Citizen Klansmen*, 60–70. Since the 1920 census does not provide breakdowns for local communities, it is impossible to systematically compare Klan members with the wider population.

27. Arno J. Mayer, "The Lower Middle Class as Historical Problem," *Journal of Modern History* 47 (Sept. 1975), 409–36; Rudy Koshar, ed., *Splintered Classes: Politics and the Lower Middle Class in Interwar Europe* (New York, 1990); Kramnick, *Republicanism and Bourgeois Radicalism*, 23–34; Benjamin DeMott, *The Imperial Middle: Why Americans Can't Think Straight About Class* (New York, 1990).

28. Rudy Koshar, "On the Politics of the Splintered Classes: Introductory Essay," in Koshar, *Splintered Classes*, 10–11; Hal Draper, *Karl Marx's Theory of Revolution*, vol. 2, *The Politics of Social Classes* (New York, 1978), esp. 288–91, and chaps. 11–18.

29. Mayer, "Lower Middle Class," 415–22; Draper, *Politics of Social Classes*, 291–300.

30. Mayer, "Lower Middle Class," 434, 423; Draper, *Politics of Social Classes*, 300–305. On the most extreme such example, Nazi anti-Semitism, see Arno J. Mayer, *Why Did the Heavens Not Darken? The "Final Solution" in History* (New York, 1990), esp. 90–109; Thomas Childers, *The Nazi Voter: The Social Foundations of Fascism in Germany, 1919–1933* (Chapel Hill, 1983); Leon Trotsky, *The Struggle Against Fascism in Germany* (New York, 1971).

31. For a view of the modern petite bourgeoisie as ineluctably conservative in times of acute crisis, see Mayer, "Lower Middle Class," esp. 416, 436. For emphasis on contingency, see Jonathan M. Weiner, "Marxism and

the Lower Middle Class: A Reply to Arno Mayer," *Journal of Modern History* 48 (1976), 666–71; Draper, *Politics of Social Classes*, esp. 305–16; Trotsky, *Struggle Against Fascism in Germany*; Nord, *Paris Shopkeepers*; and the essays in Koshar, *Splintered Classes*.

32. One study thus reported two key political blocs, one of larger merchants, professional men, and the rich generally; the other of smaller businesspeople, joined by carpenters, machinists and a smattering of mill workers, who were " 'the knockers,' or those who habitually tend to obstruct [the] progress" promoted by the first group. Rhyne, *Southern Cotton Mill*, 50–51. See also Morland, *Millways of Kent*, 223.

33. *ADB*, 11 Jan. 1918, p. 8; *ADB*, 17 Oct. 1919, p. 4; *ADB*, 13 July 1921, p. 6; CCL, *Annual Report*, 1916 to 1926.

34. H. G. Smith to Hoke Smith, 28 Dec. 1920, box 5, HS; Wright, *Old South, New South*, 149–50; Range, *A Century of Georgia Agriculture*, 213, 262.

35. Murray, *Red Scare*, 7; Kocka, *White-Collar Workers*, 155–56, 158–60, 178–80; *Kourier*, March 1926, p. 10.

36. Kocka, *White-Collar Workers*, 179, 186–89. "Office Rules," mimeo, box 3, CDF. The petition is contained in series I, frames 725–27, reel 7, CIC.

37. Bernstein, *Lean Years*, 113.

38. *Census* (1920), vol. 9, *Manufactures*, 271; Mrs. James Cornelison interview, transcript, p. 3, ACHF; see also "Life of a Retired Mill Worker," passim.

39. Tindall, *Emergence of the New South*, 164; Coleman, *History of Georgia*, 281.

40. Many instances of both are reported by Evans, who describes the GFL's more conservative wing as comprising men who were "Democratic, perhaps even Protestant and Ku Klux Klannish, and have maintained economic and social 'respectability.' " See Evans, "History of Organized Labor," 215.

41. GFL quoted in Rowland T. Berthoff, "Southern Attitudes toward Immigration, 1865–1914," *JSH* 17 (Feb. 1951), 348; L. R. Driskell to Thomas E. Watson, 6 Sept. 1922, box 15, TEW.

42. *Searchlight*, 11 Nov. 1922, p. 6; *ADH*, 21 April 1914. For instances of union collaboration with the Klan elsewhere, see Goldberg, *Hooded Empire*, 80, 134; Lay, *War, Revolution and the Ku Klux Klan*, 92, 98; Alec Dennis interview by Cliff Kuhn, transcript, p. 19, box 36, LAC.

43. Note the way office workers are pointedly described as "ladies" and their blue-collar counterparts as "females" in CCL, *Annual Report*, 1920, p. 60. For the sexual conservatism of craft union officials, see Hall, "Disorderly Women," 376.

44. Turner and Howell, "Condition of Farmers," 11, 13, 24, 29. For the freedom of renting tenants and even some croppers from direct supervision, see Range, *Century of Georgia Agriculture*, 261.

45. Hall et al., *Like a Family*, esp. chap. 1; MacDonald, *Southern Mill Hills*, 18–19, 21, 149; Rhyne, *Southern Cotton Mill*, 71–74; Morland, *Millways of Kent*, 4, 14, 22, 36; DeNatale, "Traditional Culture," 17, 51.

46. "Life on 'Happy Top,' " transcript, pp. 2, 9; also "A Visit to a Laundry and Dry Cleaning Plant," interview by S. B. Hornsby, 9–10 March 1939, transcript, box 11, FWP; "From Farm to Filling Station," loc. cit.; "I've Been Drifting," loc. cit.

47. Morland, *Millways of Kent*, 53; J. H. Blackman to TEW, 7 April 1917, box 11, TEW; Carlton, *Mill and Town*, 224, also 161.

48. See Hall, "Disorderly Women," 360–61; Hall et al., "Cotton Mill People," 273, 275; Morland, *Millways of Kent*, 23–24; MacDonald, *Southern Mill Hills*, 49, 52, 54, 86; Rhyne, *Southern Cotton Mill*, 71–77, 176, 199; also "The Red, White, and Blue Barber Shop," transcript, p. 12; "Life on 'Happy Top,' " transcript, pp. 7, 9, 17; "Life of a Retired Mill Worker," transcript, pp. 6, 14.

49. Quoted in MacDonald, *Southern Mill Hills*, 44, 107. See also Rhyne, *Southern Cotton Mill*, 18; Morland, *Millways of Kent*, 42, 61–62.

50. "Life on 'Happy Top,' " transcript, p. 6. On the exclusion of blacks from the mills, see Dittmer, *Black Georgia*, 33. In 1900, ninety-eight percent of Southern mill workers were native-born whites of native parentage, as against only seven percent in the mills of New England. Layne, *Cotton Mill Worker*, 76. Significantly, however, close observers noted that mill workers were less often extreme racists than middle-class whites, and had friendlier, more respectful relations with individual blacks. See, e.g., Morland, *Millways of Kent*, 185–87; Davis et al., *Deep South*, 50, 52.

51. Mary Ann Clawson, *Constructing Brotherhood: Class, Gender, and Fraternalism* (Princeton, 1989), 16, 95–106.

52. Ibid., 14–15, 45–51, 82–83, 145–77. See also Mark C. Carnes, *Secret Ritual and Victorian Manhood* (New Haven, 1989).

53. Clawson, *Constructing Brotherhood*, 15, also 110. For the Klan's insistence that class should not distinguish men from one another within its ranks, see Knights of the Ku Klux Klan, *The Practice of Klannishness*, Imperial Instructions, Document No. 1 (Atlanta, 1922), 1.

Chapter 4

1. *ABH*, 21 March 1930, p. 1; *ABH*, 27 April 1930, p. 14; *ABH*, 5 March 1930, p. 1; *ABH*, 4 May 1930, p. 14; "Americanism," 1926 typescript speech, box 1, AK. For Rivers' appointment as Great Titan, see Klan, Georgia, *Official Document*, 15 Feb. 1927, box 3, loc. cit.

2. *ABH*, 5 March 1930, p. 1; *ABH*, 4 May 1930, p. 14; *ABH*, 6 Aug. 1930, pp. 1, 5.

3. "Americanism," 1926 typescript, box 1, AK; also *Searchlight*, 29 March 1924, p. 8; William Joseph Simmons, *The Klan Unmasked* (Atlanta, 1923), 104.

4. The local press sought to refute beliefs, apparently common by 1921, that Jews were parasitic "plutocrats." See *ADB*, 5 Nov. 1921, p. 2. While information on Athens Jewry is lacking, a study of Atlanta Jews at the turn of the century concluded that they had achieved "astonishing [economic] success" relative to native-born gentile whites, and "a level of integration that their northern counterparts could well envy." Steven Hertzberg, *Strangers within the Gate City: The Jews of Atlanta, 1845–1915* (Philadelphia, 1978), 153, 155. Hertzberg also notes that the attitudes of gentile Atlantans towards Jews "underwent a steady erosion as the nineteenth century passed into the twentieth," with particular hostility directed toward Jewish businesspeople. Ibid., 172.

5. Phrase from the "Honorable J. O. Wood," mimeo, p. 2, AK.

6. For the concept of a "structure of feeling," see Raymond Williams, *Marxism and Literature* (Oxford, 1978), 128–35. For liberalism and republicanism in American political culture, see Daniel T. Rodgers, "Republicanism: The Career of a Concept," *JAH* 79 (1992), 11–38.

7. C. B. MacPherson, *The Political Theory of Possessive Individualism* (New York, 1964), 3, 6, 266; Harry Fuller Atwood, *Safeguarding American Ideals* (Chicago, 1921), 22, 67–68. See also *Searchlight* 4 Nov. 1922, p. 1; *Imperial Night Hawk*, 16 Jan. 1924, p. 5.

8. Simmons, *Klan Unmasked*, 142, also 248–53; Hiram Wesley Evans, *The Rising Storm: An Analysis of the Growing Conflict over the Political Dilemma of the Roman Catholics in America* (Atlanta, 1930), 175; Charles W. Gould, *America: A Family Matter* (New York, 1922), 70–71; Atwood, *Safeguarding American Ideals*, 46.

9. Simmons, *Klan Unmasked*, 165; *Imperial Night Hawk*, 16 May 1923, p. 2; Simmons, *Klan Unmasked*, 52, 190–91. See also *Kourier*, Oct. 1926, p. 23; Evans, "Ballots Behind the Ku Klux Klan," 246.

10. Simmons, *Klan Unmasked*, 163; Gould, *America*, 25–26; *Kourier*, Dec. 1927, p. 26.

11. Curry, *Klan Under the Searchlight*, 72, 229; *Searchlight*, 18 March 1922, p. 6; ibid., 2 Sept. 1922, p. 4; William Joseph Simmons, *America's Menace, or the Enemy Within* (Atlanta, 1926), 14.

12. Simmons, *Klan Unmasked*, 26–27, 146, 193, 207; Hiram Wesley Evans, "The Klan: Defender of Americanism," *Forum* 74 (Dec. 1925), 809. For the same theme, see idem, *The Public School Problem in America: Outlining Fully the Policies and the Program of the Knights of the Ku Klux Klan Toward the Public School System* (n.p., 1924), 18.

13. *Searchlight*, 5 Aug. 1922, p. 4; Curry, *Klan Under the Searchlight*, chap. 5; *Kourier*, March 1925, p. 29; Charles E. Jefferson, *Roman Catholicism and the Ku Klux Klan* (New York, 1925), 143; Simmons, *Klan Unmasked*, 204–5, also 185; *Kourier*, March 1925, p. 29; *Searchlight*, 15 April 1922, p. 4.

14. Samuel Saloman, *The Red War on the Family* (New York, 1922), 117; *Searchlight*, 5 May 1923, p. 2; *Imperial Night Hawk*, 9 May 1923, p.

8; Samuel Green, later Imperial Wizard, quoted in Lipset and Raab, *Politics of Unreason*, 135.

15. *ADB*, 11 July 1920, p. 4; Inaugural Address, 30 June 1923, Senate *Journal* (1923), 115, 126; *ABH*, 6 Aug. 1930, pp. 1, 5; Klan, Georgia, *Official Document*, Jan.–Mar. 1927, p. 4; ibid., Oct.–Nov. 1927, p. 5.

16. Klan, Georgia, *Official Document*, Nov. 1926, p. 4; Alma White, *Heroes of the Fiery Cross* (Zarepath, 1928), 85. For analysis of such "condensation symbols," see Edelman, *Symbolic Uses of Politics*, esp. 6–11, 198–99.

17. *Searchlight*, 30 Sept. 1922, p. 1; ibid., 7 Oct. 1922, p. 1; Saloman, *Red War on the Family*, v; *Imperial Night-Hawk*, 20 June 1923, pp. 2–3; Simmons, *Klan Unmasked*, 185, 124.

18. White, *Heroes of the Fiery Cross*, 85; Curry, *Klan Under the Searchlight*, 204; Simmons, *Klan Unmasked*, 192; Saloman, *Red War*, 45.

19. Simmons, *Klan Unmasked*, 113, 115; *Searchlight*, 7 Oct. 1922, p. 1; ibid., 2 Dec. 1922, p. 5; ibid., 8 Dec. 1923, p. 2; Evans, "Fight for Americanism," 42; *Imperial Night Hawk*, 20 June 1923, pp. 2–3.

20. Simmons, *Klan Unmasked*, 51, 116–117, 167, 174–75, 180; *Kourier*, Sept. 1927, p. 16. See also Hiram W. Evans, *The Attitude of the Knights of the Ku Klux Klan Toward the Jew* (Atlanta, 1923), 4.

21. *Searchlight*, 9 Dec. 1922, p. 4; ibid., 5 April 1924, p. 8; Simmons, *Klan Unmasked*, 104, 129; Curry, *Klan Under the Searchlight*, 48, 101–2. See also Evans, "Ballots Behind the Ku Klux Klan," 252; Klan, *Official Message of the Emperor*, 10; *Imperial Night Hawk*, 25 April 1923, p. 5.

22. *Birmingham News*, 7 May 1918, in series C, box 316, NAACP; *NYT*, 1 Sept. 1918, sec. IV, p. 5; *Searchlight*, 4 March 1922, p. 7; *ADB*, 25 Sept. 1921, p. 2; *ADB*, 11 Jan. 1922, p. 5; *ADB*, 17 March 1922, p. 1; Atwood, *Safeguarding American Ideals*, 16, 18; *Searchlight*, 30 Sept. 1922, p. 4. Punitive vagrancy measures we often used to compel African Americans to work for whites on the latter's terms. See Evans, "History of Organized Labor," 316–20. On producerism, see Daniel T. Rodgers, *The Work Ethic in Industrial America, 1850–1920* (New York, 1987).

23. Evans, "Ballots Behind the Ku Klux Klan," 245; *ABH*, 28 July 1926, p. 5; *Macon Telegraph*, 15 March 1928, p. 17 (Clarke was then promoting a new organization, but one with the same ideology); *Kourier*, May 1925, p. 24; ibid., Oct. 1925, p. 18.

24. "A lecture to the K.K.K.," box 3, AK; also Evans, *Klan of Tomorrow*, 19; N. B. Forrest to JPM, 11 Nov. 1925, box 1, AK; Minutes, 2 June 1925, loc. cit.; JNJ to the "K. K. Klan," 11 Aug. [no year], loc. cit.; Mrs. GJB to Mr. A., n.d., loc. cit.

25. Industrial Home, Minutes, 2 Feb. 1914, box 6, NSMC; "Welfare Work in Athens," loc. cit.; "Officers and Various Committees," loc. cit.; Minutes, 2 Dec. 1901, 5 Dec. 1910, 5 Oct. 1914, 3 June 1901, respectively, loc. cit. For the gender, race, and class moorings of privatism, see Rosalind Pollack Petchesky, *Abortion and Woman's Choice: The State, Sexuality,*

and Reproductive Freedom (Boston, 1985), 248–50; and the essays in Linda Gordon, ed., *Women, the State, and Welfare* (Madison, 1990).

26. Evans, *Public School Problem*, 5; *ABH*, 21 March 1930, p. 1; "Americanism," 1926 typescript speech, box 1, AK; Simmons, *America's Menace*, 15; Klan, *Kloran*, 34–35; E. F. Stanton, *Christ and Other Klansmen, or Lives of Love; The Cream of the Bible Spread Upon Klanism* (Kansas City, 1924), 59; Clarke quoted in Jones, *Story of the Ku Klux Klan*, 82.

27. Curry, *Klan Under the Searchlight*, 81; "Life on 'Happy Top,'" transcript. For like complaints, see *Searchlight*, 12 Aug. 1922, p. 4; Stanton, *Christ and Other Klansmen*, 35. For variants of the ideal, see Hahn, *Roots of Southern Populism*; Bruce Palmer, *"Man over Money": The Southern Populist Critique of American Capitalism* (Chapel Hill, 1980); Clawson, *Constructing Brotherhood*.

28. Klan, *Practice of Klanishness*, 2, 8. Studies of mill communities found that foremen and skilled workers joined fraternal lodges and churches in part to obtain prestige and demonstrate to uptown men their difference from ordinary operatives. See Rhyne, *Southern Cotton Mill*, 162–64, 166; Morland, *Millways of Kent*, 31, 164, 208, 213; Herring, *Welfare Work*, 147. See also Samuel Taylor Moore, "How the Kleagles Collected the Cash," *Independent* (13 Dec. 1924), 518. For local practice, see Minutes, 27 April 1925, box 2, AK; N. B. Forrest to JPM, 4 May 1926, box 1, loc. cit.; AJW to JPM, 16 July 1926, loc. cit.; WFD to L. G. Hardman, 23 July 1927, box 274, EDC.

29. Quoted in *Searchlight*, 13 May 1922, p. 1. See also Moseley, "Invisible Empire," 58; Mugleston, "Julian Harris," 287–88. On the politics of municipal reform in the South, see Brownell, *Urban Ethos in the South*, 54, 157–89; nationally, see Weinstein, *Corporate Ideal in the Liberal State*; Hays, *American Political History*.

30. *ADB*, 4 May 1914, 4; Georgia, House, *Journal* (1923), 574; *ABH*, 23 Nov. 1925; *ADH*, 16 Sept. 1914; *ADH*, 28 July 1915, p. 1; *ABH*, 12 June 1924, 1; *ABH*, 29 July 1926, p. 8.

31. Letter in *ABH*, 10 Aug. 1926, p. 8. Arnold's hostility to Democratic élites ultimately outweighed his loyalty to the party of white supremacy. In 1930, he would join a black doctor as a Clarke County's representative to the state Republican convention. *ABH*, 6 April 1930, 1.

32. *Searchlight*, 29 March 1924, p. 8; ibid., 5 April 1924, p. 4; obituary, *ABH*, 20 Nov. 1944, in Georgia Biography File, s.v. "Michael," UGA.

33. Minutes, 19 Jan. 1926, box 2, AK; *Kourier*, Feb. 1927, p. 2; Klan, *Papers Read at the Meeting*, 49; *Kourier*, Dec. 1925, p. 27; *Searchlight*, 22 March 1922; also Evans, "Ballots Behind the Ku Klux Klan," 245–46; Curry, *Klan Under the Searchlight*, 100, 105. Opposition to the League, of course, spread far beyond the ranks of the Klan.

34. Sam H. Campbell to TEW, 16 Sept. 1922, TEW, SHC; *NYT*, 11 Oct. 1920, p. 1; *Searchlight*, 8 April 1922, p. 4; ibid., 11 March 1922, p. 4;

ibid., 17 June 1922, p. 4; ibid., 25 Feb. 1922, p. 4; ibid., 16 Sept. 1922, p. 4.

35. Tom Watson to Robert L. Rodgers, 3 Jan 1921, box 5, RLR; *Searchlight*, 29 July 1922, p. 4.

36. *ABH*, 30 July 1926, p. 5; also, on the state's credit system, Range, *Century of Georgia Agriculture*, 247–48. Benton spoke in the wake of a bank failure that had cost many local residents "their earnings of years." See "Your Devoted Mama" to "My Precious Guy," 2 May 1925, folder 6, CNM Collection, UGA; also December 1925 YMCA circular, box 1, Frances Long Taylor Papers, UGA.

37. *ABH*, 6 Aug. 1930, pp. 1, 5; *Searchlight*, 25 Feb. 1922, p. 4. For other criticisms of "international banking," often anti-semitic, see ibid., 21 Oct. 1922, p. 4; *Kourier*, Feb. 1927, p. 2; ibid., Dec. 1925, p. 27.

38. *Searchlight*, 12 Aug. 1922, p. 1; Klan, *Papers Read at the Meeting*, 129. See also *Searchlight*, 5 May 1923, p. 2; *Imperial Night Hawk*, 28 May 1924, p. 2. In one of the only instances I have come across of the Klan threatening a wealthy white man with flogging, it was largely because, in their words, he did "not employ any help except Japs and Chinese." *Indianapolis News*, in box 10, William Dudley Foulke Papers, Library of Congress.

39. *ABH*, 14 Oct. 1924, p. 1; Oscar Ameringer, *If You Don't Weaken: The Autobiography of Oscar Ameringer* (New York, 1940), 371; Weaver, "Klan in Wisconsin," 233; Gerlach, *Blazing Crosses*, 32; Moore, "How the Kleagles Collected the Cash," 1.

40. *Searchlight*, 15 Dec. 1923, p. 4; ibid., 12 Aug. 1922, p. 4; Curry, *Klan Under the Searchlight*, 81.

41. George M. Marsden, *Fundamentalism and American Culture: The Shaping of Twentieth-Century Evangelicalism, 1870–1925* (New York, 1980), 119, also 141–95; Sydney E. Ahlstrom, *A Religious History of the American People*, vol. 2 (Garden City, 1975), 359–407.

42. *Kourier*, Oct. 1927, pp. 20–22. What is being described here is an ideological affinity for fundamentalism. In practice, the Klan was ecumenical. It hoped to surmount the barriers between different conservative Protestant denominations so as to unite their members behind a common secular program. See, for example, Klan, *Klan Building*, 11; Ku Klux Klan, "The University of America" (Atlanta, [1924]), in box 27, DCB. For a suggestive parallel, see the discussion of religion and the New Right in Petchesky, *Abortion and Woman's Choice*, 245. For the way evangelical Protestantism served as a court of last resort in the defense of inequality in this period, see MacDonald, *Southern Mill Hills*, 77, 144; Raper, *Preface to Peasantry*, 164–65.

43. *Searchlight*, 4 Nov. 1922, pp. 1, 4; *Kourier*, April 1925, p. 12; *Searchlight*, 20 June 1923, p. 2; ibid., 4 Nov. 1922, pp. 1, 4; Klan, *Monthly Bulletin*, Jan. 1927, p. 3. The Klan was not alone in such thinking, of course. See *AB*, 27 May 1921, p. 7.

44. Klan, *Message of the Emperor*, 12; *Kourier*, May 1925, p. 26;

Georgia, *Official Bulletin*, 1 Feb. 1926, box 3, AK (italics in original); *Kourier*, Oct. 1926, p. 21. See also H. W. Evans et al., *Is the Ku Klux Klan Constructive or Destructive?* (Girard, Kansas, 1924), 21.

45. Evans, "Fight for Americanism," 54, also 36; Evans, "Defender of Americanism," 805; Evans, *Rising Storm*, 166. See also Curry, *Klan Under the Searchlight*, 50; Wood, *Are You a Citizen?* 43; *Kourier*, Dec. 1926, p. 4. For other interpretations along these lines of the fundamentalism in the decade, see Robert A. Garson, "Political Fundamentalism and Popular Democracy in the 1920s," *South Atlantic Quarterly* 76 (Spring 1977), 219–33; Lawrence W. Levine, *Defender of the Faith: William Jennings Bryan: The Last Decade, 1915–1925* (New York, 1965). For a contemporary parallel from the anti-abortion movement, see Kristin Luker, *Abortion and the Politics of Motherhood* (Berkeley, 1984), esp. 207–10.

46. Wood, *Are You a Citizen?* 42; Blaine Mast, *K.K.K. Friend or Foe: Which?* (n.p., 1924), 64–65; *ABH*, 29 Jan. 1925, p. 1. For other examples, see *Imperial Night Hawk*, 9 March 1923; ibid., 6 June 1923, p. 1.

47. Klan, *Thirty-three Questions Answered*, p. 16, box 3, AK; Mast, *K.K.K. Friend or Foe*, 94–95.

48. Wood, *Are You a Citizen?* 48; *Imperial Night Hawk*, 16 May 1923; Curry, *Klan Under the Searchlight*, 79; Georgia, *House Journal* (1923), 1,001–2. See also Evans, "The Klan: Defender of Americanism," *Forum* 74 (1925), 813.

49. *Kourier*, April 1925, p. 9; *Searchlight*, 4 Nov. 1922, pp. 1, 4; Klan, *Papers Read at the Meeting*, 55, 125; Evans, *Rising Storm*, 175; Klan, "University of America," 2. Not surprisingly, the Protestant publications and bodies associated with the Social Gospel were also the ones who most consistently criticized the Klan. See Miller, "Protestant Churches," 357–63. For the Georgia Grand Dragon's enthusiasm about the death penalty for Communists, see *Kourier*, Sept. 1930, pp. 11–12.

50. *Kourier*, June 1925, p. 1; *ABH*, 13 Sept. 1925, p. 3; JHC to "Knights of Klu Kluck Klan," 9 Sept. 1925, box 1, AK; NBF to JPM, 31 Aug. 1925, loc. cit.; N. B. Forrest to JPM, 22 March 1926, loc. cit.; letter in Minutes, 23 March 1926, box 2, loc. cit.

51. Lillian Smith, *Killers of the Dream*, rev. ed. (New York, 1978), 103; *ABH*, 4 April 1926, p. 1; N. B. Forrest to J. P. McCall, 10 April 1926, box 1, AK; *Imperial Night Hawk*, 11 July 1923; ibid., 30 April 1924, p. 1; William G. McLoughlin, Jr., *Billy Sunday Was His Real Name* (Chicago, 1955), esp. 274–87.

52. Klan, Georgia, *Official Document*, Nov. 1926, p. 4. For earlier episodes of anti-Catholicism, see Higham, *Strangers in the Land*. For the view that anti-Catholicism was the driving force of the second Klan, see Mecklin, *Ku Klux Klan*, 28, 157–58; Fry, *Modern Klan*, 107, 128–47.

53. For example, the leading nineteenth-century anti-Catholic organization, the American Protective Association, had admitted both Jews and

African Americans. See Donald Kinzer, *An Episode in Anti-Catholicism: The American Protective Association* (Seattle, 1964), 46–47.

54. *Searchlight,* 18 Aug. 1923, 2; "A lecture to the K.K.K.," box 3, AK; Evans, *Rising Storm,* 252. For comparable fears, see Klan, Georgia, *Official Document,* May–June 1927, p. 3; *NYT,* 12 Jan. 1923, p. 10; Estes, *Roman Katholic Kingdom and the Ku Klux Klan,* 12.

55. Klan, Georgia, *Official Document,* Oct. 1926, p. 4; David Brion Davis, *The Fear of Conspiracy: Images of Un-American Subversion from the Revolution to the Present* (Ithaca, 1971), xviii; *Kourier,* Sept. 1925, p. 10. The Klan also recommended the "patriotic journalism" of the viciously anti-Catholic newspaper *The Menace;* Athens Klansmen voted to purchase bulk subscriptions to its fraternalist counterpart, the *Fellowship Forum.* See *Searchlight,* 19 May 1923; Minutes, 1 Dec. 1925, box 2, AK.

56. Evans, *Public School Problem,* 16; Estes, *Roman Katholic Kingdom,* 6, 16–17, 19, 22, 25, 29; Simmons, *Klan Unmasked,* 75; "A lecture to the K.K.K.," box 3, AK; "Oaths" in loc. cit.; Dever, *Masks Off!* 8. For more specific complaints, see Evans, *Public School Problem,* 13–14; idem, *Rising Storm,* 98; *Kourier,* April 1927, p. 21.

57. White, *Heroes of the Fiery Cross,* 86–87; idem, *The Ku Klux Klan in Prophecy* (Zarepeth, N.J., 1925), 82; Simmons, *Klan Unmasked,* 124; *Searchlight,* 21 April 1923, p. 1; *Kourier,* Oct. 1927, pp. 20–22.

58. Estes, *Roman Katholic,* 31; Klan, Alabama, *Official Document,* Oct. 1928, p. 3; also *Kourier,* Aug. 1929, pp. 20–21. For defenses of employer prerogatives, see Simmons, *Klan Unmasked,* 205–6; Atwood, *Safeguarding American Ideals,* 73–75, 78; *Kourier,* April 1929, p. 5. In many towns where the Klan built, the local class structure divided along lines of religion and ethnicity, with Catholics disproportionately working-class. See, e.g., "Memoirs of Father Peter Minwegan," Father Peter Minwegan Manuscript, SHSW; Bohn, "Klan Interpreted," 406.

59. See *Kourier,* March 1925, pp. 4–5, 29; Evans, "Ballots Behind the Klan"; Klan, *Papers Read at the Meeting,* 116; *Searchlight,* 22 July 1922, pp. 1, 2; ibid., 29 July 1929, p. 4. Population figures from Cashin, "Thomas E. Watson," 123.

Chapter 5

1. *NYT,* 18 Oct. 1923, p. 23; *ABH,* 19 Dec. 1923, p. 1; depositions (2) of W. V. Guerard, July 1922, box 48, Legal Series, JWB; Jackson, *Klan in the City,* 13–17, 35; Monteval, *Klan Inside Out,* 13–15; *Imperial Night Hawk,* 12 March 1924.

2. Fry, *Modern Klan,* 24; Wood, *Are You a Citizen?* 38; *Searchlight,* 25 Feb. 1922, p. 5. On efficacy, see correspondence in series C, box 127, and series B, boxes 48 and 49, HJA; Mecklin, *Ku Klux Klan,* 40–41, 78; Loucks, *Klan in Pennsylvania,* 39; Frost, *Challenge of the Klan,* 168; Gold-

berg, *Hooded Empire*, 7; Lila Lee Jones, "The Ku Klux Klan in Eastern Kansas during the 1920's," *Emporia State Research Studies* 23 (Winter 1975), 29; William D. Jenkins, "The Ku Klux Klan in Youngstown, Ohio: Moral Reform in the Twenties," *Historian* 61 (Nov. 1978), 86–87, 93; Lay, *War, Revolution*, 75–76; Moore, "How the Kleagles Collected the Cash," 518; Toy, "Klan in Oregon," 85.

3. For Klansmen's own self-consciousness in this regard, see *Kourier*, Oct. 1926, p. 23; also ibid., Dec. 1927, p. 26; ibid., Oct. 1925, p. 10; *Imperial Night Hawk*, 28 May 1924, p. 2. More generally, see George L. Mosse, *Nationalism and Sexuality: Respectability and Abnormal Sexuality in Modern Europe* (New York, 1985); Mayer, "Lower Middle Class."

4. Atwood, *Safeguarding American Ideals*, 24; *ADB*, 28 Sept. 1921, p. 4. For earlier efforts in Georgia and Athens, see Coleman, *History of Georgia*, 184–85, 305; *ADH*, 18 Sept. 1914, p. 1; *ADH*, 1 March 1915, p. 4; *ADH*, 14 Sept. 1915, p. 1; *ADH*, 30 Sept. 1916, p. 1; *ADH*, 19 March 1915, p. 8; *ADH*, 10 March 1914, p. 1.

5. *ADB*, 17 July 1920, p. 1. For northern cities, see Kathy Lee Peiss, *Cheap Amusements: Working Women and Leisure in New York City, 1880–1920* (Philadelphia, 1986); Joanne J. Meyerowitz, *Women Adrift: Independent Wage Earners in Chicago, 1880–1930* (Chicago, 1988).

6. A. J. Woodruff to the Hon. R. B. Russell, Jr., 12 May 1922, box 1, Pre-Gub. series, RBR; Patrick, "Cultural Analysis," 337; also Morland, *Millways of Kent*, 168.

7. Printed in *ADB*, 5 Jan. 1921, p. 2.

8. See, for example, Martin F. Amorous to Gov. L. G. Hardman, 30 June 1928, box 66, WAC; Amorous to Bishop Warren Akin Candler, 1 June 1928, loc. cit.; George Edward Drayton to James Robert Hamilton, 21 May 1922, folder 3, JRH; I. Heard to Hamilton, 14 May 1922, loc. cit.; Evans, "Organized Labor," 251; Ronald Morris Benson, "American Workers and Temperance Reform, 1866–1933" (Ph.D. diss., University of Notre Dame, 1974). Bootlegging and political radicalism came from the same districts in some areas, according to Garin Burbank, "Agrarian Radicals and Their Opponents: Political Conflict in Southern Oklahoma," *JAH* 58 (June 1971), 19. The mutual antagonisms are conveyed in the pro-Klan novel by Egbert Brown, *The Final Awakening: A Story of the Ku Klux Klan* (Brunswick, Ga., 1923), 10–29.

9. *ADB*, 7 June 1921, p. 1; *ADB*, 21 Sept. 1921, p. 8; *ABH*, 19 Feb. 1921, in series C, box 355, NAACP.

10. *ADB*, 23 Sept. 1921, p. 6; *ADB*, 8 Oct., 1921, p. 1.

11. CCSC, Minutes, Book 47 (1921), 322–23; *ADB*, 18 Oct. 1921, p. 8; *ADB*, 26 Oct. 1921, p. 1; *ADB*, 24 Nov. 1921, p. 4; *ADB*, 27 Nov. 1921, p. 1.

12. *ADB*, 24 March 1922, p. 4. Copies of the *Daily News* have not survived, so the account here comes from the established press and Klan sources.

13. *Searchlight,* 4 March 1922, p. 7; case no. 1975, State *vs.* Abe Farbstein (1922), CCSC, Office of County Clerk, Athens; *AB,* 28 March 1922, p. 2. For other complaints about dances, see Dyer, *University of Georgia,* 177–82.

14. *ADB,* 13 April 1922, p. 1; quotes from CCSC, Minutes, Book 47 (1922), 493–98.

15. CCSC, Minutes, Book 47 (1922), 494, 498, 499, 500; *ADB,* 22 April 1922, 1.

16. CCSC, Minutes (1922), 500. For the prejudices of earlier social purity efforts, see *ADH,* 16 March 1914, p. 8; Woofter, *Negroes of Athens,* 10, 54; and, more generally, Paul S. Boyer, *Urban Masses and Moral Order, 1820–1920* (Cambridge, 1978); Egal Feldman, "Prostitution, the Alien Woman and the Progressive Imagination," *American Quarterly* 19 (1967), 192–206; Barbara L. Epstein, *The Politics of Domesticity: Women, Evangelism and Temperance in Nineteenth-Century America* (Middletown, Conn., 1981); David Pivar, *Purity Crusade: Sexual Morality and Social Control, 1868–1900* (Westport, 1973).

17. *Searchlight,* 20 May 1922, p. 6.

18. *ADB,* 22 April 1922, pp. 1, 5; *ADB,* 25 April 1922, p. 4; *ADB,* 26 April 1922, p. 4.

19. *ADB,* 22 April 1922, pp. 1, 5; *ADB,* 25 April 1922, p. 4; *ADB,* 26 April 1922, p. 4.

20. *ADB,* 26 April 1922, p. 4; *ADB,* 21 Nov. 1922; *ADB,* 13 Dec. 1922, p. 4; CCSC, Minutes, Book 48 (1922), 19.

21. *Searchlight,* 22 Dec. 1923, pp. 1, 4; ibid., 29 March 1924, p. 8; University of Georgia, Senior Class, *Pandora,* s.v. "Kappa Kappa Kappa" (Athens, 1925). Leaders of student organizations also issued a statement asserting that "there has been exaggeration as to conditions here"—an oblique criticism of the Klan's use of sensationalist tactics to build its own ranks. Statement of campus leaders, 24 April 1924, University of Georgia, Board of Trustees, Correspondence and Reports, UA; also report of Board of Trustees, 12 June 1924, p. 4, loc. cit. For a similar example from the Midwest of moral issues as the hook for vicious racism, see *New York World,* 20 July 1924, in vol. 252, American Civil Liberties Union Archives, Seeley G. Mudd Manuscript Library, Princeton University, Princeton, N.J. (hereafter, ACLU).

22. CCSC, Minutes, Book 48 (1924), 571–72.

23. Anon. to Rebecca Latimer Felton, 19 Dec. 1924, box 8, RLF; *ABH,* 20 May 1924, p. 1. See also Judge M. C. Tarver to Rebecca L. Felton, 3 Sept. 1924, loc. cit.; Anon. to Judge Clifford H. Walker, 8 Jan. 1925, box 5, J. J. Brown Papers, SHC. On gender and temperance more generally, see Epstein, *Politics of Domesticity;* Ruth Bordin, *Woman and Temperance: The Quest for Power and Liberty, 1873–1900* (Philadelphia, 1981).

24. *Searchlight,* 7 June 1924, p. 3; *ABH,* 26 May 1924, pp. 1, 6; *ABH,* 28 May 1924, p. 6; *ABH,* 28 June 1923, p. 1; *NYT,* 15 May 1924, p. 3.

25. *Searchlight,* 21 June 1924, pp. 1, 8; *Imperial Night Hawk,* 28 May 1924, p. 5; CCSC, Minutes, Book 49 (1924), 103. See also *Imperial Night-Hawk,* 18 June 1924; ibid., 2 July 1924, p. 5. For rally preparations, see box 1, AK.

26. *ABH,* 25 Jan. 1925, p. 1; G. L. Johnson *vs.* Knights of the Ku Klux Klan, case no. 4776, April Term 1925, CCSC, Office of the County Clerk; Mrs. E. T. Eppes *vs.* Knights of the Ku Klux Klan et al., case no. 4761, loc. cit.; JPM to Nathan Bedford Forrest, 29 Nov. 1925, box 1, AK.

27. *ABH,* 25 Jan. 1925, p. 1; *ABH,* 21 Jan. 1925, p. 1; *ABH,* 30 June 1925, p. 2.

28. *ADB,* 28 April 1922, p. 2; *ABH,* 3 July 1930, pp. 1, 6; *ABH,* 13 June 1927, p. 1; *ABH,* 25 May 1924, p. 1.

29. See, for example, Klan, *Klan Building,* 5; *ADB,* 25 April 1922, p. 4; *ABH,* 15 Jan. 1925, pp. 1, 8; *ABH,* 18 Jan. 1925, pp. 1, 5.

30. See *ABH,* 8 Jan 1925, pp. 1, 5; *ABH,* 21 Jan 1925, pp. 1, 8; *ABH,* 13 Feb. 1925, pp. 1, 6. For the governor's involvement, see *ABH,* 25 Jan. 1925, p. 1; "Exhibit A," Mrs. E. T. Eppes *vs.* Knights of the Ku Klux Klan, op. cit. For the denials, suspension, and cover-up, see *ABH,* 9 Jan. 1925, p. 1; *ABH,* 15 Jan. 1925, pp. 1, 8; *ABH,* 18 Jan. 1925; *ABH,* 22 Jan. 1925, p. 1; *ABH,* 19 Jan. 1925, p. 4.

31. *ABH,* 8 Jan. 1925, p. 1; *ABH,* 21 Jan. 1925, p. 1.

32. *ABH,* 23 Jan. 1925, pp. 1, 5.

33. *ABH,* 23 Jan. 1925, pp. 1, 5; also 23 Jan. 1925, p. 1, and 25 Jan. 1925, p. 1. On means and ends, see, e.g., *Searchlight,* 18 March 1922, p. 1; ibid., 4 March 1922, 3.

34. *ABH,* 30 Jan. 1925, p. 1; *ABH,* 8 Feb. 1925, pp. 1, 8.

35. Minutes, 18 March 1925, box 2, AK; JPM to N. B. Forrest, 29 Nov. 1925, box 1, loc. cit.

36. M. B. Miller to Athens Klan No. 5, n.d., box 1, loc. cit.; JPM to N. B. Forrest, 29 Nov. 1929, loc. cit. N. B. Forrest to JPM, 30 Nov. 1925, loc. cit.; N. B. Forrest to JPM, 15 Feb. 1926, loc. cit.; N. B. Forrest to JPM, 23 Feb. 1926, loc. cit.; CCSC, Minutes, Book 51 (1928), 487.

37. *Kourier,* Oct. 1926, p. 23. See also ibid., May 1925, p. 23; ibid., Dec. 1927, p. 26; ibid., Oct. 1925, p. 10; *Searchlight,* 15 April 1922, p. 7; *Imperial Night-Hawk,* 28 May 1924, p. 2.

38. *Kourier,* Aug. 1925, p. 17; ibid., March 1927, pp. 9–11; White, *Heroes of the Fiery Cross,* 41–42; *Searchlight,* 18 March 1922, p. 6; Evans, *Rising Storm,* 161, 172, 178; also idem, *Public School Problem,* 6; Saloman, *Red War,* 136–37.

39. Stanton, *Christ and Other Klansmen,* 35; Saloman, *Red War,* 5, 93–113, 128; *Kourier,* June 1926, p. 16; *Searchlight,* 2 Sept. 1922, p. 4; ibid., 20 May 1922, p. 4; A. D. Kean to Thomas E. Watson, 23 March 1922, box 14, TEW; "Life on 'Happy Top,' " transcript, pp. 7, 18.

40. "Grocery Store," transcript, p. 6; "The Red, White and Blue Barber Shop," transcript, pp. 12, 14; "Oldest Barber in Town," interview by S. B.

Hornsby, 28 March 1939, box 13, FWP, 14; Minutes, 28 March 1928, box 1, Women of the Ku Klux Klan, Klan 14, Chippewa Falls, Wisc., Papers, SHSW.

41. *Searchlight*, 20 May 1922, p. 4; *ABH*, 27 June 1923, p. 1.

42. Evans, "Fight for Americanism," 39; *Searchlight*, 18 March 1922, p. 4; ibid., 15 April 1922, p. 7; Evans et al., *Constructive or Destructive*, 21; *Kourier*, April 1926, p. 11.

43. *Kourier*, Aug. 1925, p. 3; Simmons, *America's Menace*, 37; *Searchlight*, 15 April 1922, p. 7; *Kourier*, Nov. 1925, p. 13. Ideas of Pollution and purity, according to anthropologist Mary Douglas, give ritual enactment to hierarchical social divisions that have become weakened and open to challenge. Mary Douglas, *Purity and Danger: An Analysis of the Concepts of Pollution and Taboo* (London, 1966), esp. 4, 139, 141–42.

44. Atwood, *Safeguarding American Ideals*, 88; Gould, *America*, 88.

45. *Searchlight*, 4 March 1922. Also ibid., 12 Aug. 1922, p. 5; Gould, *America*, 135; Simmons, *Klan Unmasked*, 274, 277; *ABH*, 6 Aug. 1930, pp. 1, 5.

46. *Kourier*, Oct. 1926, p. 23.

47. *Searchlight*, 16 Jan. 1922, p. 5; Brown, *Final Awakening*, 102; Wood, *Are You a Citizen?* 44; "A Lecture to the K.K.K.," box 3, AK. See also Curry, *Klan Under the Searchlight*, 206. For the relationship between female chastity and male honor, see Julian Pitt-Rivers, "Honor," *International Encyclopedia of the Social Sciences*, vol. 6, ed. Daniel L. Sills (New York, 1968), 506, 510.

48. *ADH*, 9 June 1914, p. 1; *ABH*, 6 June 1927, p. 1; *ABH*, 4 Oct. 1923, p. 1; Bessie Mell Industrial Home, Minutes, 5 Feb. 1906, p. 94, NSMC; Clarke County, Office of the Ordinary, *Marriages*, Book M, p. 197, GDAH; CCSC, Minutes, Book 46 (1919), 30; ibid., Book 49 (1925), 284, 329, 376–77; *Kourier*, Nov. 1927, p. 18. See also Curry, *Klan under the Searchlight*, 108; Klan, *Papers Read at the Meeting*, 63.

49. Stanton, *Christ and Other Klansmen*, 35; *Searchlight*, 2 Sept. 1922, p. 4; also ibid., 18 March 1922, p. 4; ibid., 7 Oct. 1922, p. 4; ibid., 17 June 1922, p. 4; also ibid., 6 May 1922, p. 4. For the more permissive dating practices then developing in Athens, see "How Many Days Have I Regretted," interview by S. B. Hornsby, 23 June 1939, transcript, p. 10, box 12, FWP.

50. *Searchlight*, 29 July 1922, p. 4; also ibid., 12 Aug. 1922, p. 4; ibid., 30 Sept. 1922, p. 4; ibid., 11 March 1922, p. 4. Stanton, *Christ and Other Klansmen*, 35; Gould, *America: A Family Matter*, 101; Klan, *Papers Read at the Meeting*, 82, 90; White, *Klan in Prophecy*, 129; *Kourier*, Nov. 1926, pp. 11–21; ibid., April 1925, p. 11; Saloman, *Red War*, 119–20.

51. *Bigger's Magazine*, May 1922, p. 6, in Scrapbook, Don Hampton Biggers Papers, Southwest Collection, Texas Tech University, Lubbock, Texas; Simmons, *Klan Unmasked*, 280–82; Saloman, *Red War*, 119–20; *NYT*, 27 Nov. 1922, p. 1.

52. *Searchlight*, 24 March 1923, p. 4; *Kourier*, Dec. 1924, p. 29; also *Imperial Night Hawk*, 14 May 1924, p. 7; Klan, *Papers Read at the Meeting*, 92.

53. *Searchlight*, 8 April 1922, p. 4; *Kourier*, Oct. 1928, pp. 32–33; Mary Latimer McLendon to Richard B. Russell, Jr., 23 June 1921, Pre-Gub. Series, box 1, RBR; Clement C. Moseley, "Political Influence of the Ku Klux Klan in Georgia, 1915–1925," *GHQ* 57 (1973), 235–55.

54. *ABH*, 26 May 1924, pp. 1, 6; *Kourier*, Sept. 1928, p. 40; ibid., Jan. 1929, p. 17. See also Hux, "Klan in Macon," 60; Jenkins, "Klan in Youngstown"; Toy, "Klan in Oregon," 89.

55. White, *Heroes of the Fiery Cross*, 51; *Searchlight*, 27 May 1922, p. 2; ibid., 15 Sept. 1923, p. 1; *Kourier*, Dec. 1924, p. 21; ibid., April 1926, p. 21; ibid., Aug. 1925, p. 9.

56. Cocoltchos, "Invisible Government," 371; Lay, *War, Revolution and the Klan*, 85, 39; *Kourier*, Feb. 1928, pp. 22–23; *Searchlight*, 9 Sept. 1922, p. 3; Toy, "Klan in Oregon," 35, 71, 89, 129; Gerlach, *Blazing Crosses*, 42, 51; Hux, "Klan in Macon," 60. The Klan's rendition of the motto was "For God, and Home and Country." See *Searchlight*, 15 April 1922, p. 7.

57. For the convergence, see *ABH*, 1 Dec. 1927, p. 1; *ABH*, 8 July 1928, p. 1. For women's clubs, see *Searchlight*, 20 May 1922, p. 8; ibid., 4 March 1922, p. 3; *Kourier*, June 1925, p. 26; *Imperial Night Hawk*, 16 May 1923, p. 8; Lay, *War, Revolution*, 45, 78; *ABH*, 20 Nov. 1927, pp. 1, 4; ABH, 11 Nov. 1923, p. 1.

58. See Steven Hahn, "Honor and Patriarchy in the Old South," *American Quarterly* 31 (Spring 1984), 145–53; Ayers, *Vengeance and Justice*, 10–29, 274–75; John Mack Faragher, "History from the Inside-Out: Writing the History of Women in Rural America," *American Quarterly* 33 (Winter 1981), 537–57; McCurry, "The Two Faces of Republicanism."

59. *Imperial Night Hawk*, June 13, 1923, p. 5; Curry, *Klan Under the Searchlight*, 220. For examples of the fiction and the reality, see *Imperial Night Hawk*, 23 April 1923, p. 7; ibid., 20 June 1923, p. 8; *NYT*, 7 Nov. 1923, p. 15. These tensions led to open conflict in the late 'twenties, as described in Loucks, *Klan in Pennsylvania*, 155–60, and, more recently, in Blee, *Women of the Klan*.

60. Saloman, *Red War on the Family*, 50–53, 118, 126–28, also 132. See also *Kourier*, April 1926, p. 24; Curry, *Klan Under the Searchlight*, 32. Saloman's book was recommended to Klan members by the *Kourier*, Feb. 1927, p. 27.

61. Saloman, *Red War*, 22–23, 29–30, 50–51, 68, 71, 83–84, 90–91, 105; *Imperial Night Hawk*, 20 June 1923, p. 1. See also *Searchlight*, 28 Oct. 1922, p. 1.

62. Saloman, *Red War*, 18, 28, 32, 46, 47, 84. See also White, *Heroes of the Fiery Cross*, 90.

63. Saloman, *Red War*, 73. "The bourgeois sees in his wife a mere

instrument of production," Marx and Engels wrote. "He hears that the instruments of production are to be exploited in common, and, naturally, can come to no other conclusion than that the lot of being common to all will likewise fall to the women. He has not even a suspicion that the real point being aimed at is to do away with the status of women as mere instruments of production." Karl Marx and Frederick Engels, *Manifesto of the Communist Party* (Peking: Foreign Languages Press, 1972), 54.

64. Thomas E. Watson, *The Inevitable Crimes of Celibacy* (Thomson, Ga., 1917), box 11, TEW.

65. Evans, *Rising Storm*, 154; "A lecture to the K.K.K.," box 3, AK.

66. Quote from Fred D. Regan, "Obscenity or Politics? Tom Watson, Anti-Catholicism, and the Department of Justice," *GHQ* 70 (Spring 1986), 35; William Lloyd Clark, *The Devil's Prayer Book, or an Exposure of Auricular Confession as Practiced by the Roman Catholic Church: An Eye-Opener for Husbands, Fathers, and Brothers* (Milan, Illinois, 1922). Copy in papers of Klan member Robert L. Rodgers, box 3, RLR.

67. Quoted in Cashin, "Thomas E. Watson," 19, 46, and 135; Moseley, "Invisible Empire," 44.

68. White, *Heroes of the Fiery Cross*, 171, 176; *Searchlight*, 27 May 1922, p. 8. Among the order's favorite tactics to incite anti-Catholicism was thus speaking tours by phony ex-nuns, and the circulation of publications by such women recounting their imprisonment and humiliation. See, for example, the Klan favorite by Helen Jackson, *Convent Cruelties, or, My Life in a Convent* (Toledo, 1919). These accounts, while presented with proper Victorian outrage, often verged on the pornographic. Of the speech of one Klan-sponsored "ex-nun," the Denver press said: "for utter foulness . . . [she] simply could not be surpassed. It was the vomit not of the red light district, but of hell's depths." Quoted in Goldberg, *Hooded Empire*, 31. Titillation no doubt augmented their power.

69. "A Lecture to the K.K.K.," box 3, AK. In total, sixteen Athens Klansmen were divorced in the decade, the records of which appear in CCSC, Minutes, Books 46–52.

70. *Georgia Free Lance*, 17 Dec. 1925, p. 1; Frost, *Challenge of the Klan*, 31. See also Hux, "Klan in Macon," 33, 68–69; Confession of Luke Trimble, pp. 3–4, box 60, REM; Confession of W. C. Bishop, pp. 7–8, 13, loc. cit.; N.Y. (Pete) Rutherford interview by Cliff Kuhn, transcript, pp. 52, 54, box 10, LAC; Weaver, "Klan in Wisconsin," 115; *Searchlight*, 1 April 1922, p. 4; ibid., 4 March 1922, p. 6; *ADB*, 12 Jan. 1923, p. 1. The letters will be cited as appropriate below. Destitute women also requested aid from the Klan, while deserted wives asked it to locate their husbands. For an example of the latter, see N. B. Forrest to J. W. Arnold, 14 Dec. 1923, box 1, AK.

71. B. T. P., to "Members of the Ku Klux Klan," 19 July 1928, box 1, AK.

72. Ibid.

73. "Yours Truly" to "Klu Klux Klan," 3 Nov. 1925, box 1, AK. See also Mrs. A. L. O. to Mr. B., 3 Oct. 1926, loc. cit.

74. See divorce proceedings involving A. G. A., C. B. B., C. M. H., K. A. O., and Dr. B. B. C., as recorded in CCSC, Minutes, Book 46 (1920), 487; ibid., Book 48 (1923), 393; ibid., Book 49 (1925), 259; ibid., Book 50 (1926), 152; ibid, Book 52 (1928), 178–80, respectively. For mention of wife-beating in another Klan chapter, see Cocoltchos, "Invisible Government," 37–38.

75. *ADB*, 26 March 1922, p. 1. The case will be discussed in chapter 7. For complaints about "rough" people in the Southern Mill neighborhood from the wife of a local Klansman, see "Life on 'Happy Top,' " transcript, p. 17.

76. For the concept of moral economy, see Thompson, "Moral Economy of the English Crowd," esp. 78–79.

77. Mrs. B. T. P. to "Members of the Ku Klux Klan," box 1, AK; "Yours in this great work" to "The Klansman," 31 March 1927, loc. cit. See also Mrs. T. M. P. to "Dear Sir," July 1926, loc. cit. Whether from sincerity or desperation, some recipients of Klan aid lavished the order's members with praise. See Mrs. S. H. to "My Dear Friends in White," 11 Jan. 1925, loc. cit.; Mrs. S. K. to idem, n.d., loc. cit.; Mrs. W. P. and Children to "the Kind friends of the Klu Klux Klan," 22 Nov. [1925], loc. cit.

78. *Kourier*, Oct. 1926, p. 23; also ibid., Dec. 1927, p. 26; ibid., Oct. 1925, p. 10; *Imperial Night Hawk*, 28 May 1924, p. 2.

Chapter 6

1. Hugh M. Dorsey, *A Statement from Governor Hugh M. Dorsey as to the Negro in Georgia, April 22, 1921* (New York, 1921), in series I, reel 45, CIC; *Atlanta Independent*, March 3, 1921, p. 1; M. Ashby Jones (for the CIC) to "Dear Brother," 24 May 1921, box 3, RLR. The CIC went by different names in its early years; sometimes, as here, the Committee on Race Relations. For the sake of clarity, I refer to it always as the CIC.

2. *Atlanta Independent*, April 7, 1921, p. 6; Dorsey, *Statement from Governor Hugh M. Dorsey*, 22–23; Sosna, *In Search of the Silent South*, 26; correspondence between Marion M. Jackson and Will Alexander in series I, frames 475–76, reel 7, CIC; *ADB*, 9 Aug. 1921, p. 4; *Atlanta Independent*, March 24, 1921, p. 1.

3. Quoted in "Governor Dorsey Stirs Up Georgia," *Literary Digest* 69 (4 June 1921), 19. See also *Atlanta Journal*, 15 May 1921, in Vertical File, s.v. "Hugh Manson Dorsey," UGA. The backlash is reported in *ADB*, 21 May 1921, p. 1; "Georgia's Indictment," *Survey*, 7 May 1921, pp. 183–84. For the cast of characters, see Hux, "Klan in Macon"; Carl F. Hutcheson to Thomas E. Watson, 22 June 1921, box 12, TEW; Edward Young Clarke to "Dear Sir," n.d., box 3, Pre-Gub., RBR; Miss A. Benton to R. L. Rodgers, 14 Nov. 1921, box 3, RLR.

4. A. Benton to R. L. Rodgers, 14 Nov. 1921, box 3, RLR.

5. *The Montgomery Bus Boycott and the Women Who Started It: The Memoir of Jo Ann Gibson Robinson*, ed. David J. Garrow (Knoxville, 1987), xv.

6. For examples of such language and thought, see *Searchlight*, 26 April 1924, p. 8. My thinking about the Klan's racialism has benefitted from Barbara Fields, "Ideology and Race in American History"; Evelyn Brooks Higgenbottom, "African-American Women's History and the Meta-language of Race"; Benedict R. Anderson, *Imagined Communities: Reflections on the Origin and Spread of Nationalism* (London, 1983).

7. Dixie Defense Committee, *The "Negro in Georgia"* (n.p., 1921), 3, 6, 8, 9, 11, 12. See also *Searchlight*, 4 June 1921, p. 1.

8. Dixie Defense Committee, *"Negro in Georgia,"* 4, 6, 9. "The race question," Ridley himself stated, "has become a controversy between the two factions of the white race, and the negro question is no longer the serious one for consideration." (Ibid., 9). From the other side of the divide, Dorsey supporters recognized as a potential boon the salutary effect his statement was having on outside opinion of Georgia. (See "Governor Dorsey Stirs Up Georgia," 19).

9. Dixie Defense Committee, *"Negro in Georgia"*, 11, 4.

10. DuBois, "Shape of Fear," 301; "The Ku Klux Are Riding Again," *The Crisis* 17 (March 1919), 231.

11. Klan, *Practice of Klanishness*, 5; Klan, Georgia, *Official Document*, May–June 1927, pp. 5, 7; *Kourier*, Nov. 1929, p. 23. Evans, *Klan of Tomorrow*, 9; Klan, Georgia, *Official Document*, May–June 1927, p. 7 (italics mine).

12. Simmons, *Klan Unmasked*, 172; *Searchlight*, 5 Aug. 1922, p. 4; Klan, *Message of the Emperor*, 12–13; *Kourier*, Oct. 1926, p. 20. On Stoddard, see Carey McWilliams, *A Mask for Privilege: Anti-Semitism in America* (Boston, 1948), 63–67.

13. "Notes Taken from the Speech of Col. Wm. J. Simmons," series C, box 312, NAACP; see also "Notes from 24 Jan. 1921 Speech of Col. William J. Simmons," series I, frames 522–23, reel 7, CIC; West Virginia leader quoted in Trotter, *Coal, Class, and Color*, 128; Klan, *Official Monthly Bulletin*, 1 June 1927, p. 3; Klan, Georgia, *Official Document*, May–June 1927, p. 9; ibid., April 1927, pp. 3, 4. For illustrations of such race-baiting from Georgia, see Kuhn et al., *Living Atlanta*, 324; Angelo Herndon, *Let Me Live* (New York, 1969), 144; *ABH*, 5 Nov. 1928, p. 1.

14. Klan, *Practice of Klannishness*, 6; Klan, Georgia, *Official Document*, May–June 1927, p. 8; Simmons, *Klan Unmasked*, 22; also Wood, *Are You a Citizen?* 47; Klan, *Message of the Emperor*, 15–16; *Atlanta Independent*, 1 Feb. 1919, in series C, box 313, NAACP; *Kligrapp's Quarterly Report* forms, box 3, AK; Klan, *Constitution and Laws*, 16.

15. See "W. J. Simmons," p. 3 (affidavit by W. V. Guerard, July 1922), box 48, JWB; *ABH*, 29 July 1926, p. 5.

16. The best example from within the Klan is the lionization of the Reconstruction Klan as a force for national salvation at the center of the K-UNO degree ritual, the beginning of the Klan's fraternal ladder. See Klan, *Kloran*, 5th ed., 48–51. On the Dunning school, see W. E. B. Du Bois, *Black Reconstruction in America* (1935; reprint, Cleveland, 1964), 711–28.

17. Evans, "Fight for Americanism," 36, 41, 51; idem, "Defender of Americanism," 802–3; *ABH*, 5 Nov. 1928, p. 1. See also Guy B. Johnson, "The Race Philosophy of the Ku Klux Klan," *Opportunity* 1 (Sept. 1923), 269; Klan, Georgia, *Official Document*, Oct.–Nov. 1927, p. 7.

18. Evans, "Fight for Americanism," 42, 38; Evans, "Defender of Americanism," 803, 808–9.

19. Evans, "Defender of Americanism," 804; *Searchlight*, 5 Aug. 1922, p. 4; Klan, *Kloran*, 5th ed., 2; also Dixie Defense Committee, *"Negro in Georgia,"* 3.

20. *Kourier*, Jan. 1926, p. 19. Gould, *America*, 27, 49, 57, 124–25; *Atlanta Independent*, 13 Jan. 1921, p. 1; Evans, *Klan of Tomorrow*, 8–9, 13; Simmons, *Klan Unmasked*, 49; Evans, *Rising Storm*, 152, 154. For a contemporary's recognition of the racial element in Klan anti-Catholicism, see Andre Siegfried, *America Comes of Age*, trans. H. H. Heming and Doris Heming (New York, 1927), 137.

21. Gould, *America*, 163; also, Klan, *Official Monthly Bulletin*, 1 Nov. 1926, p. 1; Simmons, *Klan Unmasked*, 143; Evans, "Defender of Americanism," 806–7. For how racial exclusivity became part of earlier republicanism, see Morgan, *American Slavery, American Freedom*; Winthrop D. Jordan, *White over Black: American Attitudes Toward the Negro, 1550–1812* (Chapel Hill, 1968).

22. Curry, *Klan Under the Searchlight*, 137; Klan, *Message of the Emperor*, 9; Evans, *Klan of Tomorrow*, 13, 16; Grady quote from *Kourier*, Sept. 1925, p. 7; Coolidge and Hoover from Lipset and Raab, *Politics of Unreason*, 142. For the Klan's use of such statements, see *Kourier*, April 1925, pp. 28, 32; ibid., Feb. 1925, p. 12; ibid., Dec. 1927, p. 15.

23. I. A. Newby, *Jim Crow's Defense: Anti-Negro Thought in America, 1900–1930* (Baton Rouge, 1965), quote from xi; Evans, *Rising Storm*, 151; idem, "Defender of Americanism," 810; also, idem, "Fight for Americanism," 43, 49; and idem, *Klan of Tomorrow*, 13; Klan, "Thirty-Three Questions Answered," 7, box 3, AK. On eugenics, see Linda Gordon, *Woman's Body, Woman's Right: A Social History of Birth Control in America* (New York, 1977), 116–58. On anti-Semitism, see McWilliams, *Mask of Privilege*, 57–67 and passim; Donald S. Strong, *Organized Anti-Semitism in America: The Rise of Group Prejudice During the Decade 1930–1940* (1941; reprint, Westport, 1979).

24. Klan, Georgia, *Official Document*, May–June 1927, p. 4; Klan, *Kloran*, 5th ed., 49; Klan, *Practice of Klanishness*, 5; Monteval, *Klan Inside Out*, 65; Klan, *Message of the Emperor*, 12, 13.

25. Klan, *Message of the Emperor*, 9; Simmons, *Klan Unmasked*, 33;

Evans, "Fight for Americanism," 53. See also Sweeney, "Great Bigotry Merger," 9; Klan, Georgia, *Official Document*, May–June 1927, p. 8; the report in *New York World*, 20 July 1924, in vol. 252, p. 104, ACLU Papers; Dixie Defense Committee, *"Negro in Georgia,"* 12; Johnson, "Race Philosophy of the Ku Klux Klan," 269.

26. *Kourier*, Feb. 1925, p. 9; ibid., Sept. 1925, p. 27; quotes from Charles. S. Bryan to Henry Ford, 13 Nov. 1920, box 31, Charles S. Bryan Papers, SHC (hereafter, CSB). See also *Searchlight*, 30 June 1923, p. 2.

27. McWilliams, *Mask for Privilege*, 179 and passim; Abram Leon, *The Jewish Question: A Marxist Interpretation*, 2d ed. (New York, 1970), 226–27, 233; John Higham, "American Anti-Semitism Historically Reconsidered," in *Jews in the Mind of America*, ed. Charles H. Stember et al. (New York, 1966), 248. On Henry Ford, see Lipset and Raab, *Politics of Unreason*, 135–38.

28. Strong, *Organized Anti-Semitism*, 4, also 5–6, 14–15; Jean-Paul Sartre, *Anti-Semite and Jew*, trans. George J. Becker (1948; reprint, New York, 1965), 149. See also McWilliams, *Mask for Privilege*, 88, 94, 103; Morton Rosenstock, *Louis Marshall, Defender of Jewish Rights* (Detroit, 1965), 110–15, 119.

29. For religious themes, see White, *Klan in Prophecy*, 27, 49–50; Stanton, *Christ and Other Klansmen*, 40; *Imperial Night Hawk*, 27 June 1923, p. 7. For economic themes, see Charles S. Bryan to Henry Ford, 13 Nov. 1920, box 31, CSB; *Searchlight*, 30 Dec. 1922, p. 4; *Kourier*, Oct. 1925, p. 32; Curry, *Klan Under the Searchlight*, 161, 163; Klan, *Papers Read at the Meeting*, 125; *Imperial Night Hawk*, 27 June 1923, pp. 6–7; *Searchlight*, 30 June 1923, p. 2. See also Rosenstock, *Louis Marshall*, 204–5.

30. Curry, *Klan Under the Searchlight*, 161; Evans, *Attitude . . . Toward the Jew*, 6; "Americanism," typescript speech, 11 May 1926, box 1, AK; White, *Klan in Prophecy*, 53–54; idem, *Fiery Cross*, 34.

31. For a European parallel, see Leon, *The Jewish Question*, esp. 236–37.

32. Evans, *Attitude . . . Toward the Jew*, 4–5; *Kourier*, Oct. 1926, p. 23. John Higham argued that American anti-Semitism was usually "pseudo-agrarian" and bound up with a producerist ideology, in "American Anti-Semitism Historically Reconsidered," 247–48.

33. Samuel H. Campbell to TEW, 16 Sept. 1922, box 15, TEW; *Imperial Night Hawk*, 27 June 1923, pp. 6–7; also Campbell, *The Jewish Problem in America* (Atlanta, 1923). See also Evans' claim that "the Jew" acted as a parasitic "middle-man," in Evans, *Attitude . . . Toward the Jew*, 5–6.

34. *Searchlight*, 30 June 1923, p. 2; ibid., 28 Oct. 1922, p. 4; *Imperial Night Hawk*, 1 Aug. 1923, p. 7; Klan, Georgia, *Official Document*, Oct. 1926, p. 4. On anti-Semitism more generally in these years, see Rosenstock, *Louis Marshall*, 110–15, 119, 127; McWilliams, *Mask for Privilege*, 41; Strong, *Organized Anti-Semitism*, 6, 149–50.

35. For the "magical" quality of anti-Semitic thought, see Sartre, *Anti-Semite and Jew,* esp. 39. For the class basis of anti-Semitic groups, see Strong, *Organized Anti-Semitism,* 169, 172–74. On the Janus face, see McWilliams, *Mask for Privilege,* 92; Draper, *Politics of Social Classes,* 300–302, 413.

36. Klan, Georgia, *Official Document,* Oct. 1926, p. 4; *Searchlight,* 28 Oct. 1922, p. 4; *Kourier,* Feb. 1927, p. 2; ibid., Jan. 1926, 15; *Imperial Night-Hawk,* 27 June 1923, p. 7; Georgia, *Official Document,* Oct. 1926, p. 4.

37. White, *Heroes of the Fiery Cross,* 85; *Searchlight,* 5 May 1923, p. 2; ibid., 30 Sept. 1922, p. 1; *Imperial Night Hawk,* 20 June 1923, p. 1.

38. Klan, *Message of the Emperor,* 10; Simmons, *Klan Unmasked,* 69, also 132, 252; *Searchlight,* 9 Dec. 1922, p. 4. See also *Imperial Night Hawk,* 25 April 1923, p. 5; Curry, *Klan Under the Searchlight,* 101–2; *Kourier,* Oct. 1925, p. 32.

39. *Searchlight,* 19 Aug. 1922, p. 4; Evans, "Defender of Americanism," 810; *Imperial Night Hawk,* 2 May 1923, p. 7; Jones, *Story of the Klan,* 84–85; Klan, Georgia, *Official Document,* Jan.–Mar. 1927, p. 8. See also Evans, "Fight for Americanism," 40–41, 51; Curry, *Klan Under the Searchlight,* 101–2, 115; Simmons, *Klan Unmasked,* 138; *Searchlight,* 27 May 1922, p. 4; ibid, 5 May 1923, p. 2.

40. Klan, *Papers Read at the Meeting,* 129; *Imperial Night Hawk,* 28 May 1924, p. 2; *Kourier,* Dec. 1926, p. 5; Curry, *Klan Under the Searchlight,* 54, 115; *Kourier,* Oct. 1925, p. 10; Goldberg, *Hooded Empire,* 61; Bohn, "Klan Interpreted," 399; Weaver, "Klan in Wisconsin," 139–42; Gerlach, *Blazing Crosses,* 41–42; Angle, *Bloody Williamson,* 139, 145, 150–51.

41. See, for example, *Kourier,* Jan. 1926, pp. 18–19; Dixie Defense Committee, "Negro in Georgia," 3.

42. Julius S. Goetchius, in Judia C. Jackson Harris, *Race Relations* (Athens, 1925), 9; George C. Thomas, in *Race Relations,* 9; also Dixie Defense Committee, "Negro in Georgia," 6; Simmons, "Notes Taken from Speech," series C, box 312, NAACP.

43. *Searchlight,* 30 Sept. 1922, p. 2; *Imperial Night Hawk,* 18 April 1923, pp. 1, 7; *Kourier,* Nov. 1929, p. 24. For the Communist Party's novel commitment, see Robert L. Allen, *Reluctant Reformers: Racism and Social Reform Movements in the United States* (Washington, D.C., 1974). For later fears Southern Klansmen had about Communist influence on blacks, see Mary C. Brown to Hugh Gladney Grant, 30 Nov. 1932, box 3, HGG.

44. *Searchlight,* 12 Aug. 1922, p. 6; *Searchlight,* 1 April 1922, p. 1; Klan, Georgia, *Official Document,* Jan.–Mar. 1928, p. 7. A 1921 city council resolution aimed at the Catholic Church to prohibit interracial worship helped win Klansman Walter Sims the mayoralty of Atlanta. *ADB,* 20 Sept. 1921, p. 1.

45. Simmons, *Klan Unmasked*, 141, 142, 193, 248, 251, 253–255. See also *Atlanta Independent*, Jan. 13, 1921, p. 1; *Kourier*, Jan. 1926, p. 18; *Searchlight*, 22 March 1922, p. 3; Evans, *Klan of Tomorrow*, 15; *New York World*, 20 July, 1924, vol. 252, p. 104, ACLU Papers.

46. Evans, "Fight for Americanism," 40–41; *Imperial Night Hawk*, 2 May 1923, p. 7.

47. *Kourier*, March 1927, p. 11; White, *Klan in Prophecy*, 27, 49; Klan, *Papers Read at the Meeting*, 125; Klan, *Message of the Emperor*, 10, 13; Evans et al., *Constructive or Destructive?* 9, 16; *Kourier*, Nov. 1926, p. 21; *Imperial Night-Hawk*, 14 May 1924, p. 7.

48. Klan, Alabama, *Official Document*, Oct. 1928, pp. 1, 2, 4; *Kourier*, Nov. 1929, pp. 23–24; Georgia, *Official Document*, May–June 1927, p. 5; White, *Heroes of the Fiery Cross*, 144; *Kourier*, Aug. 1925, p. 10; ibid., Dec. 1929, p. 14.

49. Walker, "Inaugural Address," in Georgia, *Senate Journal* (1923), 118–19; Smith, *Killers of the Dream*, 124. On the larger pattern, see Arthur F. Raper, *The Tragedy of Lynching* (1933; reprint, New York, 1969), 20, 50; Newby, *Jim Crow's Defense*, 135; Hall, *Revolt Against Chivalry*, 145–47; Tindall, *Emergence of the New South*, 170–71; James Weldon Johnson, *Along This Way: The Autobiography of James Weldon Johnson*, (1933; reprint, New York, 1968), 170, 311–13, 365–66, 391; White, *Rope and Faggot*, 17, 54–55, 82; Ralph Ellison, *Shadow and Act* (New York, 1964), 311.

50. White, *Rope and Faggot*, 76, 82. See also White, *Rope and Faggot*, 17; Raper, *Tragedy of Lynching* 36–37; Johnson, *Along This Way*, 330; Robert L. Zangrando, *The NAACP Crusade Against Lynching* (Philadelphia, 1980), 4–8.

51. For the classic statement of this problem, see Wilhelm Reich, *The Mass Psychology of Fascism*, trans. Vincent R. Carfango (New York, 1970), 36. See also Charles Herbert Stember, *Sexual Racism: The Emotional Barrier to an Integrated Society* (New York, 1976), 13 and passim.

52. Hall, *Revolt Against Chivalry*, quote on 149, discussion on 149–53. For another attempt to make sense of the role of gender in lynching, this one more focussed on the psychology of individual white male perpetrators, see Joel Williamson, *The Crucible of Race: Black/White Relations in the American South Since Emancipation* (New York, 1984).

53. Davis et al., *Deep South*, 6, 24–25. For theoretical discussions of this kind of boundary policing, see Douglas, *Purity and Danger*, 125; Verena Martinez-Alier, *Marriage, Class and Colour in Nineteenth-Century Cuba: A Study of Racial Attitudes and Sexual Values in a Slave Society* (London, 1974), 1, 7, 123–24.

54. See, for example, Klan, *Official Monthly Bulletin*, 1 Dec. 1926, pp. 1–2; ibid., 1 April 1927, p. 2; Evans, "Defender of Americanism," 804; Klan, Georgia, *Official Document*, April 1927, p. 4; ibid., May–June 1927, p. 5. For a rare exception, see *Searchlight*, 18 Nov. 1922, p. 7.

55. Klan, Alabama, *Official Document,* Oct. 1928, p. 1.

56. Quote from account in Greensboro, N.C., *Daily News,* 18 March 1922, series C, box 312, NAACP; also *Searchlight,* 6 May 1922, p. 7; ibid., 18 Nov. 1922, p. 7; *NYT,* 25 April 1926, p. 16; *NYT,* 7 Nov. 1925, p. 7; Klan, Georgia, *Official Document,* Jan.–March 1928, p. 6; *Kourier,* Aug. 1925, p. 10.

57. White, *Klan in Prophecy,* 53; idem, *Heroes of the Fiery Cross,* 10; Klan, Georgia, *Official Document,* Oct. 1926, p. 4; *Imperial Night Hawk,* 4 April 1923, p. 3; *Searchlight,* 30 June 1923, p. 2; *Kourier,* Dec. 1926, p. 5.

58. *Imperial Night Hawk,* 30 April 1924, 6; Ridley, in Wood, *Are You a Citizen?* 44; "Americanism," box 1, AK; *Kourier,* May 1925, p. 32; *Imperial Night Hawk,* 27 June 1923, p. 7. See also White, *Heroes of the Fiery Cross,* 34, 36; idem, *Klan in Prophecy,* 53–54; also Bohn, "Klan Interpreted," 388.

59. "Americanism," box 1, AK; *Imperial Night Hawk,* 27 June 1923, p. 7; MacLean, "The Leo Frank Case Reconsidered"; Mast, *Friend or Foe?,* 19, 22.

60. *Searchlight,* 30 June 1923, p. 2; Evans et al., *Constructive or Destructive?* 14; White, *Klan in Prophecy,* 27, 49; Klan, *Papers Read at the Meeting,* 125; Simmons, *Klan Unmasked,* 55, 59; Curry, *Klan Under the Searchlight,* 162; Evans et al., *Constructive or Destructive?* 14; *Kourier,* May 1925, p. 32; Evans, *Attitude . . . Toward the Jew,* 5, 6.

61. "To the ordinary American," he wrote, "the race question at bottom is simply a matter of ownership of women; white men want the right to use all women, colored and white, and they resent the intrusion of colored men in this domain." Quoted in Giddings, *When and Where I Enter,* 61.

62. Gerda Lerner, *Black Women in White America: A Documentary History* (New York, 1973), 150, 172–73.

63. *Kourier,* Nov. 1926, p. 23; *NYT,* 7 Nov. 1925, p. 7; White, *Heroes of the Fiery Cross,* 34–35. See also Ridley, in Wood, *Are You a Citizen?* 44.

64. Evans, *Klan of Tomorrow,* 9; idem, "Fight for Americanism," 39, 44; Simmons, *Klan Unmasked,* 32–36, 274, 275, 279; *NYT,* 27 Nov. 1922, p. 1; *Searchlight,* 11 March 1922, p. 4; ibid., 29 July 1922, p. 4; ibid., 15 April 1922, p. 4; Gould, *America,* 27. On the turn-of-the-century "race suicide" panic, see Gordon, *Woman's Body, Woman's Right.*

65. *Kourier,* Oct. 1926, p. 22.

Chapter 7

1. *Atlanta Constitution,* 19 Sept. 1917, frame 552, reel 221, TINCF; *New York Herald Tribune,* 11 Sept. 1919, in series C, box 355, NAACP; *[Albany, Ga.] Supreme Circle News,* [undated], loc. cit.

2. *Atlanta Independent,* undated 1917 clippings, frames 631–32, reel 221, TINCF.

3. For evidence of Klan involvement in lynchings in Georgia and South Carolina, see *New York Call,* 23 March 1921, in vol. 190, ACLU; Walter White to Governor Thomas G. McLeod, 26 Oct. 1926, Correspondence file, Walter Francis White Papers, Yale Collection of American Literature, Beinecke Rare Book and Manuscript Library, Yale University, New Haven. For illustrations of the social roots and political character of collective violence, see Tilly, "Collective Violence in European Perspective," 4–42; and Ayers, *Vengeance and Justice,* 239–64. On the need for cultural sanction, see Thompson, "Moral Economy of the English Crowd," esp. 78–79.

4. Quotes from *Atlanta Constitution,* 20 Feb. 1921, frame 462, reel 222, TINCF; *Chicago Defender,* [date illegible], loc. cit. See also *Atlanta Independent,* 24 Feb. 1921, p. 4; W. H. Harris to Walter White, 19 Feb. 1921, series C, box 355, NAACP; *Atlanta Constitution,* 8 Feb. 1921, loc. cit.; *Atlanta Constitution,* 17 Feb. 1921, loc. cit.; *Atlanta Constitution,* 27 Feb. 1921, loc. cit.; *Macon Chronicle,* 17 Feb. 1921, loc. cit.; *ADB,* 19 Feb. 1921.

5. Quote from *Atlanta Constitution,* 6 Dec. 1921, frame 309, reel 222, TINCC; see also *Atlanta Constitution,* 26 Jan. [1922], frame 927, loc. cit.; *Atlanta Constitution,* 11 Dec. 1921, frame 461, loc. cit.; *ADB,* 6 Dec. 1921, p. 6.

6. William H. Harris to Mr. Walter White, 19 Feb. 1921, series C, box 355, NAACP; *ADB,* 19 Feb. 1921, series C, box 355, NAACP; *ADB,* 23 Sept. 1921, pp. 1, 2, 8; *ADB,* 11 April 1922, p. 1; *Atlanta Independent,* 24 Feb. 1921, frame 460, reel 222, TINCF.

7. Anon. to the Hon. Warren G. Harding, frames 289–90, reel 14, FSAA. At least one national Klan leader was arrested for robbing the mails. See U.S. Congress, Senate, *Senator from Texas: Hearings before a Subcommittee of the Committee on Privileges and Elections,* 68th Congress, 1st session (Washington, 1924), 282.

8. *ADB,* 8 March 1921, p. 1; *Atlanta Independent,* 20 Jan. 1921, p. 4. For white uneasiness about black institutions, see Jones, *Labor of Love,* 133; Brooks, "Local Study," 218. For evidence of Klan perpetration of similar cases of arson, see Monteval, *Klan Inside Out,* 65; J. Q. Jett to Rebecca Latimer Felton, [1924], box 8, RLF. Perhaps this is why some African Americans referred to the KKK as the "Knights of the Kerosene Kan." *Atlanta Independent,* 8 Sept. 1921, p. 1.

9. *ADB,* 8 March 1921, p. 1; quote from *ADB,* 13 Dec. 1921, p. 4. See also *Atlanta Constitution,* 11 Dec. 1921, frame 461, reel 222, TINCF; *NYT,* 23 Nov. 1921, p. 23. For other threats, see *Athens Herald,* clipping in series C, box 355, NAACP. Although "aimed ostensibly at blacks," Jacquelyn Hall observed in another context, lynching "also operated as a

means of indoctrination and social control over whites." See Hall, *Revolt Against Chivalry*, 142.

10. *ADB*, 26 March 1922, p. 1; *ADB*, 28 March 1922, pp. 1, 8; *ADB*, 25 March 1922, p. 5.

11. *ADB*, 24 March 1922, p. 1; *ADB*, 28 March 1922, pp. 1, 8; *ADB*, 24 March 1922, p. 1; *ADB*, 25 March 1922, p. 5; *ADB*, 26 March 1922, pp. 1–2.

12. *ADB*, 28 March 1922, pp. 1, 8; *ADB*, 24 March 1922, p. 1; *ADB*, 25 March 1922, p. 5; *ADB*, 26 March 1922, p. 1. The other case was that of a man from nearby Winterville, who was first warned by the Klan and then taken out twice and whipped. He and his wife were fighting to the point where it appeared likely that one would kill the other. *ADB*, 21 July 1922, p. 1; *ADB*, 22 July 1922, p. 1; CCSC, Minutes, Book 48 (1922), 21.

13. On the minister, see Case file 311.4113, RG 59, frames 197–245, reel 18, FSAA. For the McClusky/Peters case, see T. J. Woofter, Jr., to Roger N. Baldwin, 30 June 1922, vol. 215, 358–61, ACLU; Dykeman and Stokely, *Seeds of Southern Change*, 104–5.

14. T. J. Woofter, Jr., to Roger N. Baldwin, 30 June 1922, vol. 215, 358–61, ACLU Papers; Dykeman and Stokely, *Seeds of Southern Change*, 106–8; "Affidavit of Odessa Peters," August 1922, series 7, frames 1627–29, reel 45, CIC.

15. *ADH*, 18 Nov. 1922, p. 5; *ADB*, 26 Nov. 1922; *Atlanta Independent*, 20 Jan. 1921, pp. 1, 4; *NYT*, 23 Nov. 1922, p. 23. In Clarke County, several people complained to the grand jury "that pistol firing [had] become a nightly occurrence" in the city and out, but whether Klansmen were involved is impossible to determine. See CCSC, Minutes (1922), 391.

16. *ABH*, 16 Dec. 1922, p. 1; "Report Negroes Driven from Georgia Towns," [press release], 2 March 1923, series C, box 313, NAACP; *Atlanta Independent*, 16 Aug. 1923, p. 1; Donald Dewey Scarborough, "An Economic Study of Negro Farmers as Owners, Tenants, and Croppers," *BUGA* 25, n. 2a (Sept. 1924), 9; Fanning, "Negro Migration," 7, 9, 12–13, 15; *ABH*, 11 Oct. 1923, p. 1; Harris, *Race Relations*, esp. 4, 8, 9.

17. For interesting similarities, see A. James Hammerton, "The Targets of 'Rough Music': Respectability and Domestic Violence in Victorian England," *Gender & History* 3 (Spring 1991), 23–44.

18. K.K.K. to A.M., n.d., in Ku Klux Klan Miscellany Collection, UGA.

19. Exalted Cyclops, Bartow, Florida, Klan, to Exalted Cyclops, Athens, Georgia, 27 April 1927, box 1, AK.

20. *NYT*, 11 March 1926, p. 24; W. G. Brock to JPM, 26 July 1926, box 1, AK.

21. *ABH*, 7 June 1926, 4; *ADB*, 26 Sept. 1922, p. 1; *ADB*, 25 April 1922, p. 4.

22. B. T. P. to "Members of the Ku Klux Klan," 19 July 1928, box 1,

AK; W. A. L. [to Ku Klux Klan], 22 March 1926, loc. cit.; C. W. H. to "Dear Clammans," 28 Oct. 1925, loc. cit.

23. *Searchlight,* 16 Jan. 1922, p. 5. See also Klan, *Papers Read at the Meeting,* 136; Glavis, "When the Klan Tells the Truth," 28.

24. Brown, *The Final Awakening,* esp. 45, 62–63, 88–99, 104–8, 113, 120, 147; Simmons quoted in Albert De Silver, "The Ku Klux Klan—'Soul of Chivalry,'" *Nation* 113 (14 Sept. 1921), 285; Georgia, *Official Document,* Dec. 1926, 1. The Klan press, for its part, maintained that the order was "impelled by the same chivalric impulses as the original Knights of the Ku Klux Klan." *Searchlight,* 16 Jan. 1922.

25. One person who had driven around the Georgia countryside speaking with white farmers reported that when asked about Klan activities, "they unfailingly answer: 'The Ku Klux Klan are organized to enforce law. The time has come when a poor man can't get justice in the courts or a rich man punishment.'" See Helen Dortch Longstreet to Mrs. W. H. Felton, 20 Oct. 1923, RLF.

26. Simmons, as quoted in Jones, *Story of the Klan,* 65, and "Notes Taken from Speech of Wm. J. Simmons," 24 Jan. 1921, series C, b;x 312, NAACP; Paul M. Winter, *What Price Tolerance?* (New York, 1928), iii, vi.

27. Richard Maxwell Brown, "The American Vigilante Tradition," in Hugh Davis Graham and Ted Robert Burr, eds., *The History of Violence in America: Historical and Comparative Perspectives* (New York, 1969), 168–70, 179–83. For illustrations, see Nall, *Tobacco Night Riders;* Crozier, *White Caps.* The preservationist character of vigilante activity and its consistent association with the right wing of the political spectrum are stressed in Andrew A. Karmen, "Vigilantism," *Encyclopedia of Crime and Justice,* ed. Sanford H. Kadish, vol. 4 (New York, 1983), 1,616–18.

28. Gould, *America,* 25–26; Evans, "Fight for Americanism," 49; Siegfried, *America Comes of Age,* 134; "Knights of the Ku Klux Klan" to "Dear Sir," loose one-page lecture admission card, box KY/KU in main reading room, SHSW; *Kourier,* March 1925, p. 29; Curry, *Klan Under the Searchlight,* 169–70; Simmons, *America's Menace,* 42, 44–45; Jefferson, *Roman Catholicism,* 145.

29. Simmons, *Klan Unmasked,* 196 (italics mine), 205–6, 209–10; Evans, "Defender of Americanism," 814. See also Klan, *Thirty-Three Questions Answered.*

30. Samuel Olive, as quoted in *Atlanta Journal,* 15 May 1921, in Vertical File, s.v. "Hugh Manson Dorsey," UGA; Klan, *Kloran,* 5th ed., 12; *New York World,* 20 July 1924, in vol. 252, p. 105, ACLU Papers; Sermon by Imperial Kludd Caleb Ridley, in Wood, *Are You a Citizen?* 43. For evidence of local cross-burnings, never reported by the press, see the reimbursements of members for materials in Athens Klan, Minutes, 15 Sept. 1925 and 23 March 1926, box 2, AK.

31. *Searchlight,* 3 May 1924, p. 8; Evans, *Klan of Tomorrow,* 23, 26; Evans, "Defender of Americanism," 805; *Kourier,* Oct. 1926, p. 21. See

also George Estes, *The Roman Katholic Kingdom and the Ku Klux Klan* (Troutdale, Ore., 1923), 31; Ridley sermon in Wood, *Are You a Citizen?* 48; "A lecture to the K.K.K.," box 3, AK; Klan, *Message of the Emperor*, 15.

32. Wood, *Are You a Citizen?* 6, 43; Evans, *Constructive or Destructive*, 10; *Kourier*, Oct. 1926, p. 21; *Imperial Night Hawk*, 8 Aug. 1923, p. 6; Klan, *Message of the Emperor*, 3, 5, 8, 12, 13; Jones, *Story of the Klan*, 82; Evans, *Klan of Tomorrow*, 7, 17, 25, 27.

33. Mr. and Mrs. W. B. Romine, *Story of the Ku Klux Klan* (Pulaski, Tenn., 1924), 14–15; Wood, *Are You a Citizen?* 47; *Searchlight*, 18 March 1922, p. 1; also ibid., 4 March 1922, p. 3; ibid., 11 March 1922, p. 7. Forrest's comment was a defensive parry against Judge James K. Hines, who had recently censured "secret organizations banded together for the purpose of regulating the morals of a community without respect for the law or courts." *ABH*, 7 June 1926, p. 4; Realm of Georgia, "Special Bulletin," 27 July 1926, box 3, AK. On the fear of vice in republicanism, see Wood, *Creation of the American Republic*, 68, 91–124; Ayers, *Vengeance and Justice*, 45.

34. See the Klan letterhead in N. B. Forrest to J. P. M., 11 Nov. 1925, box 1, AK; Wood, *Are You a Citizen?* 45; Klan, *Practice of Klannishness*, 2. See also *Searchlight*, 16 Jan. 1922, p. 5; Jones, *Story of the Klan*, 83–84; Brown, *The Final Awakening*, 40, 51, 442. On honor, see Ayers, *Vengeance and Justice*, 13, 26; Bertram Wyatt-Brown, *Southern Honor: Ethics and Behavior in the Old South* (New York, 1982); Hahn, "Honor and Patriarchy."

35. *Searchlight*, 18 March 1922, p. 6; Anon. to Judge J. R. Hamilton, 31 March 1922, JRH. On the imperative to violence, see Ayers, *Vengeance and Justice*, 13, 15–19; Pitt-Rivers, "Honor," 506, 509; Lynwood Montell, *Killings: Folk Justice in the Upper South* (Lexington, 1986), 29, 146, 149; John Dollard, *Caste and Class in a Southern Town* (New York, 1949), 331–32.

36. *Kourier*, April 1924, p. 18. See also *Imperial Night Hawk*, 18 June 1924, p. 5; Simmons, *Klan Unmasked*, 80.

37. For examples, see "Confession of W. C. Bishop," 6, 10, box 60, REM; *Atlanta Independent*, 20 Jan. 1921, p. 1; Grady R. Kent, *Flogged by the Ku Klux Klan* (Cleveland, Tenn., 1942), 19; NAACP, Press Release, 18 July 1921, series C, box 312, NAACP; *NYT* 16 April 1923, p. 8; *The Madisonian*, 30 March 1928, in box 317, NAACP; W. C. Witcher, *The Reign of Terror in Oklahoma* (Fort Worth, n.d.), 14; Case file 311.4113, reel 18, frames 208, 233–34, FSAA. For related interpretations of crowd violence, see Hall, *Revolt Against Chivalry*, 149–52; Natalie Zemon Davis, "The Rites of Violence," in *Society and Culture in Early Modern France* (Palo Alto, 1975); Roger Kent Hux, "The Ku Klux Klan and Collective Violence in Horry County, South Carolina, 1922–1925," *South Carolina Historical Quarterly* 85 (July 1984), 218.

38. CIC, "Race Relations in 1927" (Annual Report), series I, frame

552, reel 4, CIC. See also Jonathan A. Manget to Editor, *Columbus Enquirer-Sun*, 21 Feb. 1923, in box 5, JLH. One might infer the frequency of local flogging raids from the cases that occasionally came to light. The Klan vigilantism uncovered in a series of Georgia counties comprised scores of separate incidents, conducted sometimes more than once a week in the periods about which evidence emerged. For Bibb County, see Hux, "Klan in Macon"; for Toombs County, *NYT*, 14 Feb. 1927, p. 10; *NYT*, 14 March 1927, p. 1; for Fulton County, the typed confessions of three Klansmen in box 60, REM; *Nation*, 6 April 1940, pp. 445–46; for Barrow County, *ADB*, 13 Feb. 1922, p. 1. For a similar frequency in Alabama, see *Outlook*, 2 Nov. 1927, p. 261; in Texas, see article reprinted from *Houston Press*, series I, frames 533–36, reel 7, CIC. It is possible, however, that the zeal of these chapters contributed to their eventual prosecution.

39. For examples, see *Literary Digest*, 26 March 1927, p. 8; *NYT*, 8 Jan. 1925, p. 12; *NYT*, 12 Jan. 1923, p. 10; *NYT*, 31 Jan. 1924, p. 30; *NYT*, 30 Nov. 1926, p. 31; *NYT*, 27 Aug. 1924, p. 1; *NYT*, 16 June 1922, p. 17; N. B. Forrest to J. W. Arnold, 14 Dec. 1923, box 1, AK; idem to J. P. M., 19 March 1926, loc. cit.; Klan, Georgia, *Official Document*, March 1927, p. 3; ibid., Jan.–Mar. 1928, p. 3; ibid., April 1927, p. 4; *Searchlight*, 1 April 1922, p. 4.

40. For examples, see NAACP Press Release, 18 July 1921, series C, box 312, NAACP; *The Madisonian*, 30 March 1928, in series C, box 317, NAACP; *Augusta Chronicle*, 15 June 1927; *NYT*, 26 April 1923, p. 8; *NYT*, 15 Dec. 1924, p. 14; NYT, 22 Nov. 1923, p. 3; *NYT*, 27 Feb. 1927, p. 5; *NYT*, 17 Jan. 1923, p. 8; *NYT*, 18 Jan. 1923, p. 9; Lay, *War, Revolution and the Ku Klux Klan*, 108.

41. For examples, see Kuhn et al., *Living Atlanta*, 317–19; *NYT*, 25 April 1926, p. 16; *NYT*, 7 Nov. 1925, p. 7; *NYT*, 30 June 1922, in series C, box 344, NAACP; Greensboro *Daily News*, 18 March 1922, in loc. cit., box 312; Confession of W. C. Bishop, 24 March 1940, 5, box 3, REM. Even here, the alleged offense was sometimes a fabrication to cover economic jealousy. See, for example, Roger Baldwin to N. G. Romey, 5 July 1922; and N. G. Romey to Roger Baldwin, 7 July 1922, vol. 215 (1922), pp. 323–25, ACLU.

42. See, for example, T. J. Woofter, Jr., to Roger N. Baldwin, 30 June 1922, vol. 215, 358–59, ACLU; Mary McLeod Bethune to William Alexander, 12 Jan. 1921, series I, frame 520 of reel 7, CIC; Mrs. R. E. B. Lake to Grace Abbott, 26 Aug. 1923, box 162, U.S. Children's Bureau Papers, National Archives, Washington, D.C. For such gendered racial terror in slavery and after Emancipation, see Gerda Lerner, *Black Women in White America*, 149, 163–64, 172–73.

43. Monteval, *Klan Inside Out*, 79, 81; *ABH*, 31 July 1930, p. 1. For other examples, see *Kourier*, Sept. 1930, pp. 8–12; *NYT*, 31 Oct. 1920, p. 12; Case file BS 213410, RG 65, frames 924–65, reel 7, FSAA; Case file BS 212117, frame 921, loc. cit.; *NYT*, 7 Oct. 1920, p. 2; *NYT*, 9 July 1920, p.

3; *NYT,* 6 Sept. 1922, p. 19; *NYT,* 14 March 1922, p. 1; *NYT,* 6 Dec. 1922, p. 4; Kuhn et al., *Living Atlanta,* 42–45, 316.

44. See, e.g., *New York Herald Tribune,* 14 March 1927, in series C, box 354, NAACP; *NYT,* 14 March 1927, loc. cit.; *NYT,* 26 Dec. 1920, loc. cit. 2; *NYT,* 6 Sept. 1921, p. 17; *NYT,* 4 July 23, p. 1; *Literary Digest,* 26 March 1927, p. 8; *Baltimore Afro-American,* 10 Feb. 1922, in series C, box 313, NAACP; CIC, *Race Relations in 1927,* series I, frame 552, reel 4, CIC.

45. James B. Clark to William Pickens, 19 July 1923, series C, box 3313, NAACP.

46. For examples, see *NYT,* 9 Jan. 1918, IV, 5; *Searchlight,* 25 Feb. 1922, p. 7; ibid., 11 Aug. 1923, p. 1; *ABH,* 18 July 1930, p. 1; "For and Against the Ku Klux Klan," *Literary Digest* 70 (24 Sept. 1921), 38; De Silver, "The Ku Klux Klan—'Soul of Chivalry,' " 286; Jackson, *Klan in the City,* 80, 192, 209; *Federated Press Bulletin,* [Chicago], 28 Jan. 1922, in series C, box 313, NAACP; ACLU, *Annual Report* (1923), 20; Toy, "Klan in Oregon," 143.

47. Duffus, "Klan in the Middle West," 365–66; *NYT,* 2 Oct. 1923, p. 9; also Mecklin, *Ku Klux Klan,* 97–98; Toy, "Klan in Oregon," 125. This is not to deny that some local unions and members supported or collaborated with the Klan. For examples, see Toy, "Klan in Oregon," 91; Lipset and Raab, *Politics of Unreason,* 127–28; Goldberg, *Hooded Empire,* 80; Layne, *Cotton Mill Worker,* 222.

48. See, for example, Goldberg, *Hooded Empire,* vii; Cocoltchos, 4; Moore, *Citizen Klansmen,* 1–2.

49. Herbert G. Gutman, *Slavery and the Numbers Game: A Critique of Time on the Cross* (Urbana, 1975), 17–20. For an illustration, see the poignant inquiry from the wife of a black Virginia minister: Mrs. R. E. B. Lake to Miss Grace Abbott, 26 Aug. 1923, box 162, Children's Bureau Papers. I am grateful to Linda Gordon for this source.

50. For examples, see *Imperial Night Hawk,* 12 Sept. 1923, p. 1; *Kourier,* March 1925, p. 13; ibid., August 1925, pp. 19, 31; Simmons, *Klan Unmasked,* 43.

51. See, for example, Simmons quoted in *NYT,* 1 Nov. 1920, 27; Simmons, *Klan Unmasked,* 12, 15, 20–22, 29, 38. Georgia Governor Clifford Walker and Grand Dragon Nathan Bedford Forrest, for example, maintained publicly that they opposed people taking the law into their own hands. Yet behind the scenes, both men had collaborated in the extralegal entrapment effort and cover-up that came out of the Athens anti-vice crusade.

52. Klan, *Klan Building,* 14–15 (italics mine); Klan, *Constitution,* 11–12 (italics mine). For a declaration of opposition to vigilantism, followed by approval in cases where the law or courts acted negligently or inadequately, see *Searchlight,* 4 March 1922, 3. For similar qualifications, see Klan, *Kloran,* 2; Klan, *Papers Read at the Meeting,* 133; Klan, "University of America," [4]; Oregon Klan leader quoted in Glavis, "When the

Klan Tells the Truth," 28. The Klan also encouraged study of the Constitution, "American courts," and the "history of law"—presumably to equip
members for safe, efficacious action. See Klan, *Klan Building*, 9–10.

53. *New York World*, 20 July 1924, in vol. 252, p. 104, ACLU. On the
assumptions, see Simmons, *Klan Unmasked*, 156–57; quotes in Johnson,
"Race Philosophy of the Ku Klux Klan," 268, 269.

54. *New York World*, 20 July 1924, in vol. 252, p. 104, ACLU; Klan,
Georgia, *Official Document*, March 1927, p. 4. See also ibid., Jan–Mar.
1928, p. 4; Klan, *Official Monthly Bulletin*, 1 Dec. 1926, pp. 2, 3; ibid.,
Jan. 1927, 2; ibid., Feb. 1927, p. 2.

55. Klan, *Klan Building*, 9–10; Petition, G. L. Johnson *vs.* Knights of
the Ku Klux Klan, p. 2, Case no. 4776, filed 24 Feb. 1925, CCSC; "Application for Citizenship" forms, box 2, AK; Klan, Georgia, *Official Document*,
Nov. 1926, p. 4; Witcher, *Reign of Terror*, 27; *Atlanta Independent*, 15
Nov. 1923, p. 1; *NYT*, 2 Sept. 1924, p. 21; *NYT*, 11 April 1925, p. 9.

56. Monteval, *Klan Inside Out*, 66, also 69, 70, 83–84; W. V. Guerard,
deposition on "W. J. Simmons," box 48, JWB, 1, 3; Duffus, "Klan in the
Middle West," 368; *Atlanta Independent*, 3 Jan. 1924, p. 4; also J. Q. Jett
to R. L. Felton, [1924], box 8, RLF. For similar testimony from Pennsylvania, see *NYT*, 10 April 1928, in series C, box 317, NAACP; from Indiana,
see *New York Evening World*, in loc. cit.; *NYT*, 2 April 1928, p. 9; *NYT*,
10 April 1928, p. 1; *NYT*, 11 April 1928, p. 1; *NYT*, 14 April 1928, p. 1.

57. This description is compiled from several kinds of overlapping accounts: internal documents, court cases, members' confessions, victims'
complaints, and reports by contemporaries. See, in addition to the Athens
material discussed earlier; Klan, *Klan Building*, 8, 14–15; Hux, "Klan in
Macon"; Sweeney, "Great Bigotry Merger," 9; Monteval, *Klan Inside Out*,
65, 76–78; Duffus, "How the Klan Sells Hate," 177; Loucks, *Klan in Pennsylvania*, 60, 78–79; Kent, *Flogged by the Ku Klux Klan*, 19; Klan, *Official
Monthly Bulletin*, 1 April 1927, p. 1; confessions of Atlanta Klan floggers
in box 60, REM.

58. Confession of F. I. Lee, 28 March 1940, pp. 1–2, box 4, REM; Confession of Luke Trimble, 1, loc. cit.; Confession of W. C. Bishop, 1–2, loc.
cit.; also Loucks, *Klan in Pennsylvania*, 79. For the claim that ministers
were kept "in the dark" and then "used to hold up the cross & defend"
the Klan after it was charged with illegal acts, see J. Q. Jett to Mrs. W. H.
Felton, [1924], pp. 3–4, box 8, RLF. For similar patterns in the flogging
operations of the post–World War II Klan, see Stetson Kennedy, *I Rode
with the Ku Klux Klan* (London, 1954), 95–119.

59. Monteval, *Klan Inside Out*, 86–90, 99–100; Kuhn et al., *Living
Atlanta*, 320; G-1-NJ to E. P. Banta, 13 June 1923, Edwin P. Banta Papers,
Rare Books and Manuscripts Division, New York Public Library, New
York City; Hux, "Klan in Macon," 43; Dykeman and Stokely, *Seeds of
Southern Change*, 107; "Sincerely yours" to J. R. Hamilton, 31 March
1922, folder 3, JRH; *ADB*, 12 Dec. 1922, p. 1; *ADB*, 13 Dec. 1922, 1; *NYT*,

14 March 1927, series C, box 354, NAACP; *NYT,* 28 Oct. 1923, 1. On jury-stacking, see Klan, *Klan Building,* 10; John V. Baiamonte, Jr., *Spirit of Vengeance: Nativism and Louisiana Justice, 1921–1924* (Baton Rouge, 1984), 120, 125. On attempts to intimidate journalists, see Shepherd, "Fighting the K.K.K. on Its Home Grounds," esp. 509; Henry Louis Suggs, ed., *The Black Press in the South, 1865–1979* (Westport, 1983). On willful destruction of evidence, see *ABH,* 13 Feb. 1925, pp. 1, 6; Minutes, 29 Dec. 1925, AK; W.G.B. to J.P.M., 26 July 1926, box 1, loc. cit.

60. Vagts, *History of Militarism,* 66–74, 175–79; Hux, "Klan in Macon," 24. As illustrated by the Athens social purity entrapment scandal, this tactic was generally successful.

61. Howard A. Tucker, *History of Governor Walton's War on [the] Ku Klux Klan* (Oklahoma City, 1923), esp. 26, 64. For an account of the 1987 case, see Morris Dees, with Steve Fiffer, *A Season for Justice: The Life and Times of Civil Rights Lawyer Morris Dees* (New York, 1991).

62. See, for example, the controls on the use of robes in Klan, *Klan Building,* 14.

63. For examples, see *NYT,* 25 July 1926, II, 2; *NYT,* 4 Oct. 1921, p. 5; *NYT,* 1 Sept. 1923, p. 5; *NYT,* 5 June 1923, p. 4; *NYT,* 20 Oct. 1924, p. 18; Loucks, *Klan in Pennsylvania,* 128 ff.

64. On anti-mask efforts, see Klan, *Official Monthly Bulletin,* Jan. 1927, p. 2; H. G. Connor, Jr. to E. C. Branson, 8 March 1923, box 6, ECB; Jackson, *Klan in the City,* 79.

65. For reports to this effect, see Johnson, "Sociological Interpretation of the New Ku Klux Klan," 444. The most extensive and sensational episodes were also largely concentrated in the South. See, for example, Witcher, *Reign of Terror.*

66. Hux, "Klan in Macon," 52–53; Case file 311.4113, RG 59, frames 199 and 219, reel 18, FSAA; *New York World,* 12 Dec. 1926, series C, box 317, NAACP; Patton, "Klan Reign of Terror," 54; Witcher, *Reign of Terror,* 30.

67. Fry, *Modern Klan,* 7; Tucker, *Governor Walton's War,* 26; Monteval, *Klan Inside Out,* 93–94, 127; Ben B. Lindsey, "My Fight with the Ku Klux Klan," *Survey* 54 (1 June 1925), 274; *New York World,* 12 Dec. 1926, in ACLU; *NYT,* 29 Oct. 1926, p. 48; Glavis, "When the Klan Tells the Truth," 26; Baiamonte, *Spirit of Vengeance,* 16, 62, 120.

68. Dollard, *Caste and Class,* quote on 333; see also 211–12, 328, 339; "Notes Taken from Speech of Wm. J. Simmons," 24 Jan. 1921, series C, box 312, NAACP. For the pattern in Georgia, see W. C. Bishop Confession, box 60, REM, 13; Kuhn et al., *Living Atlanta,* 313–14; Hux, "Klan in Macon," 34. Elsewhere, see Grand Jury Report to Judge J. R. Hamilton, 5 Dec. 1921, folder 2, JRH; "Yours for Law & Order," [anon.] to J. R. Hamilton, 28 June 1921, folder 1, loc. cit.; *NYT,* 26 April 1923, p. 40; *NYT,* 18 March 1928, p. 2; *NYT,* 13 April 1928, p. 27; Duffus, "Ancestry and End," 534; Crowell, "Collapse of Constitutional Government," 334; Harold W.

Stephens, "Mask and Lash in Crenshaw," *North American Review* 225 (April 1928), 400, 438; Toy, "Klan in Oregon," 75.

69. N.Y. "Pete" Rutherford interview by Cliff Kuhn, transcript, pp. 49, 55, box 10, LAC. For examples, see James A. McDermott to Gov. H. J. Allen, 12 Oct. 1922, series B, box 49, HJA; Hux, "Klan in Macon," 34; Grand Jury Report to Judge J. R. Hamilton, JRH.

70. For examples, see Stephens, "Mask and Lash in Crenshaw," 437; Toy, "Klan in Oregon," 75; Moseley, "Invisible Empire," 40–41; Baiamonte, *Spirit of Vengeance*, 141; CIC, "Race Relations in 1927," series I, reel 4, frame 552, CIC. See also Arthur Raper interview by Cliff Kuhn, transcript, p. 24, box 39, LAC.

71. See, for example, the incidents recounted in Dorsey, *Statement of Governor*, esp. p. 12; also *New York Tribune*, 14 March 1927, in series C, box 354, NAACP; Mrs. R. E. B. Lake to Grace Abbott, 26 Aug. 1923, box 162, Children's Bureau Papers; N. G. Romey to ACLU, 7 July 1922, vol. 215, p. 326, ACLU; *Atlanta Independent*, 23 Aug. 1923, p. 4; *ADB*, 27 Aug. 1921, p. 1; *ADB*, 23 Oct. 1921, p. 1.

72. *NYT*, 14 March 1927, in series C, box 354, NAACP; Hux, "Klan in Macon," 20, 39–40; James A. McDermott to Gov. H. J. Allen, 12 Oct. 1922, series B, box 49, HJA.

73. Quoted in *New York Herald Tribune*, 14 March 1927, in series C, box 354, NAACP; Walter Horace Carter interview, 17 Jan. 1976, transcript, p. 23, Southern Oral History Program Collection, SHC.

74. Walter White to Thomas G. McLeod, 26 Oct. 1928, folder 136, Walter White Papers; also White, *Rope and Faggot*, 29–33; Zangrando, *NAACP*, 92–93.

75. *Atlanta Independent*, 6 Dec. 1923, p. 1.

Chapter 8

1. Thomas Mann, "Germany and the Germans," speech at the Library of Congress after the defeat of Germany and the fall of the Nazis, May 1945, quoted in the exhibit *Degenerate Art: The Fate of the Avant-Garde in Nazi Germany*, ed. Stephanie Barron (Los Angeles, 1991).

2. *NYT*, 17 Aug. 1924, VIII, 4; *NYT*, 21 Feb. 1926, VIII, 1; *NYT*, 5 Feb. 1928, II, 1; ACLU, *Annual Report* (1925), 4; Lynd and Lynd, *Middletown*, 484; Loucks, *Klan in Pennsylvania*, 162; Weaver, "Klan in Wisconsin," 43–44, 142; Toy, "Klan in Oregon," 33; Gerlach, *Blazing Crosses*, xvi–xvii. For a dissenting view, see Cocoltchos, "Invisible Government," 567.

3. Jackson, *Klan in the City*, 43, 253–54; Rice, *Klan in American Politics*, 92.

4. N. B. Forrest, "Special Bulletin," 2 Sept. 1926, box 3, AK; Klan, Georgia, *Official Document*, Dec. 1926, p. 3; Coleman, *History of Georgia*,

293; *Kligrapp's Quarterly Report* forms in box 3, AK; Samuel Green to H.K.B., 24 June 1930, box 1, loc. cit.

5. Hux, "Klan in Macon," 21; Loucks, *Klan in Pennsylvania,* 164; Moore, *Citizen Klansmen,* 184–86; Goldberg, *Hooded Empire,* 58, 94–95, 178–79; Gerlach, *Blazing Crosses,* xvi-xvii, 8; Jenkins, *Steel Valley Klan,* 153; Alexander, *Crusade for Conformity,* 27.

6. The Klan's "ultimate weakness," maintained Kenneth Jackson, "was its lack of a positive program and a corresponding reliance upon emotion rather than reason." "The genuine American sense of decency," he concluded, "finally asserted itself and consigned the once mighty Klan to obscurity." Jackson, *Klan in the City,* 254–55. For similar, if less cheerful, views, see Gerlach, *Blazing Crosses,* 8, 83.

7. Gerlach, *Blazing Crosses,* 154; Loucks, *Klan in Pennsylvania,* 163; Blee, *Women of the Klan,* 5–6.

8. To the extent that scholars of the Klan make the comparison, they tend to dispose of it quickly and speciously. See, for example, Cocoltchos, "Invisible Government," 626. Exceptions to the prevailing provincialism are Robert Moats Miller, "The Ku Klux Klan," in *Change and Continuity in Twentieth-Century America: The 1920's,* ed. John Braeman, Robert H. Bremner, and David Brody (n.p., 1968), 215–55; Victor C. Ferkiss, "Populist Influences on American Fascism," *Western Political Quarterly* 10 (June 1957), 350–73. Both of these works are now quite dated; Ferkiss's work is particularly flawed by its one-sided caricature, taking off from Richard Hofstadter, of the Populism of the 1890s and its assertion of a direct, unmediated link between it and the fascisms of the twentieth century.

9. Mayer, *Why Did the Heavens Not Darken?* 31. For discussion of the international retreat in recent years from comparative study of these movements, see Tim Mason, "Whatever Happened to Fascism?" *Radical History Review* 49 (1991), 89–98. For the differences between counterrevolution, reaction, and conservatism and why they matter, see Arno J. Mayer, *Dynamics of Counterrevolution in Europe, 1870–1956: An Analytic Framework* (New York, 1971).

10. Quoted in James R. Green, *Grass-Roots Socialism: Radical Movements in the Southwest, 1895–1943* (Baton Rouge, 1978), 401–5; Siegfried, *America Comes of Age,* 134; Arthur Corning White, "An American Fascismo," *Forum* 72 (1924), 636–42. See also *Atlanta Independent,* 21 Dec. 1922, p. 1; "Our Own Secret Fascisti," *Nation* 115 (15 Nov. 1922), 514; Burbank, "Agrarian Radicals and Their Opponents"; see also DuBois, "The Shape of Fear," 293; Bohn, "The Klan Interpreted," 397. The discussion that follows concentrates on the commonalities between the Klan and fascist movements, not fascist governments in power, given the changes in composition, ideology, and object that occurred once fascist leaders assumed direction of the state.

11. *Searchlight,* 4 Nov. 1922, p. 4; *Imperial Night Hawk,* 4 April

1923, p. 2; *Kourier,* June 1925, p. 10; Jefferson, *Roman Catholicism and the Ku Klux Klan,* 145.

12. Quotes from Charles H. Martin, "White Supremacy and Black Workers: Georgia's 'Black Shirts' Combat the Great Depression." *LH,* 18 (1977) 366, also 369; and John Roy Carlson [pseud.], *Under Cover: My Four Years in the Nazi Underworld of America* (Philadelphia, 1943), 152–53, 198. See also John Hammond Moore, "Communists and Fascists in a Southern City: Atlanta, 1930," *South Atlantic Quarterly* 57 (1968), 446–47, 460–61; Toy, "Klan in Oregon," 45–46; *ABH,* 31 Aug. 1930, p. 1. For Bell's Klan involvement, see White, *Heroes of the Fiery Cross,* 197.

13. The discussion of Nazi ideology in this and ensuing paragraphs builds heavily on Mayer, "The Syncretism of *Mein Kampf,*" chap. 4 of *Why Did the Heavens Not Darken?* 90–109. Hereafter, it will only be cited when quoted; other sources will be cited as appropriate.

14. See William Sheridan Allen, *The Nazi Seizure of Power: The Experience of a Single German Town, 1930–1935* (Chicago, 1965), 26–27; also Mayer, *Dynamics of Counterrevolution,* 68, 70; Childers, *The Nazi Voter,* 268; Geoff Eley, "What Produces Fascism: Preindustrial Traditions or a Crisis of the Capitalist State?" in Michael N. Dobkowski and Isidor Walliman, eds., *Radical Perspectives on the Rise of Fascism in Germany, 1919–1945* (New York, 1989), 91; Reich, *Mass Psychology of Fascism,* 81; Mayer, *Dynamics of Counterrevolution,* 63–66.

15. Quotes from Mayer, *Why Did the Heavens Not Darken?* 9, 99. See also idem, *Dynamics of Counterrevolution,* 6–66; George L. Mosse, *Nazi Culture: Intellectual, Cultural, and Social Life in the Third Reich* (New York, 1966), 3, 95–96, 133–34.

16. Quotes from Allen, *Nazi Seizure of Power,* 22; Mayer, *Why Did the Heavens Not Darken?* 98. See also Mosse, *Nazi Culture,* 319–22; Mayer, *Dynamics of Counterrevolution,* 64; Gunter W. Remmling, "The Destruction of the Workers' Mass Movements in Nazi Germany," in Dobkowski and Walliman, *Radical Perspectives on the Rise of Fascism,* 215–30; Kurt Patzold, "Terror and Demagoguery in the Consolidation of the Fascist Dictatorship in Germany, 1933–34," in loc. cit., 231–46.

17. The phrase "revolutionary counterrevolutionism" comes from Mayer, who uses it to capture Hitler's self-representation as "a revolutionary against revolution." (Mayer, *Why Did the Heavens Not Darken?* 94). On Nazi racialism, see Mosse, *Nazi Culture,* 2, 57–60.

18. For the way this conservative sexual politics was rooted in Hitler's disdain for the rise of a concentrated urban working class, see Mayer, *Why Did the Heavens Not Darken?* 103. My understanding of the sexual politics of European fascism comes from Reich, *Mass Psychology of Fascism;* the essays in Renate Bridenthal et. al., eds., *When Biology Became Destiny: Women in Weimar and Nazi Germany* (New York, 1984); Claudia Koonz, *Mothers in the Fatherland: Women, the Family, and Nazi Politics* (New York, 1987); and Mosse, *Nationalism and Sexuality.*

19. For analyses along these lines of fascism as a social movement, see Allen, *Nazi Seizure of Power;* the essays in David Forgacs, ed., *Rethinking Italian Fascism: Capitalism, Populism and Culture* (London, 1986); Giampiero Carocci, *Italian Fascism,* trans. Isabel Quigly (Baltimore, 1975), esp. 7–27; Childers, *The Nazi Voter: The Social Foundations of Fascism in Germany, 1919–1933* (Chapel Hill, 1983); Mayer, *Dynamics of Counterrevolution;* Trotsky, *Struggle Against Fascism in Germany;* idem, *Whither France?* (1936; reprint, New York, 1968); Daniel Guerin, *Fascism and Big Business,* trans. Frances and Mason Merr (n.p., 1973); Felix Morrow, *Revolution & Counter-Revolution in Spain* (New York, 1974). For a dissenting view about lower-middle-class dominance in Hitler's popular following, and emphasis, instead, on upper- and upper-middle-class backing, see Richard F. Hamilton, *Who Voted for Hitler?* (Princeton, 1982).

20. Roderick Kedward, "Afterword: What Kind of Revisionism," in Forgacs, *Rethinking Italian Fascism,* 198; Mayer, *Why Did the Heavens Not Darken?* 54–55.

21. See, for example, *Searchlight,* 22 July 1922, p. 4.

22. See Simmons, *Klan Unmasked,* 208. For examples, see *Searchlight,* 11 Nov. 1922, p. 6; ibid., 19 Aug. 1922, p. 5; Evans, "History of Organized Labor," 291–92, 294–95; Toy, "Klan in Oregon," 143.

23. For illustrations, see Duffus, "Klan in the Middle West," 365; Gladys L. Palmer, *Union Tactics and Economic Change* (Philadelphia, 1932), 38–44; Cocoltchos, "Invisible Government," 194–96, 260–61, 331.

24. See Brian Peterson, "Regional Élites and the Rise of National Socialism," in Dobkowski and Walliman, *Radical Perspectives on the Rise of Fascism,* esp. 172; Mayer, *Why Did the Heavens Not Darken?* 94–95; idem, *Dynamics of Counterrevolution,* 119. Earlier in U.S. history, such threatened élites had played important roles in anti-abolition vigilantism. See Leonard L. Richards, *"Gentlemen of Property and Standing": Anti-Abolition Mobs in Jacksonian America* (New York, 1970), 131–50.

25. J. Q. Jett to Mrs. W. H. Felton, [1924], p. 7, RLF. For examples of intramural rivalries among whites over black labor involving the Klan, see *Atlanta Independent,* 20 Jan. 1921, pp. 1, 3; *Imperial Night-Hawk,* 4 April 1922. The question of the Klan's precise role—or roles—in the class relations of the rural countryside is a complex one that merits more research. But it should be addressed in light of evidence—unrelated to the Klan—that white farm tenants and small farmers sometimes employed whitecapping to drive out their black rivals. See Mathews, "Studies in Race Relations in Georgia," 193. For an interesting discussion of how élite support for vigilantism waned with changes in the political economy in one Southern city, see Ingalls, "Lynching and Establishment Violence in Tampa," esp. 640, 643–44.

26. *Kourier,* Dec. 1929, p. 4; also *Imperial Night-Hawk,* 4 April 1923, p. 2; White, *Heroes of the Fiery Cross,* 79–84.

27. For examples from Athens, see *ABH,* 21 Jan. 1925, p. 1; J. T. Jones

to Exalted Cyclops and Klansmen of Athens, 1 June 1925, box 1, AK. Nationally, see Klan, Georgia, *Official Document*, Nov. 1926, pp. 1, 2–3; Klan, *Official Monthly Bulletin*, 1 Dec. 1926, pp. 1, 2. On churches, see Allen, *The Nazi Seizure of Power*, 272; Mosse, *Nazi Culture*, 235–40.

28. Hamilton, *Who Voted for Hitler?* 3; Jackson, *Klan in the City*, 137–38. On conditions in Germany at the time of the election of 1932, see Childers, *Nazi Voter*, 192–93. In terms of social setting, the American experience of these years was more like that of England. Victors in the war, both of these countries developed similar movements of the far right, yet the impasse was less severe and their systems more able to accommodate the strain.

29. See, e.g., *ABH*, 5 April 1927, p. 1; *ABH*, 20 Jan. 1930, p. 1.

30. Leo Wolman, *The Growth of American Trade Unions*, 26, 33–37; Barnett, "American Trade Unionism"; Montgomery, *Fall of the House of Labor*, 453–54; Bernstein, *Lean Years*; and Dunn, *Americanization of Labor*. The rout of labor in the South was especially thorough. See Layne, *Cotton Mill Worker*, 203–6; Evans, "History of Organized Labor," 90; Yabroff and Herlihy, "History of Work Stoppages," 368–70. On the Left, see Draper, *American Communism and Soviet Russia*, 187, 513; Stein, *World of Marcus Garvey*, 130–31; Hamilton, *Who Voted for Hitler?* 301.

31. Stein, *World of Marcus Garvey*, 162; McMillen, *Dark Journey*, 314–16.

32. For the rise and demise of black militancy in the postwar U.S., see Stein, *World of Marcus Garvey*, esp. 129–32. For Garvey's accommodation to the Ku Klux Klan in the South, see ibid., 153–54, 159–60; Zangrando, *NAACP Crusade Against Lynching*, 91.

33. Rebecca Latimer Felton to Editor, *Walton News*, 16 Sept. 1924, box 8, RLF; A. B. Martin to J. R. Hamilton, 7 April 1922, folder 3, JRH; *Searchlight*, 22 July 1922, p. 4. See also J. M. Robinson to James W. Johnson, 20 Oct. 1921, series C, box 312, NAACP; W. H. Woodhouse to Henry J. Allen, 8 July 1922, series B, box 48, HJA.

34. White, *Rope and Faggot*, 120.

35. Allen, *Nazi Seizure of Power*, esp. 276; David Forgacs, "The Left and Fascism: Problems of Definition and Strategy," in Forgacs, ed., *Rethinking Italian Fascism*; Mabel Berezin, "Created Constituencies: The Italian Middle Classes and Fascism," in loc. cit., esp. 158; Hamilton, *Who Voted for Hitler?* esp. 422, 441; Rudy Koshar, "On the Politics of the Splintered Classes: An Introductory Essay," in Koshar, ed., *Splintered Classes*, esp. 6, 15; and the tragically prescient commentary by Leon Trotsky, *Struggle Against Fascism in Germany*.

36. The phrase comes from DeMott, *Imperial Middle*. For growing militarism and antipathy to labor struggle in the larger, multi-million-member fraternal tradition, see Clawson, *Constructing Brotherhood*, chap. 8, 26–28; Lynn Dumenil, *Freemasonry and American Culture, 1880–1930* (Princeton, 1984), 147; and chap. 4. For American racialist traditions, see,

for example, Richard Hofstadter, *Social Darwinism in American Thought,*
rev. ed. (Boston, 1955); Stephen Steinberg, *The Ethnic Myth: Race, Ethnicity, and Class in America* (Boston, 1981); Higham, *Strangers in the
Land;* Gordon, *Woman's Body, Woman's Right,* 136–58; Thomas G. Dyer,
Theodore Roosevelt and the Idea of Race (Baton Rouge, 1980). On the
race-coding of class, see Jordan, *White Over Black;* Morgan, *American
Slavery, American Freedom;* Ronald T. Takaki, *Iron Cages: Race and Cul-
ture in Nineteenth-Century America* (New York, 1979); Alexander Saxton,
*The Rise and Fall of the White Republic: Class Politics and Mass Culture
in Nineteenth-Century America* (London, 1990); David R. Roediger, *The
Wages of Whiteness: Race and the Making of the American Working
Class* (London, 1991). On gender and middle-class consciousness, see, for
a start, Mary P. Ryan, *Cradle of the Middle Class: The Family in Oneida
County, New York, 1790–1865* (Cambridge, Mass., 1981); Christine
Stansell, *City of Women: Sex and Class in New York, 1789–1860* (New
York, 1986).

37. James Weldon Johnson, testimony before United States Senate, *To
Prevent and Punish the Crime of Lynching,* Hearing before a Subcommittee of the Committee on the Judiciary, U.S. Senate, 69th Congress, 1st
session, 16 February 1926 (Washington, D.C., 1926), 24. For the cultural
legitimacy of such violence, see, in particular, Richard Maxwell Brown,
*Strain of Violence: Historical Studies of American Violence and Vigilan-
tism* (New York, 1975), esp. chap. 6, "Lawless Lawfulness: Legal and Be-
havioral Perspectives on American Vigilantism."

38. See Leo P. Ribuffo, *The Old Christian Right: The Old Protestant
Right from the Great Depression to the Cold War* (Philadelphia, 1983);
Alan Brinkley, *Voices of Protest: Huey Long, Father Coughlin, and the
Great Depression* (New York, 1983).

39. Anne Braden, "Lessons from a History of Struggle," *Southern Exposure* 8, n. 2 (Summer 1980), 56. There are even some reports of former
Klansmen in the South crossing over to the interracial Southern Tenant
Farmers Union and the Socialist and Communist Parties. See Foner, *Organized Labor and the Black Worker,* 207; Green, *Grass-Roots Socialism,*
414; Robin D. G. Kelley, *Hammmer and Hoe: Alabama Communists During the Great Depression,* (Chapel Hill, 1990), 28, 61. For an extraordinary,
suggestive recent example of such a turn, see " 'Why I Quit the Klan';
Studs Terkel Interviews C. P. Ellis," *Southern Exposure* 8, n. 2 (Summer
1980), 95–98.

▲ ▲ ▲

Bibliography

Primary Sources

Manuscript Collections, by Repository

Atlanta Historical Society:
  Brown, Joseph Mackey, Papers
  Living Atlanta Collection.
Clark Atlanta University, Robert W. Woodruff Library, Special Collections
    and Archives:
  Commission on Interracial Cooperation, Papers. Microfilm ed. Ann
    Arbor: University Microfilms International, 1984.
Duke University (Durham, N.C.), Special Collections Library:
  Bailey, Josiah William, Papers.
  Etheridge, Paul Sharp, Jr., Papers.
  Grant, Hugh Gladney, Papers.
Emory University, Special Collections, Robert W. Woodruff Library:
  Candler, Warren Akin, Papers.
  Harris, Julian LaRose, Papers.
  McGill, Ralph Emerson, Papers.
*Federal Surveillance of Afro-Americans (1917–1925): The First World
    War, the Red Scare, and the Garvey Movement.* Ed. Theodore Korn-
    weibel. Black Studies Research Sources: Microfilms from Major Archi-
    val and Manuscript Collections. Frederick, Maryland: University Pub-
    lications of America, 1985.
Georgia, Department of Archives and History, Atlanta:
  Records of the Department of Archives and History, Record Group 4,
    File II, Clarke County.

Records of the Executive Department, Record Group 1, Incoming Correspondence, 1927–1931.

Rodgers, Robert L., Collection, 1872–1927. AC 27–101.

Library of Congress, Washington, D.C., Manuscript Division:

Allen, Henry J., Papers.

Foulke, William Dudley, Papers.

National Association for the Advancement of Colored People, Papers.

New York Public Library, Astor, Lenox, and Tilden Foundations, Rare Books and Manuscripts Division:

Banta, Edwin P., Papers.

Princeton University, Seeley G. Mudd Manuscript Library:

American Civil Liberties Union Archives. Quoted with permission of Princeton University Libraries.

State Historical Society of Wisconsin, Madison, Manuscripts Section:

Blaine, John James, Papers.

Minwegen, Father Peter, Manuscript.

Women of the Ku Klux Klan, Klan No. 14, Chippewa Falls, Wisconsin, Papers.

Texas Tech University, Lubbock, Southwest Collection:

Biggers, Don Hampton, Papers.

Tuskegee Institute, Hollis Burke Frissell Library:

Tuskegee Institute News Clippings File.

University Archives, Main Library, University of Georgia, Athens:

Barrow, David Crenshaw, Papers.

Tate, William, Archives.

University of Georgia, Board of Trustees, Correspondence and Reports.

——, Board of Trustees, Minutes.

——, Demosthenian Society, Minutes.

——, Prudential Committee, Minutes and Miscellaneous Papers.

University of Georgia Libraries, Athens, Hargrett Rare Book and Manuscript Library:

Athens-Clarke County Heritage Foundation. Oral History Tapes.

Brooks, Robert Preston, Papers.

Brown, Joseph Mackey, Papers.

Carlton-Newton-Mell Collection.

Cobb, Howell, Papers.

Coulter, E. Merton, Pamphlet Collection.

Felton, Rebecca L., Collection.

Flanigen, Cameron Douglas, Papers.

Jett, John Quincy, Papers.

Ku Klux Klan, Athens Chapter No. 5, Papers.

Ku Klux Klan, Miscellany.

Scudder, Nina, Memorial Collection.

Taylor, Frances Long, Papers.

University of Georgia Libraries, Athens, Richard B. Russell Memorial Library:

Russell, Richard B., Jr., Collection.

Smith, Hoke, Papers.

University of North Carolina at Chapel Hill, Southern Historical Collection:

Branson, Eugene Cunningham, Papers.

Brown, J. J., Papers.

Bryan, Charles S., Papers and Books.

Federal Writers' Project, Papers.

Hamilton, James Robert, Papers.

Robinson, W. D., Papers.

Southern Oral History Program, Collection.

Watson, Thomas Edward, Papers.

Yale University, Beinecke Rare Book and Manuscript Library, Yale Collection of American Literature:

White, Walter Francis, Papers.

State and Federal Government Documents and Publications

Georgia. Department of Archives and History. *Georgia's Official Register,* 1923–1931. Atlanta, 1923–1931.

———. Commissioner of Commerce and Labor. *Annual Report of the Commissioner of Commerce and Labor,* 1915–1929/30. Atlanta, 1916–1931.

———. House of Representatives. *Journal of the House of Representatives of the State of Georgia,* 1920–1929. Atlanta, 1920–1929.

———. Senate. *Journal of the Senate of the State of Georgia,* 1920–1929. Atlanta, 1920–1929.

———. State Tax Commissioner. *Annual Report,* 1917–1929/30. Atlanta, 1918–1931.

Turner, Howard A., and L. D. Howell. "Condition of Farmers in a White-Farmer Area of the Cotton Piedmont, 1924–1926." U. S. Department of Agriculture. Circular No. 78. Washington, D.C., 1929.

United States Congress. House of Representatives. Committee on Rules.

Hearings on the Ku Klux Klan. 67th Congress, 1st session. Washington, D.C., 1921.

———. Senate. *To Prevent and Punish the Crime of Lynching.* Hearing before a Subcommittee of the Committee on the Judiciary, U.S. Senate, 69th Congress, 1st session, 16 February 1926. Washington, D.C., 1926.

———. Senate. *Punishment for the Crime of Lynching.* 73rd Congress, 2nd session. Washington, D.C., 1934.

———. Senate. *Senator from Texas, Hearings Before a Subcommittee of the Committee on Privileges and Elections.* 68th Congress, 1st session. Washington, D.C., 1924.

———. Senate. *Testimony Taken by the Joint Select Committee to Inquire into the Condition of Affairs in the Late Insurrectionary States.* Georgia, vols. 1 and 2. 42nd Congress, 2nd Session. Washington, D.C., 1872.

United States Department of the Interior. Census Office. *Tenth Census of the United States: 1880.* Washington, D.C., 1883.

———. Census Office. *Eleventh Census of the United States: 1890.* Washington, D.C., 1895.

———. Census Office. *Twelfth Census of the United States: 1900.* Washington, D.C., 1901–1904.

United States Department of Commerce. Bureau of the Census. *Census of Religious Bodies, 1926,* Part I, Summary and Detailed Tables. Washington, D.C., 1930.

———. Bureau of the Census. *Marriage and Divorce, 1916.* Washington, D.C., 1919.

———. Bureau of the Census. *Marriage and Divorce, 1922.* Washington, D.C., 1925.

———. Bureau of the Census. *Thirteenth Census of the United States: 1910.* Washington, D.C., 1912–1914.

———. Bureau of the Census. *Fourteenth Census of the United States: 1920.* Washington, D.C., 1921–1926.

———. Bureau of the Census. *Fifteenth Census of the United States: 1930.* Washington, D.C., 1931–1933.

———. Bureau of the Census. *Census of Agriculture: 1925,* Pt. II, *The Southern States.* Washington, D.C.

County Records

Note: Owing to the extensive demographic research on local Klan members conducted for this study, county sources are too numerous to list individually here. They included the federal manuscript census; city directories; tax digests; court records; miscellaneous church minutes, histories, records, registers, directories, and cemetery files; histories of local civic and social organizations; county Servicemen's Books and Veteran's Discharge records. All of these records can be found in either the Hargrett Rare Book and Manuscript Library in Athens or the Georgia Department of Archives and History in Atlanta. The citations that follow refer only to sources cited in the notes of this work.

Clarke County. Office of the Ordinary. Marriages. Georgia Department of Archives and History, Atlanta.

Clarke County. Office of the Ordinary. Minutes, 1889–1946. Georgia Department of Archives and History, Atlanta.

Clarke County. Superior Court. Minutes, 1914–1929. Georgia Department of Archives and History, Atlanta.

Clarke County. Superior Court. Registration of Charters, 1890–1929. Georgia Department of Archives and History, Atlanta.

Newspapers, Periodicals, and Annuals

American Civil Liberties Union. *Annual Reports,* vol. I, Jan. 1920–May 1930. Reprint, New York: Arno Press, 1970.

Athens Banner Herald (Athens, Ga.)

Athens Daily Banner (Athens, Ga.)

Athens Daily Herald (Athens, Ga.)

Atlanta Independent (Atlanta, Ga.)

Columbus Enquirer-Sun (Columbus, Ga.)

Knights of the Ku Klux Klan. Department of the Imperial Klaliff. *Official Monthly Bulletin* (Atlanta).

———. *Imperial Night Hawk* (Atlanta).

———. *Kourier* (Atlanta).

———. Realm of Georgia. Office of the Grand Dragon. *Official Document* (Atlanta).

New York Times

Searchlight (Atlanta)

Contemporary Books, Articles, and Pamphlets

Aikman, Duncan. "Prairie Fire." *American Mercury* 6 (Sept.–Dec. 1925), 209–14.

"Alabama Aroused." *Outlook* 147 (2 November 1927), 261.

American Red Cross, *Report of Survey: Athens and Clarke County.* Athens: n.p., 1920.

Atwood, Harry Fuller. *Safeguarding American Ideals.* Chicago: Laird & Lee, 1921.

Bohn, Frank. "The Klan Interpreted." *American Journal of Sociology* 30 (January 1925), 385–407.

Boyd, Thomas. "Defying the Klan." *Forum* 76 (July–Dec. 1926), 48–56.

Brooks, Robert Preston. *The Agricultural Revolution in Georgia, 1865–1912. Bulletin of the University of Wisconsin*, no. 639 (Madison, 1914).

———. "A Local Study of the Race Problem." *Political Science Quarterly* 26 (June 1911), 193–221.

———. *Georgia Studies; Selected Writings.* Athens: University of Georgia Press, 1952.

Brown, Egbert. *The Final Awakening; A Story of the Ku Klux Klan.* Brunswick, Georgia: Overstreet & Company, 1923.

Brunner, Edmund D. *Church Life in the Rural South: A Study of the Opportunity of Protestantism Based upon Data from Seventy Counties.* 1923; reprint, New York: Negro Universities Press, 1969.

Carlson, John Roy. *Under Cover: My Four Years in the Nazi Underworld of America.* Philadelphia: Blakiston Co., 1943.

Clark, William Lloyd. *The Devil's Prayer Book, or an Exposure of Auricular Confession as Practiced by the Roman Catholic Church: An Eye-Opener for Husbands, Fathers, and Brothers.* Milan, Ill., 1922.

Crowell, Chester T. "The Collapse of Constitutional Government." *Independent* 109 (9 Dec. 1922), 333–34.

Curry, Leroy Amos. *The Ku Klux Klan Under the Searchlight: An Authoritative, Dignified and Enlightened Discussion of the American Klan.* Kansas City, Mo.: Western Baptist Publishing Company, 1924.

Dau, W. H. T. *Weighed, and Found Wanting.* Fort Wayne, Ind.: American Luther League, n.d.

De Silver, Albert. "The Ku Klux Klan—'Soul of Chivalry.'" *Nation* 113 (14 Sept. 1921), 285–86.

Dever, Lem A. *Masks Off! Confessions of an Imperial Klansman.* 2nd ed. N.p: n.p., 1925.

Dixie Defense Committee. *The "Negro in Georgia."* N.p.: n.p., 1921.

Dorsey, Hugh M. *A Statement from Governor Hugh M. Dorsey as to the Negro in Georgia, April 22, 1921.* New York: NAACP, 1921.

Du Bois, W. E. Burghardt. "Georgia: Invisible Empire State." In *These United States: A Symposium*, vol. I, ed. Ernest Gruening. New York: Boni & Liveright, 1924.

———. "The Shape of Fear." *North American Review* 223 (1926), 291–304.

Duffus, Robert L. "Ancestry and End of the Ku Klux Klan." *World's Work* 46 (1923), 527–36.

———. "Counter-Mining the Ku Klux Klan." *World's Work* 46 (1923), 275–84.

———. "How the Klan Sells Hate." *World's Work* 46 (1923), 174–83.

———. "The Ku Klux Klan in the Middle West." *World's Work* 46 (1923), 363–72.

Dunn, Robert W. *The Americanization of Labor: The Employers' Offensive Against the Trade Unions.* New York: International Publishers, 1927.

Estes, George. *The Roman Katholic Kingdom and the Ku Klux Klan.* Troutdale, Oregon: George Estes, 1923.

Evans, Hiram Wesley. *The Attitude of the Knights of the Ku Klux Klan Toward the Jew.* Atlanta: n.p., 1923.

———. "The Ballots Behind the Ku Klux Klan." *World's Work* 55 (1927/28), 243–52.

———. "The Klan: Defender of Americanism." *Forum* 74 (Dec. 1925), 801–14.

———. *The Klan of Tomorrow and the Klan Spiritual.* N.p.: Knights of the Ku Klux Klan, 1924.

———. "The Ku Klux Klan's Fight for Americanism." *North American Review* 223 (1926), 33–63.

———. *The Public School Problem in America: Outlining Fully the Policies and the Program of the Knights of the Ku Klux Klan Toward the Public School System.* N.p: Ku Klux Klan, 1924.

———. *The Rising Storm: An Analysis of the Growing Conflict over the Political Dilemma of the Roman Catholics in America.* Atlanta: Buckhead Publishing Co., 1930.

——— et al. *Is the Klan Constructive or Destructive? A Debate Between Imperial Wizard Evans, Israel Zangwill, and Others.* Girard, Kansas: Haldeman-Julius Company, 1924.

Evans, Mercer G. "Are Southern Cotton Mills Feudalistic?" *American Federationist* 35 (Nov. 1928), 1,354–59.

"For and Against the Ku Klux Klan." *Literary Digest* 70 (24 Sept. 1921), 34–40.

Fowler, C. Lewis. *The Ku Klux Klan: Its Origin, Meaning and Scope of Operation.* Atlanta: n.p. 1922.

Frank, Glenn. "Christianity and Racialism: Has the Klan the Right to Celebrate Christmas?" *Century* 109 (1924/25), 277–84.

Frost, Stanley. *The Challenge of the Klan.* Indianapolis: Bobbs-Merrill, 1924.

Fry, Henry P. *The Modern Ku Klux Klan.* Boston: Small, Maynard & Company, 1922.

"Georgia's Indictment." *Survey,* 7 May 1921, pp. 183–84.

Glavis, Louis R. "The Ku Klux Klan Invades Two Capitals." *Hearst's International Magazine* (Feb. 1924), 10.

———. "When the Klan Tells the Truth." *Hearst's International Magazine* (March 1924), 26–29.

Googe, George L. "Organizing the Workers of Georgia." *American Federationist* 35 (Nov. 1928), 1,326–27.

Gould, Charles W. *America: A Family Matter*. New York: Charles Scribner's Sons, 1922.

"Governor Dorsey Stirs Up Georgia." *Literary Digest* 69 (4 June 1921), 19.

Harris, Judia C. Jackson. *Race Relations*. Athens: n.p., 1925.

Herndon, Angelo. *Let Me Live*. New York: Arno Press, 1969.

Jackson, Helen. *Convent Cruelties; or, My Life in a Convent*. Toledo, Ohio: John & Helen Jackson, 1919.

Jefferson, Charles E. *Roman Catholicism and the Ku Klux Klan*. New York: Fleming H. Revell Company, 1925.

Johnson, Guy B. "The Race Philosophy of the Ku Klux Klan." *Opportunity* 1 (1923), 268–70.

———. "A Sociological Interpretation of the New Ku Klux Klan Movement." *Social Forces* 1 (May 1923), 440–45.

Johnson, M. K. "School Conditions in Clarke County, Georgia, with Special Reference to Negroes." *Bulletin of the University of Georgia* 16, n. 11a (August 1916).

Jones, Winfield. *Story of the Ku Klux Klan*. Washington, D.C.: American Newspaper Syndicate, 1921.

Kennedy, Stetson. *I Rode with the Ku Klux Klan*. London: Arco Publishers, 1954.

Kent, Grady R. *Flogged by the Ku Klux Klan*. Cleveland, Tenn.: White Wing Publishing House, 1942.

Knights of the Ku Klux Klan. *Constitution and Laws of the Knights of the Ku Klux Klan*. Atlanta: 1921.

———. *Kloran*. 5th ed. Atlanta: Imperial Palace, 1916.

———. *Official Message of the Emperor of the Invisible Empire of the Knights of the Ku Klux Klan to the Initial Session of the Imperial Klonvocation*. Atlanta: n.p., 1922.

———. *Papers Read at the Meeting of the Grand Dragons of the Knights of the Ku Klux Klan, At their First Annual Meeting held at Asheville, North Carolina, July 1923, Together With Other Articles of Interest to Klansmen*. Atlanta: Knights of the Ku Klux Klan, 1923.

———. *The Practice of Klannishness*. Imperial Instructions, Document No. 1. Atlanta: Imperial Palace, 1924.

———. *Thirty-Three Questions Answered*. Atlanta: Knights of the Ku Klux Klan, n.d.

———. Department of Realms. *Klan Building: An Outline of Proven Klan*

Methods for Successfully Applying the Art of Klancraft in Building and Operating Local Klans. Atlanta: n.p., n.d.

"The Ku Klux Are Riding Again." *The Crisis* 17 (March 1919), 231.

"The Ku Klux Klan." *Catholic World* 116 (January 1923), 433–43.

Kuczynski, Jurgen. "The Merchant and His Wage Earning Customers in the South and in the North." *American Federationist* 35 (Nov. 1928).

Lindsey, Ben B. "My Fight with the Ku Klux Klan." *Survey* 54 (1 June 1925), 271–74.

Lynd, Robert S., and Helen Merrell Lynd. *Middletown: A Study in Contemporary American Culture.* New York: Harcourt, Brace, and World, 1929.

Mast, Blaine. *K.K.K. Friend or Foe: Which?* N.p.: Blaine Mast, 1924.

Mecklin, John Moffat. *The Ku Klux Klan: A Study of the American Mind.* 1923; reprint, New York: Russell & Russell, 1963.

Monteval, Marion. *The Klan Inside Out.* 1924; reprint, Westport: Negro Universities Press, 1970.

Moore, Samuel Taylor. "How the Kleagles Collected the Cash." *Independent* (13 Dec. 1924).

Odum, Howard Washington. *An American Epoch: Southern Portraiture in the National Picture.* New York: Henry Holt and Company, 1930.

O'Kelley, H. S. "Sanitary Conditions among the Negroes of Athens,Georgia." *Bulletin of the University of Georgia* 18, n. 7 (July 1918).

"Our Own Secret Fascisti." *Nation* 115 (15 Nov. 1922), 514.

Patton, R. A. "A Klan Reign of Terror." *Current History* 28 (April 1928), 51–5.

"The Reign of the Tar Bucket." *Literary Digest* 70 (27 August 1921), 12–13.

Robertson, William Joseph. *The Changing South.* New York: Boni and Liveright, 1927.

Rogers. J. A. *The Ku Klux Spirit.* New York: Messenger Publishing Co., 1923.

Romine, Mr., and Mrs. W. B. *A Story of the Ku Klux Klan.* Pulaski, Tenn.: Pulaski Citizen, 1924.

Saloman, Samuel. *The Red War on the Family.* New York: J. J. Little & Ives, 1922.

Scarborough, Donald Dewey. "An Economic Study of Negro Farmers as Owners, Tenants, and Croppers." *Bulletin of the University of Georgia* 25, n. 2a (Sept. 1924).

Siegfried, André. *America Comes of Age.* Trans. H. H. Heming and Doris Heming. New York: Harcourt, Brace and Company, 1927.

Shepherd, William G. "Fighting the K.K.K. on Its Home Grounds." *Leslie's* 133 (15 Oct. 1921).

Simmons, William Joseph. *America's Menace, or the Enemy Within.* Atanta: Bureau of Patriotic Books, 1926.

———. *The Klan Unmasked.* Atlanta: William E. Thompson Publishing Company, 1923.

Stanton, E. F. *Christ and Other Klansmen, or Lives of Love; The Cream of the Bible Spread upon Klanism.* Kansas City, Mo.: Stanton & Harper, 1924.

Stephens, Harold W. "Mask and Lash in Crenshaw." *North American Review* 225 (April 1928), 435–42.

Sweeney, Charles P. "The Great Bigotry Merger." *Nation* 115 (5 July 1922).

Taylor, Francis Long. "The Negroes of Clarke County, Georgia, During the Great War." *Bulletin of the University of Georgia* 29, n. 8 (Sept. 1919).

Tucker, Howard A. *History of Governor Walton's War on [the] Ku Klux Klan.* Oklahoma City: Southwest Publishing Company, 1923.

University of Georgia. Senior Class of 1925. *Pandora.* Athens: n.p., 1925.

Upshaw, William David. *Clarion Calls from Capitol Hill.* New York: Fleming H. Revell Company, 1923.

Watson, Thomas E. *The Inevitable Crimes of Celibacy.* Thomson, Ga.: Jeffersonian Publishing Co., 1917.

———. *The Life and Speeches of Thomas E. Watson.* Thomson, Ga.: Jeffersonian Publishing Co., 1916.

———. *Socialists and Socialism.* Thomson, Ga.: Press of theJeffersonians, 1910.

White, Alma. *Heroes of the Fiery Cross.* Zarepath, N.J.: Good Citizen, 1928.

———. *The Ku Klux Klan in Prophecy.* Zarepath, N.J.: Good Citizen, 1925.

White, Arthur Corning. "An American Fascismo." *Forum* 72 (July–Dec. 1924), 636–42.

Winter, Paul M. *What Price Tolerance?* New York: All-American Book, Lecture and Research Bureau, 1928.

Witcher, W. C. *The Reign of Terror in Oklahoma.* Fort Worth: The author, n.d.

Wood, J. O. *Are You a Citizen?* Atlanta: Searchlight Publishing Company, n.d.

Woofter, Thomas Jackson, Jr. *Negro Migration: Changes in Rural Organization and Population of the Cotton Belt.* New York: W. D. Gray, 1920.

———. "The Negroes of Athens, Georgia." *Bulletin of the University of Georgia* 14, n. 4 (Dec. 1913).

Secondary Sources

Books

Ahlstrom, Sydney E. *A Religious History of the American People.* Vol. 2. Garden City, N.Y.: Doubleday & Company, 1975.

Alexander, Charles C. *Crusade for Conformity: The Ku Klux Klan in Texas, 1920–1930.* Houston: Texas Gulf Coast Historical Association, Publication Series, vol. 6, n. 1 (August, 1962).

———. *The Ku Klux Klan in the Southwest.* Lexington: University of Kentucky Press, 1965.

Allen, Robert L. *Reluctant Reformers: Racism and Social Reform Movements in the United States.* Washington, D.C.: Howard University Press, 1974.

Allen, William Sheridan. *The Nazi Seizure of Power: The Experience of a Single German Town, 1930–1935.* Chicago: Quadrangle Books, 1965.

Ameringer, Oscar. *If You Don't Weaken: The Autobiography of Oscar Ameringer.* New York: Henry Holt & Co., 1940.

Anderson, Benedict R. *Imagined Communities: Reflections on the Origin and Spread of Nationalism.* London: Verso Books, 1983.

Angle, Paul M. *Bloody Williamson: A Chapter in American Lawlessness.* New York: Alfred A. Knopf, 1952.

Arnett, Alex Matthews. *The Populist Movement in Georgia.* New York, 1922.

Ayers, Edward L. *Vengeance and Justice: Crime and Punishment in the Nineteenth-Century American South.* New York: Oxford University Press, 1984.

Baiamonte, John V., Jr. *Spirit of Vengeance: Nativism and Louisiana Justice, 1921–1924.* Baton Rouge: Louisiana State University Press, 1986.

Barrow, David C., et al. *History of Athens and Clarke County.* Athens: H. J. Rowe, 1923.

Bartley, Numan V. *The Creation of Modern Georgia.* Athens: University of Georgia Press, 1983.

Bernstein, Irving. *The Lean Years.* Boston: Houghton Mifflin, 1960.

Blee, Kathleen M. *Women of the Klan: Racism and Gender in the 1920s.* Berkeley: University of California Press, 1991.

Boyer, Paul S. *Urban Masses and Moral Order in America, 1820–1920.* Cambridge: Harvard University Press, 1978.

Bridenthal, Renate, et al., eds. *When Biology Became Destiny: Women in Weimar and Nazi Germany.* New York: Monthly Review Press, 1984.

Brinkley, Alan. *Voices of Protest: Huey Long, Father Coughlin, and the Great Depression.* New York: Random House, 1983.

Brown, Richard Maxwell. *Strain of Violence: Historical Studies of American Violence and Vigilantism.* New York: Oxford University Press, 1975.

Brownell, Blaine A. *The Urban Ethos in the South, 1920–1930.* Baton Rouge: Louisiana State University Press, 1975.

Burnham, Walter Dean. *Critical Elections and the Mainsprings of the American Party System.* New York: W. W. Norton, 1970.

———. *The Current Crisis in American Politics.* New York: Oxford University Press, 1982.

Carby, Hazel. *Reconstructing Womanhood: The Emergence of the Afro-American Woman Novelist.* New York: Oxford University Press, 1987.

Carlton, David L. *Mill and Town in South Carolina, 1880–1920.* Baton Rouge: Louisiana State University Press, 1982.

Carnes, Mark C. *Secret Ritual and Manhood in Victorian America.* New Haven: Yale University Press, 1989.

Carocci, Giampiero. *Italian Fascism.* Trans. Isabel Quigly. Baltimore: Penguin Books, 1975.

Chalmers, David M. *Hooded Americanism: A History of the Ku Klux Klan.* Rev. ed. New York: Franklin Watts, 1981.

Childers, Thomas. *The Nazi Voter: The Social Foundations of Fascism in Germany, 1919–1933.* Chapel Hill: University of North Carolina Press, 1983.

Clark, Norman. *Deliver Us from Evil: An Interpretation of American Prohibition.* New York: W. W. Norton, 1976.

Clawson, Mary Ann. *Constructing Brotherhood: Class, Gender, and Fraternalism.* Princeton: Princeton University Press, 1989.

Coben, Stanley. *Rebellion Against Victorianism: The Impetus for Change in 1920s America.* New York: Oxford University Press, 1991.

Coleman, Kenneth, ed. *A History of Georgia.* Athens: University of Georgia Press, 1977.

——— and Charles Stephen Gurr, eds. *Dictionary of Georgia Biography.* Athens: University of Georgia Press, 1983.

Collins, Patricia Hill. *Black Feminist Thought: Knowledge, Consciousness, and the Politics of Empowerment.* New York: Routledge, 1990.

Cott, Nancy. *The Grounding of Modern Feminism.* New Haven: Yale University Press, 1987.

Crozier, E. W. *The White-Caps: A History of the Organization in Sevier County.* 1899; reprint, n.p.: Brazos Printing Co., 1963.

Daniel, Pete. *The Shadow of Slavery: Peonage in the South, 1901–1969.* Urbana: University of Illinois Press, 1972.

Davis, Allison, Burleigh B. Gardner, and Mary R. Gardner. *Deep South: A Social Anthropological Study of Caste and Class.* Chicago: University of Chicago Press, 1941.

Davis, David Brion, ed. *The Fear of Conspiracy: Images of Un-American Subversion from the Revolution to the Present.* Ithaca: Cornell University Press, 1971.

Dees, Morris, with Steve Fiffer. *A Season for Justice: The Life and Times of Civil Rights Lawyer Morris Dees.* New York: Charles Scribner's Sons, 1991.

D'Emilio, John, and Estelle B. Freedman. *Intimate Matters: A History of Sexuality in America.* New York: Harper & Row, 1988.

DeMott, Benjamin. *The Imperial Middle: Why Americans Can't Think Straight About Class.* New York: William Morrow and Company, 1990.

Dinnerstein, Leonard. *The Leo Frank Case.* New York: Columbia University Press, 1968.

Dittmer, John. *Black Georgia in the Progressive Era.* Urbana: University of Illinois Press, 1977.

Dobkowski, Michael N., and Isidor Walliman, eds. *Radical Perspectives on the Rise of Fascism in Germany, 1919–1945.* New York: Monthly Review Press, 1989.

Dollard, John. *Caste and Class in a Southern Town.* Garden City, New York: Doubleday, 1949.

Douglas, Mary. *Purity and Danger: An Analysis of the Concepts of Pollution and Taboo.* London: Routledge and Kegan Paul, 1966.

Draper, Hal. *The Politics of Social Classes*, vol. 2, *Karl Marx's Theory of Revolution.* New York: Monthly Review Press, 1978.

Draper, Theodore. *American Communism and Soviet Russia.* New York: Random House, 1986.

Du Bois, W. E. Burghardt. *Black Reconstruction in America.* 1935; reprint, Cleveland: World Publishing Company, 1964.

Dumenil, Lynn. *Freemasonry and American Culture, 1880–1930.* Princeton: Princeton University Press, 1984.

Dyer, Thomas G. *Theodore Roosevelt and the Idea of Race.* Baton Rouge: Louisiana State University Press, 1980.

———. *The University of Georgia: A Bicentennial History, 1785–1985.* Athens: University of Georgia Press, 1985.

Dykeman, Wilma and James Stokely. *Seeds of Southern Change: The Life of Will Alexander.* Chicago: University of Chicago Press, 1962.

Edelman, Murray. *The Symbolic Uses of Politics*. Urbana: University of Illinois Press, 1967.

Ellison, Ralph. *Shadow and Act*. 1964; reprint, New York: Vintage, 1972.

Epstein, Barbara. *The Politics of Domesticity: Women, Evangelism and Temperance in Nineteenth-Century America*. Middletown: Wesleyan University Press, 1981.

Escott, Paul D. *Many Excellent People: Power and Privilege in North Carolina, 1850–1900*. Chapel Hill: University of North Carolina Press, 1985.

Fite, Gilbert C. *American Farmers: The New Minority*. Bloomington: University of Indiana Press, 1981.

Flynn, Charles. *White Land, Black Labor: Caste and Class in Late Nineteenth-Century Georgia*. Baton Rouge: Louisiana State University Press, 1983.

Foner, Eric. *Reconstruction: America's Unfinished Revolution, 1863–1877*. New York: Harper & Row, 1988.

———. *Tom Paine and Revolutionary America*. New York: Oxford University Press, 1976.

Foner, Philip S. *Organized Labor and the Black Worker, 1619–1981*. 2d ed. New York: International Publishers, 1982.

Forgacs, David, ed. *Rethinking Italian Fascism: Capitalism, Populism and Culture*. London: Lawrence and Wishart, 1986.

Formisano, Ronald P. *Boston Against Busing: Race, Class, and Ethnicity in the 1960s and 1970s*. Chapel Hill: University of North Carolina, 1990.

Friedman, Jean E. *The Enclosed Garden: Women and Community in the Evangelical South, 1830–1900*. Chapel Hill: University of North Carolina Press, 1985.

Galambos, Louis, and Joseph Pratt. *Rise of the Corporate Commonwealth: U.S. Business and Public Policy in the 20th Century*. New York: Basic Books, 1988.

Garlock, Jonathan, comp. *A Guide to Local Assemblies of the Knights of Labor*. Westport: Greenwood Press, 1982.

Garrow, David J., ed. *The Montgomery Bus Boycott and the Women Who Started It: The Memoir of Jo Ann Gibson Robinson*. Knoxville: University of Tennessee Press, 1987.

Gerber, David, ed. *Anti-Semitism in American History*. Urbana: University of Illinois Press, 1986.

Gerlach, Larry R. *Blazing Crosses in Zion: The Ku Klux Klan in Utah*. Logan, Utah: Utah State University Press, 1982.

Giddings, Paula. *When and Where I Enter: The Impact of Black Women on Race and Sex in America*. New York: Bantam Books, 1985.

Goldberg, Robert Alan. *Hooded Empire: The Ku Klux Klan in Colorado.* Urbana: University of Illinois Press, 1981.

Goodwyn, Lawrence. *The Populist Moment: A Short History of Agrarian Revolt in America.* New York: Oxford University Press, 1978.

Gordon, Linda. *Heroes of Their Own Lives: The Politics and History of Family Violence; Boston, 1880–1960.* New York: Viking, 1988.

————. *Woman's Body, Woman's Right: A Social History of Birth Control in America.* New York: Penguin Books,1977.

————, ed. *Women, the State, and Welfare.* Madison: University of Wisconsin Press, 1990.

Green, James. *Grass Roots Socialism: Radical Movements in the Southwest, 1895–1943.* Baton Rouge: Louisiana State University Press, 1978.

Guerin, Daniel. *Fascism and Big Business.* Trans. Frances Merrill and Mason Merrill. New York: Monad Press, 1973.

Gutman, Herbert G. *Slavery and the Numbers Game: A Critique of Time on the Cross.* Urbana: University of Illinois Press, 1975.

Hahn, Steven. *The Roots of Southern Populism: Yeoman Farmers and the Transformation of the Georgia Upcountry, 1850–1890.* New York: Oxford University Press, 1983.

Hamilton, Richard F. *Who Voted for Hitler?* Princeton: Princeton University Press, 1982.

Hall, Jacquelyn Dowd. *Revolt against Chivalry: Jessie Daniel Ames and the Women's Campaign against Lynching.* New York: Columbia University Press, 1979.

————, James Leloudis, Robert Korstad, Mary Murphy, Lu Ann Jones, and Christopher B. Daly. *Like a Family: The Making of a Southern Cotton Mill World.* Chapel Hill: University of North Carolina Press, 1987.

Harris, J. William. *Plain Folk and Gentry in a Slave Society: White Liberty and Black Slavery in Augusta's Hinterlands.* Middletown: Wesleyan University Press, 1985.

Hatcher, Orie Latham. *Rural Girls in the City for Work.* Richmond, Va.: Garratt and Massie, 1930.

Hays, Samuel P. *American Political History as Social Analysis.* Knoxville: University of Tennessee Press, 1980.

Herring, Harriet L. *Passing of the Mill Village.* Chapel Hill: University of North Carolina Press, 1949.

————. *Welfare Work in Mill Villages.* Chapel Hill: University of North Carolina Press, 1929.

Hertzberg, Steven. *Strangers within the Gate City: The Jews of Atlanta,*

1845–1915. Philadelphia: Jewish Publication Society of America, 1978.

Higham, John. *Strangers in the Land: Patterns of American Nativism, 1860–1925.* 1963; reprint, New York: Atheneum, 1974.

Hofstadter, Richard. *The Age of Reform.* New York: Random House, 1955.

———. *Social Darwinism in American Thought,* rev. ed. Boston: Beacon Press, 1955.

Hooks, Bell. *From Margin to Center.* Boston: South End Press, 1984.

Jackson, Kenneth T. *The Ku Klux Klan in the City, 1915–1930.* New York: Oxford University Press, 1967.

Jenkins, William D. *Steel Valley Klan: The Ku Klux Klan in Ohio's Mahoning Valley.* Kent: Kent State University Press, 1990.

Johnson, James Weldon. *Along This Way: The Autobiography of James Weldon Johnson.* 1933; reprint, New York: Viking Press, 1968.

Jones, Jacqueline. *Labor of Love, Labor of Sorrow: Black Women, Work, and the Family, from Slavery to the Present.* New York: Random House, 1986.

Jordan, Winthrop D. *White Over Black: American Attitudes Toward the Negro, 1550–1812.* Chapel Hill: University of North Carolina Press, 1968.

Kelley, Robin D. G. *Hammer and Hoe: Alabama Communists During the Great Depression.* Chapel Hill, 1990.

Kerber, Linda. *Women of the Republic: Intellect and Ideology in Revolutionary America.* New York: W. W. Norton, 1986.

Kinzer, Donald. *An Episode in Anti-Catholicism: The American Protective Association.* Seattle: University of Washington Press, 1964.

Kocka, Jurgen. *White Collar Workers in America, 1890–1940.* Trans. Maura Kealey. Beverly Hills: Sage Publications, 1980.

Koonz, Claudia. *Mothers in the Fatherland: Women, the Family, and Nazi Politics.* New York: St. Martin's Press, 1987.

Koshar, Rudy, ed. *Splintered Classes: Politics and the Lower Middle Class in Interwar Europe.* New York: Holmes & Meier, 1990.

Kousser, J. Morgan. *The Shaping of Southern Politics: Suffrage Restriction and the Establishment of the One-Party South, 1880–1910.* New Haven: Yale University Press, 1974.

Kramnick, Isaac. *Republicanism and Bourgeois Radicalism: Political Ideology in Late Eighteenth-Century England and America.* Ithaca: Cornell University Press, 1990.

Kuhn, Clifford M., Harlon E. Joye, and E. Bernard West. *Living Atlanta: An Oral History of the City, 1914–1948.* Athens: University of Georgia Press, 1990.

Lay, Shaun. *War, Revolution and the Ku Klux Klan: A Study of Intolerance in a Border City.* El Paso: Texas Western Press, 1985.

———, ed. *The Invisible Empire in the West: Toward a New Historical Appraisal of the Ku Klux Klan of the 1920s.* Urbana: University of Illinois Press, 1992.

Layne, Herbert Jay. *The Cotton Mill Worker in the Twentieth Century.* New York: Farrar & Rinehart, 1944.

Leon, Abram. *The Jewish Question: A Marxist Analysis.* Second ed. New York: Pathfinder Press, 1970.

Lerner, Gerda. *Black Women in White America: A Documentary History.* New York: Vintage Books, 1973.

Levin, N. Gordon, Jr. *Woodrow Wilson and World Politics: America's Response to War and Revolution.* New York: Oxford University Press, 1968.

Levine, Lawrence W. *Defender of the Faith: William Jennings Bryan; The Last Decade, 1915–1925.* New York: Oxford University Press, 1965.

Lewis, Hylan. *Blackways of Kent.* Chapel Hill: University of North Carolina Press, 1955.

Lipset, Seymour M., and Earl Raab. *The Politics of Unreason: Right-Wing Extremism in America, 1790–1970.* New York: Harper & Row, 1970.

Loucks, Emerson H. *The Ku Klux Klan in Pennsylvania: A Study in Nativism.* Harrisburg, Pa.: Telegraph Press, 1936.

Luker, Kristin. *Abortion and the Politics of Motherhood.* Berkeley: University of California Press, 1984.

MacDonald, Lois. *Southern Mill Hills: A Study of Social and Economic Forces in Certain Mill Villages.* New York: A.L. Hillman, 1928.

MacPherson, C. B. *The Political Theory of Possessive Individualism.* New York: Oxford University Press, 1964.

Marsden, George M. *Fundamentalism and American Culture: The Shaping of Twentieth-Century Evangelicalism, 1870–1925.* New York: Oxford University Press, 1980.

Marshall, F. Ray. *Labor in the South.* Cambridge: Harvard University Press, 1967.

Martinez-Alier, Verena. *Marriage, Class and Colour in Nineteenth-Century Cuba: A Study of Racial Attitudes and Sexual Values in a Slave Society.* London: Cambridge University Press, 1974.

Marx, Karl, and Frederick Engels. *Manifesto of the Communist Party.* Peking: Foreign Languages Press, 1972.

May, Elaine Tyler. *Great Expectations: Marriage and Divorce in Post-Victorian America.* Chicago: University of Chicago Press, 1980.

Mayer, Arno J. *Dynamics of Counterrevolution in Europe, 1870–1956: An Analytic Framework.* New York: Harper & Row, 1971.

———. *Why Did the Heavens Not Darken? The "Final Solution" in History.* New York: Pantheon Books, 1990.

McLoughlin, William G., Jr. *Billy Sunday Was His Real Name.* Chicago: University of Chicago Press, 1955.

McMath, Robert C., Jr. *Populist Vanguard: A History of the Southern Farmers' Alliance.* Chapel Hill: University of North Carolina Press, 1975.

McMillen, Neil R. *Dark Journey: Black Mississippians in the Age of Jim Crow.* Urbana: University of Illinois Press, 1989.

McWilliams, Carey. *A Mask for Privilege: Anti-Semitism in America.* Boston: Little, Brown & Co., 1948.

Meyerowitz, Joanne J. *Women Adrift: Independent Wage Earners in Chicago, 1880–1930.* Chicago: University of Chicago Press, 1988.

Miller, Nora. *The Girl in the Rural Family.* Chapel Hill: University of North Carolina Press, 1935.

Mills, C. Wright. *White Collar: The American Middle Classes.* 1951; reprint, New York: Oxford University Press, 1976.

Montell, Lynwood. *Killings: Folk Justice in the Upper South.* Lexington: University of Kentucky Press, 1986.

Montgomery, David. *The Fall of the House of Labor: The Workplace, the State, and American Labor Activism, 1865–1925.* Cambridge: Cambridge University Press, 1989.

Moore, Leonard J. *Citizen Klansmen: The Ku Klux Klan in Indiana, 1921–1928.* Chapel Hill: University of North Carolina Press, 1991.

Morgan, Edmund S. *American Slavery, American Freedom: The Ordeal of Colonial Virginia.* New York: W. W. Norton, 1975.

Morland, John Kenneth. *Millways of Kent.* Chapel Hill: University of North Carolina Press, 1958.

Mosse, George L. *Nationalism and Sexuality: Respectability and Abnormal Sexuality in Modern Europe.* New York: Howard Fertig, 1985.

———. *Nazi Culture: Intellectual, Cultural, and Social Life in the Third Reich.* New York: Grosset & Dunlap, 1966.

Murray, Robert K. *Red Scare: A Study of National Hysteria, 1919–1920.* New York: McGraw-Hill Book Company, 1964.

Nall, James O. *The Tobacco Night-Riders of Kentucky and Tennessee, 1905–1909.* Louisville: Standard Press, 1939.

Neverdon-Morton, Cynthia. *Afro-American Women of the South and the Advancement of the Race, 1895–1925.* Knoxville: University of Tennessee Press, 1989.

Newby, I. A. *Jim Crow's Defense: Anti-Negro Thought in America, 1900–1930*. Baton Rouge: Louisiana State University Press, 1965.

———. *The South: A History*. N.Y.: Holt, Rinehart and Winston, 1978.

Nord, Phillip G. *Paris Shopkeepers and the Politics of Resentment*. Princeton: Princeton University Press, 1986.

Ostrander, Gilman M. *American Civilization in the First Machine Age: 1890–1940*. New York: Harper & Row, 1970.

Painter, Nell Irvin. *Standing at Armageddon: The United States, 1877–1919*. New York: W. W. Norton, 1987.

Palmer, Bruce. *"Man Over Money": The Southern Populist Critique of American Capitalism*. Chapel Hill: University of North Carolina Press, 1980.

Palmer, Gladys L. *Union Tactics and Economic Change*. Philadelphia: University of Pennsylvania Press, 1932.

Patterson, Horace C., and Gilbert C. Fite. *Opponents of War, 1917–1918*. Seattle: University of Washington Press, 1968.

Peiss, Kathy Lee. *Cheap Amusements: Working Women and Leisure in New York City, 1880–1920*. Philadelphia: Temple University Press, 1986.

Percy, William Alexander. *Lanterns on the Levee: Recollections of a Planter's Son*. 1941; reprint, Baton Rouge: Louisiana State University Press, 1973.

Petchesky, Rosalind P. *Abortion and Woman's Choice: The State, Sexuality, and Reproductive Freedom*. Boston: Northeastern University Press, 1984.

Pivar, David. *Purity Crusade: Sexual Morality and Social Control, 1868–1900*. Westport: Greenwood, 1973.

Potwin, Marjorie A. *Cotton Mill People of the Piedmont: A Study in Social Change*. New York: Columbia University Press, 1927.

Preston, William, Jr. *Aliens and Dissenters: Federal Suppression of Radicals, 1903–1933*. New York: Harper & Row, 1963.

Range, Willard. *A Century of Georgia Agriculture, 1850–1950*. Athens: University of Georgia Press, 1954.

Raper, Arthur F. *Preface to Peasantry: A Tale of Two Black Belt Counties*. Chapel Hill: University of North Carolina Press, 1936.

———. *The Tragedy of Lynching*. 1933; reprint, New York: Arno Press, 1969.

Reed, Thomas Walter. *David Crenshaw Barrow*. Athens: The author, 1935.

Reich, Wilhelm. *The Mass Psychology of Fascism*. Trans. Vincent R. Carfagno. New York: Simon & Schuster, 1970.

Rhyne, Jennings J. *Some Southern Cotton Mill Workers and Their Villages.* Chapel Hill: University of North Carolina Press, 1930.

Ribuffo, Leo P. *The Old Christian Right: The Protestant Far Right from the Great Depression to the Cold War.* Philadelphia: Temple University Press, 1983.

Rice, Arnold S. *The Ku Klux Klan in American Politics.* Washington: Public Affairs Press, 1962.

Richards, Leonard L. *"Gentlemen of Property and Standing": Anti-Abolition Mobs in Jacksonian America.* New York: Oxford University Press, 1970.

Rodgers, Daniel T. *Contested Truths: Keywords in American Politics Since Independence.* New York: Basic Books, 1987.

———. *The Work Ethic in Industrial America, 1850–1920.* Chicago: University of North Carolina Press, 1978.

Roediger, David R. *The Wages of Whiteness: Race and the Making of the American Working Class.* London: Verso, 1991.

Rose, Douglas, ed. *The Emergence of David Duke and the Politics of Race.* Chapel Hill: University of North Carolina, 1992.

Rosenstock, Morton. *Louis Marshall, Defender of Jewish Rights.* Detroit: Wayne State University Press, 1965.

Rude, George. *The Crowd in History: A Study of Popular Disturbances in France and England, 1730–1848.* Rev. ed. London: Lawrence and Wishart, 1964.

Ryan, Mary. *Cradle of the Middle Class: The Family in Oneida County, New York, 1790–1865.* New York: Cambridge University Press, 1981.

Sartre, Jean-Paul. *Anti-Semite and Jew.* Trans. George G. Becker. 1948; reprint, New York: Schocken Books, 1965.

Saxton, Alexander. *The Rise and Fall of the White Republic: Class Politics and Mass Culture in Nineteenth-Century America.* London: Verso, 1990.

Schmidt, Alvin J. *Fraternal Organizations.* Vol. 3 of *Greenwood Encyclopedia of American Institutions.* Westport: Greenwood Press, 1980.

Scott, Ann Firor. *The Southern Lady: From Pedestal to Politics, 1830–1930.* Chicago: University of Chicago Press, 1970.

Shaw, Barton C. *The Wool-Hat Boys: Georgia's Populist Party.* Baton Rouge: Louisiana State University Press, 1984.

Singal, Daniel Joseph. *The War Within: From Victorian to Modernist Thought in the South, 1919–1945.* Chapel Hill: University of North Carolina Press, 1982.

Skowronek, Stephen. *Building a New American State: The Expansion of National Administrative Capacities, 1877–1920.* Cambridge: Cambridge University Press, 1982.

Smith, Lillian. *Killers of the Dream.* New York: W. W. Norton, 1978.

Sosna, Morton. *In Search of the Silent South: Southern Liberals and the Race Issue.* New York: Columbia University Press, 1977.

Stansell, Christine. *City of Women: Sex and Class in New York, 1789–1860.* New York: Alfred A. Knopf, 1986.

Stein, Judith. *The World of Marcus Garvey: Race and Class in Modern Society.* Baton Rouge: Louisiana State University Press, 1986.

Steinberg, Stephen. *The Ethnic Myth: Race, Ethnicity, and Class in America.* Boston: Beacon Press, 1981.

Stember, Charles Herbert. *Sexual Racism: The Emotional Barrier to an Integrated Society.* New York: Elsevier, 1976.

Strong, Donald S. *Organized Anti-Semitism in America: The Rise of Group Prejudice During the Decade 1930–40.* 1941; reprint, Westport: Greenwood Press, 1979.

Suggs, Henry Louis, ed. *The Black Press in the South, 1865–1979.* Westport: Greenwood Press, 1983.

Takaki, Ronald T. *Iron Cages: Race and Culture in Nineteenth -Century America.* New York: Alfred A. Knopf, 1979.

Tindall, George. *The Emergence of the New South, 1913–1945.* Baton Rouge: Louisiana State University Press, 1967.

Tippett, Tom. *When Southern Labor Stirs.* New York: Jonathan Cape and Harrison Smith, 1931.

Trelease, Allen W. *White Terror: The Ku Klux Klan Conspiracy and Southern Reconstruction.* New York: Harper & Row, 1971.

Trotsky, Leon. *The Struggle Against Fascism in Germany.* New York: Pathfinder Press, 1971.

———. *Whither France?* 1936; reprint, New York: Merit Publishers, 1968.

Trotter, Joe William. *Coal, Class, and Color: Blacks in Southern West Virginia, 1915–1932.* Urbana: University of Illinois Press, 1990.

Vagts, Alfred. *A History of Militarism, Civilian and Military.* Rev. ed. New York: The Free Press, 1959.

Vanderwood, Paul J. *Nightriders of Reelfoot Lake.* Memphis: Memphis State University Press, 1969.

Wade, Wyn Craig. *The Fiery Cross: The Ku Klux Klan in America.* New York: Simon and Schuster, 1987.

Weinstein, James. *The Corporate Ideal in the Liberal State, 1900–1918.* Boston: Beacon Press, 1968.

White, Walter. *Rope and Faggot: A Biography of Judge Lynch.* New York: Alfred A. Knopf, 1929.

Wiebe, Robert H. *The Search for Order, 1870–1920.* New York: Hill & Wang, 1967.

Wilentz, Sean. *Chants Democratic: New York City & the Rise of the*

American Working Class, 1788–1850. New York: Oxford University Press, 1984.

Williams, Raymond. *Marxism and Literature.* Oxford: Oxford University Press, 1978.

Williamson, Joel. *The Crucible of Race: Black/White Relations in the American South Since Emancipation.* New York: Oxford University Press, 1984.

Wolman, Leo. *The Growth of American Trade Unions, 1880–1923.* New York: Bureau of Economic Research, 1924.

Wolters, Raymond. *The New Negro on Campus: Black College Rebellions of the 1920s.* Princeton: Princeton University Press, 1975.

Wood, Gordon. *The Creation of the American Republic, 1776–1787.* Chapel Hill: University of North Carolina Press, 1969.

Woodward, C. Vann. *Origins of the New South, 1877–1913.* Baton Rouge: Louisiana State University Press, 1951.

————. *Tom Watson: Agrarian Rebel.* New York: Oxford University Press, 1977.

Wright, Gavin. *Old South, New South: Revolutions in the Southern Economy Since the Civil War.* New York: Basic Books, 1986.

Wyatt-Brown, Bertram. *Southern Honor: Ethics and Behavior in the Old South.* New York: Oxford University Press, 1982.

Zangrando, Robert L. *The NAACP Crusade Against Lynching, 1909–1950.* Philadelphia: Temple University Press, 1980.

Articles

Braden, Anne. "Lessons from a History of Struggle." *Southern Exposure* 8, n. 2 (Summer 1980), 56–61.

Beatty, Bess. "Textile Labor in the North Carolina Piedmont: Mill Owner Images and Mill Worker Response." *Labor History* 25 (Fall 1984), 486–503.

Berthoff, Rowland T. "Southern Attitudes Toward Immigration, 1865–1914." *Journal of Southern History* 17 (February 1951), 328–60.

Bloch, Ruth H. "The Gendered Meanings of Virtue in Revolutionary America." *Signs* 13 (1987), 37–58.

Brown, Elsa Barkley. "Womanist Consciousness: Maggie Lena Walker and the Independent Order of Saint Luke." *Signs* 14 (Spring 1989), 251–74.

Brown, Richard Maxwell. "The American Vigilante Tradition." In *The History of Violence in America: Historical and Comparative Perspectives,* ed. Hugh Davis Graham and Ted Robert Gurr. New York: Frederick A. Praeger, 1969.

————. "Historical Patterns of Violence in America." In *The History of*

Violence in America: Historical and Comparative Perspectives, ed. Hugh Davis Graham and Ted Robert Gurr. New York: Frederick A. Praeger, 1969.

Burbank, Garin. "Agrarian Radicals and Their Opponents: Political Conflict in Southern Oklahoma, 1910–1924." *Journal of American History* 58 (June 1971), 5–23.

Davis, Natalie Zemon. "The Rites of Violence." In her *Society and Culture in Early Modern France.* Palo Alto: Stanford University Press, 1975.

Fanning, John William. "Negro Migration: A Study of the Exodus of the Negroes between 1920 and 1925 from Middle Georgia Counties." *Bulletin of the University of Georgia* 30, n. 8b (June 1930).

Faragher, John Mack. "History from the Inside-Out: Writing the History of Women in Rural America." *American Quarterly* 33 (Winter 1981), 537–57.

Feldman, Egal. "Prostitution, the Alien Woman, and the Progressive Imagination." *American Quarterly* 19 (1967), 192–206.

Ferkiss, Victor C. "Populist Influences on American Fascism." *Western Political Quarterly* 10 (June 1957), 350–73.

Fields, Barbara J. "Ideology and Race in American History." In *Region, Race, and Reconstruction,* ed. J. Morgan Kousser and James B. McPherson. New York: Oxford University Press, 1982.

Flynt, J. Wayne. "Folks Like Us: The Southern Poor White Family." In *The Web of Southern Social Relations,* ed. Walter J. Fraser, Jr., et. al. Athens: University of Georgia Press, 1985.

Garson, Robert A. "Political Fundamentalism and Popular Democracy in the 1920s." *South Atlantic Quarterly* 76 (Spring 1977), 219–233.

Gunderson, Joan. "Independence, Citizenship, and the American Revolution." *Signs* 13 (1987), 59–77.

Hahn, Steven. "Class and State in Postemancipation Societies: Southern Planters in Comparative Perspective." *American Historical Review* 95 (Feb. 1990), 75–98.

———. "Honor and Patriarchy in the Old South." *American Quarterly,* 31 (Spring 1984), 145–53.

Hall, Jacquelyn Dowd. "Disorderly Women: Gender and Labor Militancy in the Appalachian South." *Journal of American History* 73 (Sept. 1986), 354–82.

———, Robert Korstad, and James LeLoudis. "Cotton Mill People: Work, Community, and Protest in the Textile South, 1880–1940." *American Historical Review* 91 (1986), 245–86.

Hammerton, A. James. "The Targets of 'Rough Music': Respectability and Domestic Violence in Victorian England." *Gender & History* 3 (Spring 1991), 23–44.

Higginbottom, Evelyn Brooks. "African-American Women's History and the Metalanguage of Race." *Signs* 17 (Winter 1992), 51–74.

Higham, John. "American Anti-Semitism Historically Reconsidered." In *Jews in the Mind of America*, ed. Charles H. Stember et al. New York: Basic Books, 1966.

Hill, Walter B. "Rural Survey of Clarke County, Georgia, with Special Reference to the Negroes." *Bulletin of the University of Georgia* 15, n. 3 (March 1915).

Holmes, William F. "Moonshining and Collective Violence: Georgia, 1889–1895." *Journal of American History* 67 (1980), 589–611.

———. "The Southern Farmers' Alliance: The Georgia Experience." *Georgia Historical Quarterly* 72 (Winter 1988), 627–52.

———. "Whitecapping in Georgia: Carroll and Houston Counties, 1893." *Georgia Historical Quarterly* 64 (1980), 388–404.

Hux, Roger K. "The Ku Klux Klan and Collective Violence in Horry County, South Carolina, 1922–1925." *South Carolina Historical Magazine* 85 (July 1984), 211–19.

———. "The Ku Klux Klan in Macon, 1919–1925." *Georgia Historical Quarterly* 62 (1978), 155–68.

Ingalls, Robert P. "Lynching and Establishment Violence in Tampa, 1858–1935." *Journal of Southern History* 53 (Nov. 1987), 613–44.

Institute for the Study of Georgia Problems. "Survey of Athens and Clarke County, Georgia; Part I." *Bulletin of the University of Georgia* 44, n. 3 (March 1944).

Jenkins, William D. "The Ku Klux Klan in Youngstown, Ohio: Moral Reform in the Twenties." *Historian* 61 (Nov. 1978), 76–93.

Jones, Lila Lee. "The Ku Klux Klan in Eastern Kansas During the 1920's." *Emporia State Research Studies* 23 (Winter 1975).

Karmen, Andrew A. "Vigilantism." *Encyclopedia of Crime and Justice*, vol. 4. Ed. Sanford Kadish. New York: Free Press, 1983.

Klein, Rachel N. "Ordering the Backcountry: The South Carolina Regulation." *William and Mary Quarterly*, 3rd series, vol. 38 (Oct. 1981), 661–80.

Korobkin, Russell. "The Politics of Disfranchisement in Georgia." *Georgia Historical Quarterly* 74 (Spring 1990), 20–58.

MacLean, Nancy. "The Leo Frank Case Reconsidered: Gender and Sexual Politics in the Making of Reactionary Populism." *Journal of American History* 78 (Dec. 1991), 917–48.

Mann, Thomas. "Germany and the Germans." Speech at the Library of Congress after the Defeat of Germany and the Fall of the Nazis, May 1945. Quoted in the exhibit *Degenerate Art: The Fate of the Avant-Garde in Nazi Germany*. Ed. Stephanie Barron, Los Angeles County Museum of Art. New York: Larry N. Abrams, 1991.

Martin, Charles H. "White Supremacy and Black Workers: Georgia's 'Black Shirts' Combat the Great Depression." *Labor History* 18 (1977), 366–81.

Mason, Tim. "Whatever Happened to Fascism?" *Radical History Review* 49 (1991), 89–98.

Mayer, Arno J. "The Lower Middle Class as Historical Problem." *Journal of Modern History* 47 (Sept. 1975), 409–36.

McCurry, Stephanie. "The Two Faces of Republicanism: Gender and Proslavery Politics in Antebellum South Carolina." *Journal of American History* 78 (March 1992), 1,245–64.

Miller, Robert Moats. "The Ku Klux Klan." *Change and Continuity in Twentieth-Century America: The 1920's.* Ed. John Braeman, Robert H. Bremner, and David Brody. N.p.: Ohio State University Press, 1968.

———. "A Note on the Relationship Between the Protestant Churches and the Revival of the Ku Klux Klan." *Journal of Southern History* 22 (Aug. 1956), 355–68.

Montgomery, David. "On Goodwyn's Populists." *Marxist Perspectives* 1 (Spring 1978), 166–73.

———. "Violence and the Struggle for Unions in the South, 1880–1930." *Perspectives on the American South* 1 (1981), 35–47.

Moore, John Hammond. "Communists and Fascists in a Southern City: Atlanta, 1930." *South Atlantic Quarterly* 57 (1968).

Moore, Leonard J. "Historical Interpretation of the 1920's Klan: The Traditional View and the Populist Revision." *Journal of Social History* 24 (Winter 1990).

Moseley, Clement Charlton. "Political Influence of the Ku Klux Klan in Georgia, 1915–1925." *Georgia Historical Quarterly* 57 (1973), 235–55.

Mugleston, William F. "Julian Harris, the Georgia Press, and the Ku Klux Klan." *Georgia Historical Quarterly* 59 (1975), 284–95.

———. "The Press and Student Activism at the University of Georgia in the 1920s." *Georgia Historical Quarterly* 64 (Fall 1980), 241–52.

Newman, Dale. "Work and Community Life in a Southern Town," *Labor History* 19 (Spring 1978), 204–25.

Peiss, Kathy. "Dance Madness: New York City Dance Halls and Working-Class Sexuality, 1900–1920." In *Life and Labor: Dimensions of American Working-Class History,* ed. Charles Stephenson and Robert Asher. New York: SUNY Press, 1986.

Pitt-Rivers, Julian. "Honor." *International Encyclopedia of the Social Sciences,* vol. 6. Ed. Daniel L. Sills. New York: Macmillan, 1968.

Racine, Philip N. "The Ku Klux Klan, Anti-Catholicism, and Atlanta's Board of Education, 1916–1927." *Georgia Historical Quarterly* 57 (1973), 63–75.

Reed, Adolph, Jr., and Julian Bond. "Equality: Why We Can't Wait." *Nation* 253 (9 Dec. 1991), 733–37.

Regan, Fred D. "Obscenity or Politics? Tom Watson, Anti-Catholicism, and the Department of Justice." *Georgia Historical Quarterly* 70 (Spring 1986), 17–46.

Rodgers, Daniel T. "Republicanism: The Career of a Concept." *Journal of American History* 79 (1992), 11–38.

Rogin, Michael. "'The Sword Became a Flashing Vision'; D. W. Griffith's *The Birth of a Nation*." *Representations* 9 (Winter 1985), 150–95.

Smith-Rosenberg, Carroll. "The New Woman as Androgyne." *Disorderly Conduct: Visions of Gender in Victorian America*. New York: Alfred A. Knopf, 1985.

Stricker, Frank. "Affluence for Whom? Another Look at Prosperity and the Working Classes in the 1920s." *Labor History* 24 (1983), 5–33.

Taylor, A. Elizabeth. "Revival and Development of the Woman Suffrage Movement in Georgia." *Georgia Historical Quarterly* 42 (1958).

———. "The Last Phase of the Woman Suffrage Movement in Georgia." *Georgia Historical Quarterly* 43 (1959).

Terkel, Studs. "'Why I Quit the Klan': Studs Terkel Interviews C. P. Ellis." *Southern Exposure* 8, n. 2 (Summer 1980), 5–98.

Thompson, E. P. "The Moral Economy of the English Crowd in the Eighteenth Century." *Past & Present* 50 (Feb. 1971), 76–136.

Thornton, J. Mills, III. "Alabama Politics, J. Thomas Heflin, and the Expulsion Movement of 1929." *Alabama Review* 22 (April 1968).

Tilly, Charles. "Collective Violence in European Perspective." *The History of Violence in America: Historical and Comparative Perspectives*, ed. Hugh Davis Graham and Ted Robert Gurr. New York: Frederick A. Praeger, 1969.

Tyrrell, Ian R. "Drink and Temperance in the Antebellum South: An Overview and Interpretation." *Journal of Southern History* 48 (Nov. 1982), 485–510.

Wardlow, Ralph. "Negro Suffrage in Georgia, 1867–1930," *Bulletin of the University of Georgia* 33, n. 2a (Sept. 1932).

Weiner, Jonathan M. "Marxism and the Lower Middle Class: A Reply to Arno Mayer." *Journal of Modern History* 48 (1976), 666–71.

Wilson, Earnest B. "The Water Supply of the Negro." *Bulletin of the University of Georgia*, v. 31, n. 3a (March 1931).

Woodman, Harold D. "Postbellum Social Change and Its Effects on Marketing the South's Cotton Crop." *Agricultural History* 56 (Jan. 1982), 215–30.

———. "Post–Civil War Southern Agriculture and the Law." *Agricultural History* 53 (Jan. 1979), 319–37.

Yabroff, Bernard, and Ann J. Herlihy. "History of Work Stoppages in Textile Industries." *Monthly Labor Review* 76 (April 1953), 367–70.

Unpublished Dissertations, Theses, and Papers

Benson, Ronald Morris. "American Workers and Temperance Reform, 1866–1933." Ph.D. diss., University of Notre Dame, 1974.

Cashin, Edward L. "Thomas E. Watson and the Catholic Layman's Association of Georgia." Ph.D. diss., Fordham University, 1962.

Cocoltchos, Christopher Nickolas. "The Invisible Government and the Viable Community: Ku Klux Klan in Orange County, California, During the 1920s." Ph.D. diss., UCLA, 1979.

DeNatale, Douglas. "Traditional Culture and Community in a Piedmont Textile Mill Village." M.A. thesis, University of North Carolina, Chapel Hill, 1980.

Evans, Mercer Griffin. "The History of the Organized Labor Movement in Georgia." Ph.D. diss., University of Chicago, 1929.

Huffman, Frank Jackson. "Old South, New South: Continuity and Change in a Georgia County, 1850–1880." Ph.D. diss., Yale University, 1974.

Hux, Roger Kent. "The Ku Klux Klan in Macon, 1922–1925." M.A. thesis, University of Georgia, 1972.

Jones, Alton Dumar. "Progressivism in Georgia, 1898–1918." Ph.D. diss., Emory University, 1963.

Matthews, John Michael. "Studies in Race Relations in Georgia, 1890–1930." Ph.D. diss., Duke University, 1970.

Moseley, Clement C. "Invisible Empire: The History of the Ku Klux Klan in Twentieth-Century Georgia." Ph.D. diss., University of Georgia, 1968.

Mullins, Mildred Gregory. "Home-Front Activities of Atlanta Women During World War I." M.A. thesis, Emory University, 1947.

Patrick, Ralph C., Jr. "A Cultural Approach to Social Stratification." Ph.D. diss., Harvard University, 1953.

Price, Margaret Nell. "The Development of Leadership by Southern Women Through Clubs and Organizations." M.A. thesis, University of North Carolina, 1945.

Ragsdale, Annie Laurie. "The History of Co-Education at the University of Georgia, 1918–1945." M.A. thesis, University of Georgia, 1948.

Roth, Darlene Rebecca. "Matronage: Patterns in Women's Organizations, Atlanta, Georgia, 1890–1940." Ph.D. diss., George Washington University, 1978.

Sims, Anastasia. "Feminism and Femininity in the New South: White Women's Organizations in North Carolina, 1883–1930." Ph.D. diss., University of North Carolina, 1985.

Toy, Edward Vance. "The Ku Klux Klan in Oregon; Its Program and Purpose." M.A. thesis, University of Oregon, 1959.

Turcheneske, John Anthony, Jr. "The Ku Klux Klan in Northwestern Wisconsin." M.A. thesis, Wisconsin State University–River Falls, 1971.

Weaver, Norman Fredric. "The Knights of the Ku Klux Klan in Wisconsin, Indiana, Ohio and Michigan." Ph.D. diss., University of Wisconsin, 1954.

Wrigley, Steven Wayne. "The Triumph of Provincialism: Public Life in Georgia, 1898–1917." Ph.D. diss., Northwestern University, 1986.

▲ ▲ ▲

Index